The New World Species and Stemmadenia

The New World Species and Stemmadenia

A.J.M. LEEUWENBERG

Royal Botanic Gardens, Kew
Wageningen Agricultural University

First published 1994

General Editor: J.M. Lock

Special Editor for this volume: D.J. Goyder

Cover Design by Media Resources, RBG, Kew

ISBN 0 947643 74 5

Author's address:
Dr A.J.M. Leeeuwenberg, Herbarium Vadense, Wageningen Agricultural University, P.O.B. 8010, 6700 ED Wageningen, The Netherlands.

Series of Revisions of Apocynacae: XXXVI

Date of Publication 1 September 1994

Typeset by Computype, West Drayton, Middlesex

Printed and Bound in Great Britain by
Whitstable Litho, Whitstable, Kent.

CONTENTS

INTRODUCTION

The present publication is a monographic revision of all New World species of the tribe Tabernaemontaneae which, in the Americas, comprises the genera *Stemmadenia* and *Tabernaemontana*. *Woytkowskia* has been reduced to a synonym of *Tabernaemontana*. The study is based mainly on herbarium material but, in addition, the author has had the opportunity of observing living plants of 8 species of *Tabernaemontana* and 2 of *Stemmadenia* – 7 of the *Tabernaemontana* species in the wild, *T. citrifolia* and the 2 species of *Stemmadenia* in cultivation. The revision of New World *Tabernaemontana* brings the total number of species in the genus to 99, 44 of which occur in the New World. 2 of these are new to science. 10 species of *Stemmadenia* are maintained here, 3 of which are new to science.

It has been possible to trace almost all type specimens of the many names and synonyms discussed in the present revision. Besides the key to all New World species of *Tabernaemontana*, 11 keys have been added for practical reasons: for Central America inclusive of Mexico; Cuba; Jamaica; Colombia; Venezuela; Trinidad; the three Guianas; Brazil; Ecuador; Peru and Bolivia. These keys may be used for the respective Floras in due course.

An index of exsiccatae of the almost 10,000 herbarium specimens studied may be obtained from WAG on request, even on disk.

A revision of the old world species of *Tabernaemontana* was published in an earlier volume: Tabernaemontana I (Leeuwenberg 15 July 1991).

List of new species and new names

Chionanthus guianensis (Aubl.) P. Green, **comb. nov**. p. 437
Tabernaemontana cumata Leeuwenberg, **sp. nov**., p. 283
Tabernaemontana lagenaria Leeuwenberg, **sp. nov**., p. 315
Tabernaemontana lorifera (Miers) Leeuwenberg, **comb. nov**., p. 330
Stemmadenia brasiliensis Leeuwenberg, **sp. nov**., p. 401
Stemmadenia pauli Leeuwenberg, **sp. nov**., p. 428
Stemmadenia stenoptera Leeuwenberg, **sp. nov**., p. 432

GEOGRAPHICAL DISTRIBUTION

The species of *Tabernaemontana* that occupies the largest area is *T. pandacaqui* from eastern Asia, the Pacific and Australia. The New World species cover smaller areas of distribution. Of these, the most widely distributed are *T. heterophylla, T. sananho, T. siphilitica and T. undulata*, known from northern South America. *T. heterophylla* and *T. undulata* also occur in Costa Rica and Panama, while *T. siphilitica* is not known from Central America, and goes further south, even as far as Bolivia. *T. amygdalifolia* also occupies a large area, from Mexico to Venezuela and Peru. *T. alba* is the most frequently collected species of Central America inclusive of Mexico and Cuba. It has once been collected in the pinewoods near Miami, Florida. That locality is now covered by a residential area, an unfortunate situation as it was the only one in North America for the entire genus.

T. catharinensis is the only species known from Argentina and Paraguay. It occurs also in Brazil and Bolivia. Several species occupy a rather restricted area in Brazil, e.g. *T. angulata, T. hystrix, T. salzmannii* and *T. solanifolia* (see maps). The same is true for *T. palustris* in southeastern Venezuela and adjacent Colombia and Brazil, *T. cerea* in eastern Venezuela, Guyana and Suriname, *T. albiflora* and *T. lorifera* in the Guianas and adjacent Brazil. The endemics of Jamaica occupy very small areas; three of the four species represented there, *T. ochroleuca, T. ovalifolia* and *T. wullschlaegelii*, are restricted to the island, the fourth, *T. laurifolia* occurs also in Grand Cayman. *T. apoda* is endemic to Cuba, *T. oppositifolia*, to Puerto Rico and *T. cumata* and *T. muricata* have been found in a small area of the Amazon basin. *T. attenuata* has a disjunct distribution, as it is known from eastern Venezuela, Trinidad, Suriname and French Guiana. It has never been collected in Guyana.

Stemmadenia obovata is the most widespread species of that genus, it ranges from Mexico to Ecuador. *S. litoralis* and the very variable *S. macrophylla* are known from Mexico to Colombia. *S. grandiflora* occurs in Panama and northern South America. *S. donnell-smithii* goes from Mexico to Panama. *S. tomentosa* is restricted to Mexico. The three new species described in this volume are local endemics, as far as is known. *S. pauli* is restricted to the province of Puntarenas in Costa Rica while the Brazilian *S. brasiliensis* and the Mexican *S. stenoptera* are only known from the type specimens.

RELATIONSHIP WITH OTHER GENERA

Tabernaemontana and *Stemmadenia* both belong to the tribe Tabernaemontaneae, in which the interrelationship of the genera is very strong in almost all characters, as can be read in the tribe description presented below. It is even difficult to determine fruiting specimens of the tribe to genus. *Tabernaemontana* and *Stemmadenia* are the only genera in the tribe in which the corolla lobes are inflexed in bud. The constant character to distinguish *Stemmadenia* from *Tabernaemontana* is the presence of wings inside the corolla tube above the stamens. The other characters are mainly useful in combination with each other. *Stemmadenia* flowers are mostly large and showy with a thin brightly coloured corolla and large thin sepals which do not clasp the corolla base. In contrast, flowers of *Tabernaemontana* often have less showy, fleshy corollas with small thick sepals frequently clasping the corolla base. These latter differences are difficult to use in isolation, as they are easily confused. The sepals of *S. alfari* are of the same size and shape as those of *T. amygdalifolia* and *T. longipes*. In this case the size of the inflorescence is of help, as it is small and mostly 1-4-flowered in *Stemmadenia* and mostly more than 10-flowered in the two aforementioned *Tabernaemontana* species. The same is true for the length of the peduncles and pedicels, short in *Stemmadenia* and long in the two *Tabernaemontana* species. Moreover the corolla tube of *S. alfari* is much longer than of *T. amygdalifolia* and *T. longipes*.

Fruit characters are useful in distinguishing species but have limited value at the generic level. However, caudate fruits generally indicate the genus *Stemmadenia* as the character has been observed in *Tabernaemontana* only in *T. amplifolia*. Muricate fruits are present in several *Tabernaemontana* species but are unknown in *Stemmadenia*. Fruiting specimens can be named by the calyx and often also by their leaves if the fruit itself does not show clear characters.

S. brasiliensis, a species with relatively small corollas, bears a superficial resemblance to *T. debrayi* from Madagascar which has rather large sepals not clasping the corolla base. The species most difficult to place is *S. stenoptera*, the type of which has erroneously been identified as *T. amygdalifolia*. The sepals and corolla of *S. stenoptera* are thin, but small and similar size to those of *T. amygdalifolia*. The wings inside the corolla tube, albeit narrow, the size of the inflorescence and the length of peduncle and pedicels place it correctly in the genus *Stemmadenia*. Moreover, leaves of *S. stenoptera* are membranaceous and therefore thinner than those of *T. amygdalifolia*.

Although some border-line species exist, e.g. *S. stenoptera* and *T. debrayi*, and some authors erroneously placed species in the wrong genus, e.g. *S. cerea* (= *T.cerea*) and *T. alfari* (= *S. alfari*), the present author prefers to keep the two genera separate.

The other closely allied genus, *Voacanga*, has corolla lobes which are not inflexed in bud, a calyx often shed along with the corolla and a corolla tube shorter in relation to the calyx. A curious feature is that *Voacanga* has a pistil head coherent with the anthers which results in the style and pistil head being shed along with the corolla, a character unique in the Plumerioideae, but rule in the Apocynoideae. Only three species of *Tabernaemontana* shed the style and pistil head along with the corolla, but the pistil head is never coherent with the anthers. These species are *T. flavicans, T. rupicola* and *T. siphilitica*.

The genus *Woytkowskia*, the first open flower of which has recently been collected, is not maintained here as it does not have even a single distiguishing character to justify its retention. For a discussion see *T. cuspidata*.

Tabernaemontaneae G. Don, Gen. Syst. 4: 87 (1837), partly excl. *Cameraria, Vahea* (= *Landolphia*), *Plumeria, Vinca* and *Catharanthus*; Pichon in Mém. Mus. natn. Hist. Nat. II, 27: 238 (1949); in Mém. natn. Hist. Nat. Ser. B, Bot. 1: 147 (1950); Boiteau & Sastre in Phytologia 31: 247 (1975); Boiteau & Allorge in Adansonia II, 16: 9 (1981); Leeuwenberg in Journ. Ethnopharmacology 10: 4 (1984). – Type genus: *Tabernaemontana* L.

Homotypic synonyms:

Tabernaemontaninae (as subtribe) K. Schum. in Engler & Prantl, Nat.Pflanzenf. 4, 2: 145 (1895), partly, excl. *Geissospermum. Tabernaemontanoideae* (as subfamily) Stapf in Fl. Trop. Afr. 4, 1: 26 (1902); Pichon in Mém. Mus. natn. Hist. Nat. II, 27: 212 (1949); Boiteau & Sastre in Adansonia II, 15: 247 (1975), partly, as regards *Tabernaemontana* and *Voacanga*; Boiteau & Allorge, l.c.

Heterotypic synonyms:

Woytkowskieae Boiteau & Sastre in op. cit. 244. – Type genus *Woytkowskia* Woods.
Voacangeae Boiteau & Sastre in op. cit. 246. – Type genus *Voacanga* Thou.
Ervatamioideae (as subfamily) Tsiang & P.T. Li, Apocynaceae in Fl. Rep. Pop. Sinicae 63: 98 (1977). Ervatamieae (as tribe) Tsiang & P.T. Li, l.c. *Ervatamiieae* (as subtribe) Boiteau & Allorge, l.c. – Type genus *Ervatamia* Stapf (= *Tabernaemontana* L.).

Shrubs or trees, repeatedly dichotomously branched with two inflorescences (one of which is often absent) in the forks. Especially in American species, one of the branches may be missing as well but this never happens throughout the whole plant. Only *Calocrater* is seemingly unbranched as each branch bears one branchlet with a single leaf pair and one inflorescence at the apex. Leaves opposite and those of a pair equal or, especially in the apical pair of each module, frequently unequal; petiolate or less often sessile. Corolla lobes mostly overlapping to the left (to the right in *Callichilia subsessilis, Schizozygia, Tabernaemontana gamblei*, and *T. heyneana*). Anthers mostly narrowly triangular and sagittate at the base, less often oblong and cordate at the base. Pistil head mostly subglobose or lampshade-shaped or conical-truncate (see fig. 59.4, p. 243), with or without a basal ring, with a bilobed stigmoid apex. Receptive zone (stigma) at the base. Ovary superior, of two separate or less often united or partly united carpels, surrounded by a disk or not. Fruit of two separate or less often united or partly united, usually fleshy, mostly dehiscent mericarps. Seeds surrounded by a usually pulpy aril, with a deep hilar groove and ruminate endosperm; embryo straight or nearly so, spathulate.

Nine genera mainly in tropical areas, two in the New World.

KEY TO THE GENERA OF THE TRIBE TABERNAEMONTANEAE

1. Bracts large, 4-11 x 2.5-7 cm, completely covering the sepals; inflorescence 1-2-flowered with large pendulous flowers. Gabon to Angola .... **Crioceras**
   Bracts shorter to slightly longer than the sepals, never covering them 2
2. Pistil head coherent with the anthers and therefore style and pistil head shed with the corolla; calyx also shed along with the corolla (Asian species and *V. pachyceras* and *V. thouarsii*) or some time after, at least before the fruit matures; in the latter case corolla tube up to 1.4 x as long as the calyx (exceptions to this last character: *V. bracteata* and *V. chalotiana*). Old World ... . **Voacanga**
   Pistil head not coherent with the anthers and therefore pistil still complete after the corolla has been shed, (style and pistil head shed with the corolla, both not coherent with the anthers in the American *Tabernaemontana flavicans, T. rupicola* and *T. siphilitica*); calyx persistent even in fruit (shed in the American *Tabernaemontana macrocalyx*, the only species with triangular branchlets; see also *Stemmadenia*); corolla tube probably never less than twice as long as the calyx 3
3. Corolla lobes inflexed in bud and therefore seemingly much shorter in bud than in open flowers 4
   Corolla lobes not inflexed in bud and therefore not or only slightly shorter in bud than in open flowers. Africa 5
4. Inside of corolla tube with 5 wings below the anthers; inflorescence few-flowered; sepals mostly leafy and not clasping the corolla base; corolla large, thin and showy. Americas .... **Stemmadenia**
   Inside of corolla tube without wings; inflorescence few- or many-flowered; sepals mostly thick and clasping the corolla base; corolla, if large, mostly fleshy **Tabernaemontana**
5. Corolla lobes overlapping to the right; inflorescence very short; fruit capsular, veined (see also *Callichilia subsessilis*). East Africa .... **Schizozygia**
   Corolla lobes overlapping to the left; inflorescence short or long; fruit more or less fleshy, not veined 6
6. Seemingly unbranched shrub about 50 cm high with seemingly axillary inflorescences; corolla infundibuliform. Southern Cameroun and Gabon .... **Calocrater**
   Plants clearly dichotomously branched; inflorescences in the forks; corolla infundibuliform or not 7
7. Corolla large, tube (13-)20-110 mm long; lobes with 2 apices, one rounded and one acute; sepals mostly more than 5 mm long .... **Callichilia**
   Corolla much smaller, tube (3-)5-10 mm long, lobes with a single rounded or obtuse apex; sepals up to 4 mm long 8
8. Corolla tube campanulate, not contracted at the mouth, 8-10 mm long, lobes in bud not forming a head; carpels free; fruit of 2 pod-like smooth mericarps. East Africa .... **Carvalhoa**
   Corolla tube almost cylindrical, contracted at the mouth, widest just above the base and there often wider than the ovoid apical head when in bud, (3-)5-8.7 mm long; carpels fused or not; mericarps fused and smooth or lumpy or separate and with soft blunt prickles. Central Africa .... **Tabernanthe**

Callichilia Stapf, Beentje, H.J. (1978). A revision of Callichilia Stapf (Apocynaceae). Meded. Landbouwh. Wageningen 78, 7: 1-32.
Calocrater K. Schum., Hallé, N. (1965). Calocrater preussii K. Schum. Apocynacée du Gabon. Adansonia 5: 507-510.
Carvalhoa K. Schum., Leeuwenberg, A.J.M. (1985). Series of revisions of Apocynaceae XVII. A revision of Carvalhoa K. Schum. Agric. Univ. Pap. Wageningen 85, 2: 47-55.
Crioceras Pierre, Hallé, N. (1971). Crioceras dipladeniiflorus (Stapf) K. Schum. Apocynacée du Gabon et du Congo. Adansonia sér. 2. 11:301-308.
Schizozygia Baill., Barink, M.M. (1984). Series of revisions of Apocynaceae XII. A revision of Pleioceras Baill., Stephanostema K. Schum. and Schizozygia Baill. Agric. Univ. Pap. Wageningen 83, 7: 21-52(47-52).
Stemmadenia Benth., this volume.
Tabernaemontana L., Leeuwenberg, A.J.M. (1991). Tabernaemontana One. The Old World Species. Royal Botanic Gardens Kew. 223 pages and this book.
Tabernanthe Baill., Vonk, G.J.A. & A.J.M.Leeuwenberg (1989). Series of revisions of Apocynaceae XXIX, A taxonomic revision of the genus Tabernanthe and a study of the wood anantomy of T. iboga. Wageningen Agric. Univ. Pap. 89, 4: 1-18.
Voacanga Thou., Leeuwenberg, A.J.M. (1985). Series of revisions of Apocynaceae XV. Agric. Univ. Pap. Wageningen 85, 3: 1-80.

INTERRELATIONSHIP OF THE SPECIES OF TABERNAEMONTANA

In regard to its delimitation, the genus *Tabernaemontana* is the most disputed genus of the entire family of the Apocynaceae. A. De Candolle (1844) made the first segregates, while the most recent ones were proposed by Boiteau & Allorge (1976). When reading the diagnoses of the segregates *Bonafousia* A.DC. (one species) and *Peschiera* A.DC. (three species) it has been impossible to discover characters to distinguish them from *Tabernaemontana* as their author has delimited it. The confusion De Candolle started is understandable as he also mentioned all four species housed in his two segregates in *Tabernaemontana*, only under other species names. Fortunately the latter could easily be checked by studying his herbarium at Geneva.

The history is more complicated with several of the following authors, who even placed two or three species simultanously in two or three genera in the same publication, also under different species names, e.g. Miers(1878) placed *T. cymosa* under *Merizadenia, Peschiera and Taberna* and *T. flavicans* under *Anartia, Bonafousia* and *Taberna*; Markgraf (1935) placed *T. pandacaqui* under *Ervatamia*, *Pagiantha* and *Rejoua*, *T. muricata* and *T. undulata* under *Anacampta* and *Bonafousia* (Markgraf 1938), and *T. calcarea* under *Hazunta* and *Pandaca* (Markgraf 1976), and finally *T. ciliata* under *Pandaca* and *Pandacastrum*. The confusion about the relationship of the species with each other continued as the authors who published on the genus never knew enough species or had great problems in distinguishing them from each other.

The author of this publication hopes to solve this problem with the two volumes on *Tabernaemontana* and also to stabilize the nomenclature of the species as several of them are used for various purposes, especially medicinally.

A new sectional arrangement was proposed by Leeuwenberg (1990), although all of these sections remain weakly defined, but the necessary new combinations are published here. References and typification are mentioned with the genus description. The species are listed in more or less natural sequence.

1. Section **Tabernaemontana** (including *Capuronetta, Hazunta, Leptopharyngia, Oistanthera, Protogabunia* and *Taberna* Miers, as genera).

Leaves mostly thin. Inflorescence mostly lax and long. Corolla thin; tube almost cylindrical. Stamens inserted in upper half of tube, exserted or not; anthers sagittate at the base. Pistil head various, mostly with basal ring and 5 lobes around stigmoid apex. Mericarps mostly pod-like, thin-walled and dehiscent, smooth; pulp absent.
Species:
T. alba Mill.
T. citrifolia L.
T. laurifolia L.
T. wullschlaegelii Griseb.
T. ovalifolia Urb.
T. oppositifolia (Spreng) Urb.
T. amygdalifolia Jacq.
T. amplifolia Allorge
T. longipes Donn.Sm.
T. ochroleuca Urb.
T. calcarea Pichon
T. coffeoides Boj.ex A. DC.
T. mocquerysii Aug. DC.
T. persicariifolia Jacq.
T. stellata Pichon

T. capuronii Leeuwenberg
T. elegans Stapf
T. le-testui (Pellegr.) Pichon

2. Section **Peschiera** (A. DC.) Leeuwenberg, **comb. nov.** (including *Stenosolen*, as genus).

Basionym:

Peschiera A. DC., Prod. 8: 360 (1844), as genus.

Leaves mostly thin. Inflorescence mostly lax and long. Corolla thin; tube flask-shaped or almost cylindrical. Stamens inserted in lower half of tube, deeply included; anthers sagittate at the base. Pistil head various, with basal ring often composed of 10 suberect acute lobes or none, usually with 5 lobes around stigmoid apex. Mericarps rather thick-walled and dehiscent, mostly prickly or warty, obliquely ellipsoid; pulp absent.

Species:
T. heterophylla Vahl
T. vanheurckii Muell. Arg.
T. linkii A. DC.
T. salzmannii A. DC.
T. solanifolia A. DC.
T. arborea Rose ex Donn.Sm.
T. apoda C. Wright
T. cymosa Jacq.
T. laeta Mart.
T. lagenaria Leeuwenberg
T. catharinensis A. DC.
T. hystrix Steud.

3. Section **Bonafousia** (A. DC.) Leeuwenberg, **comb. nov.** (including *Anacampta, Anartia, Camerunia, Codonemma, Merizadenia, Phrissocarpus, Quadricasaea* and *Taberna* Mgf., as genera).

Basionym:

Bonafousia A. DC., Prod. 8: 359 (1844), as genus.

Leaves mostly rather thick. Inflorescence mostly congested and short. Corolla rather thin; tube almost cylindrical. Stamens inserted in upper half of corolla tube, included; anthers sagittate at the base. Pistil head similar to a lampshade or conical-truncate, with basal ring, often with 5 longitudinal grooves and 5 lobes around stigmoid apex. Mericarps rather thick-walled and dehiscent, smooth, mostly obliquely ellipsoid; pulp absent.

Species:
T. cuspidata Rusby
T. attenuata (Miers) Urb.
T. flavicans Willd. ex Roem. & Schult.
T. cerea (Woods.) Leeuwenberg
T. palustris Mgf.
T. cumatala Leeuwenberg
T. muricata Link ex Roem. & Schult.
T. angulata Mart. ex Muell. Arg.
T. disticha A. DC.
T. undulata Vahl

T. coriacea Link ex Roem. & Schult.
T. rupicola Benth.
T. lorifera (Miers) Leeuwenberg
T. albiflora (Miq.) Pulle
T. chocoensis (A. Gentry) Leeuwenberg
T. markgrafiana Macbride
T. siphilitica (L.f.) Leeuwenberg
T. sananho Ruiz & Pav.
T. penduliflora K. Schum.
T. bouquetii (Boiteau) Leeuwenberg
T. columbiensis (Allorge) Leeuwenberg
T. panamensis (Mgf., Boiteau & Allorge) Leeuwenberg
T. macrocalyx Muell. Arg.
T. maxima Mgf.

4. Section **Ervatamia** A. DC., Prod. 8: 373 (1844) (including *Muntafara, Pterotaberna* and *Taberna* Miers, as genera).

Homotypic synonym:

Ervatamia (A. DC.) Stapf in Fl. Trop. Afr. 4, 1: 126 (1902), as genus.

Leaves mostly thin. Inflorescence mostly lax and long. Corolla thin; tube almost cylindrical. Stamens inserted on various levels, included; anthers cordate at the base. Pistil head mostly subglobose, without basal ring, not grooved nor lobed. Mericarps mostly pod-like, thin-walled and dehiscent, smooth; pulp absent.

Species:
T. heyneana Wall.
T. gamblei Subramanyam & A.N. Henry
T. cordata Merr.
T. sessilifolia Bak.
T. crispa Roxb. ex Wall.
T. peduncularis Wall.
T. pauciflora Bl.
T. pandacaqui Lam.
T. inconspicua Stapf
T. antheonycta Leeuwenberg
T. corymbosa Roxb. ex Wall.
T. divaricata (L.) R.Br. ex Roem. & Schult.
T. bovina Lour.
T. bufalina Lour.
T. granulosa Pitard

5. Section **Pagiantha** (Mgf.) Leeuwenberg, **comb. nov.**

Basionym:

Pagiantha Mgf. in Notizbl. Bot. Gart. Berlin 12: 549 (1935), as genus.

Leaves mostly thick. Inflorescence rather lax, mostly long. Corolla mostly thick and fleshy; tube almost cylindrical. Stamens inserted on various levels, included; anthers cordate at base. Pistil head mostly subglobose, without basal ring, not grooved, nor lobed. Mericarps mostly thick-walled and dehiscent, smooth, mostly subglobose; pulp absent.

Species:
T. salomonensis (Mgf.) Leeuwenberg
T. thurstonii Horne ex Bak.
T. remota Leeuwenberg
T. dichotoma Roxb. ex Wall.
T. macrocarpa Jack
T. cerifera Panch. & Séb.
T. sphaerocarpa Bl.

6. Section **Pandaca** (Noronha ex Du Petit Thouars) Leeuwenberg, **comb. nov.** (including *Conopharyngia, Domkeocarpa, Gabunia, Ochronerium* and *Pandacastrum*, as genera).

Basionym:

Pandaca Noronha ex Du Petit Thouars, Gen. nov. madag. 10 (1806), as genus.

Leaves mostly thick. Inflorescence lax. Corolla thick and fleshy or not; tube almost cylindrical. Stamens mostly inserted in lower half of tube, deeply included; anthers sagittate at the base. Pistil head mostly similar to a lampshade, with basal ring, with 5 longitudinal grooves and 5 lobes around stigmoid apex. Mericarps mostly very thick-walled, subglobose or ellipsoid, mostly dehiscent, mostly smooth; pulp absent.

Species:
T. psorocarpa (Pierre ex Stapf) Pichon
T. eglandulosa Stapf
T. glandulosa (Stapf) Pichon
T. hallei (Boiteau) Leeuwenberg
T. africana Hook.
T. contorta Stapf
T. pachysiphon Stapf
T. stenosiphon Stapf
T. odoratissima (Stapf) Leeuwenberg
T. crassa Benth.
T. stapfiana Britten
T. brachyantha Stapf
T. ventricosa Hochst. ex A. DC.
T. eusepaloides (Mgf.) Leeuwenberg
T. debrayi (Mgf.) Leeuwenberg
T. phymata Leeuwenberg
T. eusepala Aug. DC.
T. crassifolia Pichon
T. retusa (Lam.) Palacky
T. sambiranensis Pichon
T. humblotii (Baill.) Pichon
T. ciliata Pichon

7. Section **Rejoua** (Gaud.) Leeuwenberg, **comb. nov.**

Basionym:

Rejoua Gaud. in Freycinet, Voy. Uran. Bot. 451 (1829), as genus.

Leaves rather thick. Inflorescence lax. Corolla thin; tube narrowly infundibuliform. Stamens inserted in lower half of tube, included; anthers sagittate at the base. Pistil head cylindrical, without basal ring, often with 5 shallow longitudinal grooves and with 5 lobes around stigmoid apex. Mericarps subglobose to pod-like, indehiscent, smooth, spongy; pulp spongy, not separable from fruit wall.

Species:

T. aurantiaca Gaud.

SYSTEMATIC PART

Tabernaemontana L., Sp. Pl. 210 (1753). – Lectotype species: *T. citrifolia* L. (designated by Britton & Wilson, Sc. Surv. Porto Rico and Virgin islands 6: 89 (1914)).

Homotypic synonym:

Tabernaemontana sect. *Taberna* A.DC., Prod. 8: 361 (1844).

Heterotypic synonyms:

Pandaca Noronha ex Du Petit Thouars, Gen. nov. madag. 10 (1806). *Conopharyngia* G. Don, Gen. Syst. 4: 94 (1837). – Type species: *P. retusa* (Lam.) Mgf. (= *C. retusa* (Lam.) G. Don = *T. retusa* (Lam.) Palacky).

Rejoua Gaud. in Freycinet Voy. Uran. Bot. 451 (1829). – Type species: *R. aurantiaca* (Gaud.) Gaud. (= *T. aurantiaca Gaud.*).

Bonafousia A.DC., Prod. 8: 359 (1844). – Type species: *B. undulata* (Vahl) A.DC. (= T. *undulata* Vahl).

Peschiera A.DC., op. cit. 360. – Lectotype species: *P. hystrix* (Steud.) A.DC. (= *T. hystrix* Steud., designated by Markgraf in Notizbl. Bot. Gart. Berlin 14: 171 (1938)).

Taberna Miers, Apoc. S. Am. 61 (1878), not of A.DC. – Lectotype species: *T. discolor* (Sw.) Miers (= *Tabernaemontana discolor* Sw., designated by Leeuwenberg in Adansonia II, 16: 390 (1976) = *T. divaricata* (L) R.Br. ex Roem. & Schult.).

Anacampta Miers, op. cit. 64. – Lectotype species: *A. congesta* Miers (= *A. coriacea* (Link ex Roem. & Schult.) Mgf. = T. *coriacea* Link ex Roem. & Schult., designated by Markgraf in op. cit. 162).

Phrissocarpus Miers, op. cit. 71. – Type species: *P. rigidus* Miers (= *T. muricata* Link ex Roem. & Schult.).

Codonemma Miers, op. cit. 72. – Type species: *C. calycina* Miers (= *T. macrocalyx* Muell. Arg.).

Merizadenia Miers, op. cit. 78. – Lectotype species: *M. sananho* (Ruiz & Pav.) Miers (= *T. sananho* Ruiz & Pav., designated by Markgraf in op. cit. 166).

Anartia Miers, op. cit. 79. – Lectotype species: *A. flavicans* (Willd. ex Roem. & Schult.) Miers (= *T. flavicans* Willd. ex Roem & Schult., designated by Markgraf in op. cit. 165).

Ochronerium Baill., Hist. Pl. 10: 199 (1889); in Bull. Soc. Linn. Paris 1: 774 (1889). – Type species: *O. humblotii* Baill. (= *T. humblotii* (Baill.) Pichon).

Gabunia K. Schum. in Engler Bot. Jahrb. 23: 224 (1896). – Lectotype species *G. crispiflora* (K. Schum.) Stapf (= *T. crispiflora* K. Schum., designated by Bullock in Kew Bull. 15: 395 (1962) (= *T. eglandulosa* Stapf).

Ervatamia (A.DC.) Stapf in Fl. Trop. Afr. 4, 1: 126 (1902); *Tabernaemontana* section *Ervatamia* A.DC., op. cit. 373. – Type species: *E. coronaria* (Jacq.) Stapf (= *T. divaricata* (L.) R. Br. ex Roem. & Schult.).

Pterotaberna Stapf, l.c. – Type species: *P. inconspicua* (Stapf) Stapf (= *T. inconspicua* Stapf).

Pagiantha Mgf. in Notizbl. Bot. Gart. Berlin 12: 549 (1935). – Type species: *P. dichotoma* (Roxb. ex Wall.) Mgf. (= *T. dichotoma* Roxb. ex Wall.).

Oistanthera Mgf. in op. cit. 550. – Type species: *O. telfairiana* (Wall.) Mgf. (= *T. telfairiana* Wall. = *T. persicariifolia* Jacq.).

Testupides Mgf., l.c. – Type species: *T. recurva* (Roxb. ex Lindl.) Mgf. (= *T. recurva* Roxb. ex Lindl. = *T. divaricata* (L.) R. Br. ex Roem. & Schult.).

Stenosolen Mgf. in Pulle (ed.), Fl. Surinam 4, 1: 344 (1937). – Type species: *S. heterophyllus* (Vahl) Mgf. (= *T. heterophylla* Vahl).
Taberna Mgf. in Notizbl. Bot. Gart. Berlin 14: 166 (1938), not of A.DC. or of Miers. – Type species: *T. albiflora* (Miq.) Mgf. (= *Tabernaemontana albiflora* (Miq.) Pulle).
Domkeocarpa Mgf. in Notizbl. Bot. Gart. Berlin 15: 421 (1941). – Type species: *D. pendula* Mgf. (= *T. ventricosa* Hochst. ex A.DC.).
Quadricasaea Woods. in Ann. Miss. Bot. Gard. 28: 271 (1941). – Type species: *Q. inaequilateralis* Woods. (= *T. macrocalyx* Muell. Arg.).
Hazunta Pichon in Not. Syst. ed. Humbert 13: 207 (1948). – Type species: *H. modesta* (Bak.) Pichon (= *T. modesta* Bak. = *T. coffeoides* Boj. ex A.DC.).
Muntafara Pichon in op. cit. 209. – Type species: *M. sessilifolia* (Bak.) Pichon (= *T. sessilifolia* Bak.).
Pandacastrum Pichon, l.c.. – Type species: *P. saccharatum* Pichon (= *T. ciliata* Pichon).
Woytkowskia Woods. in Ann. Miss. Bot. Gard. 47: 74 (1960), **syn. nov.** – Type species: *W. spermatochorda* Woods. (= *T. cuspidata* Rusby).
Capuronetta Mgf. in Adansonia II, 12: 61 (1972). – Type species: *C. elegans* Mgf. (= *T. capuronii* Leeuwenberg, not *T. elegans* Stapf).
Sarcopharyngia (Stapf) Boiteau in Adansonia II, 16: 272 (1976). *Conopharyngia* sect. *Sarcopharyngia* Stapf in op. cit. 140. *Tabernaemontana* subg. *Sarcopharyngia* (Stapf) Pichon in op. cit. 250. – Type species: *S. ventricosa* (Hochst. ex A.DC.) Boiteau (= *T. ventricosa* Hochst. ex A.DC.).
Camerunia (Pichon) Boiteau in op. cit. 274. *Tabernaemontana* subg. *Sarcopharyngia* sect. *Camerunia* Pichon in op. cit. 252. – Type species: *C. penduliflora* (K. Schum.) Boiteau (= *T. penduliflora* K. Schum.).
Leptopharyngia (Stapf) Boiteau in op. cit. 276. *Conopharyngia* sect. *Leptopharyngia* Stapf in op. cit. 141. *Tabernaemontana* subg. *Leptopharyngia* (Stapf) Pichon in op. cit. 249. – Type species: *L. elegans* Stapf (= *T. elegans* Stapf).
Protogabunia Boiteau, l.c. – Type species: *P. letestui* (Pellegr.) Boiteau (= *T. letestui* (Pellegr.) Pichon).

Shrubs or trees, repeatedly dichotomously branched from low down. Trunk terete, only in large trees rarely with buttresses. Bark pale to dark grey-brown or brown, smooth or rough, usually with large lenticels, mostly thick, with much white latex; wood rather soft, frequently pale yellowish. Branches with large lenticels, with conspicuous leaf scars, with 2 inflorescences just above each ramification (one of which sometimes absent; in several American species one of the two branchings also often absent by which the only developing inflorescence becoming seemingly axillary); ramifications sometimes umbellate and 0-4 branchlets with 4-0 inflorescences; branchlets terete (less often elliptic or triangular in section), often sulcate and angular when dried. *Leaves* opposite, those of a pair equal or unequal, petiolate or sometimes sessile; petioles channelled above, those of a pair usually connate into a conspicuous ocrea (ocrea frequently widened into intrapetiolar stipules), with colleters in the axils; blade broadly to narrowly elliptic or obovate, equal- or unequal-sided at the base, entire or sometimes sinuate or undulate. *Inflorescence* mostly distinctly pedunculate, corymbose, rather lax to congested, but rarely dense. Bracts deciduous, but still present at anthesis, scale- to sepal-like, with colleters in the axils, leaving large leaf scars. *Flowers* 5-merous, actinomorphic except for the subequal sepals, mostly fragrant. *Sepals* green or of different colour, almost free to about halfway connate, thick and fleshy or thin and leafy, erect or less often spreading or recurved, imbricate in bud, entire, inside mostly with colleters, persistent (except in the American T. *macrocalyx*) when in fruit. *Corolla* white, pale yellow, pink or mauve (some Madagascan species), with an often green or greenish or less often otherwise coloured tube and an often pale yellow throat, thin or thick and fleshy; tube mostly at least twice as long as the calyx, twisted or not; lobes in the bud overlapping to the left except in the Indian *T. gamblei* and *T. heyneana* and folded inwards, twisted, either obliquely elliptic, usually falcate and curved to the right, rounded or obtuse, often auriculate at the left side of the base (seen from inside) and sometimes with an acute lobe at the other side, or dolabriform and with two apices one of which often acute instead of rounded, frequently undulate especially near the apex, spreading and often recurved later. *Stamens* included or less often exserted; filaments short or more often reduced to ridges; anthers narrowly triangular or oblong, acuminate, apiculate or acute at the mostly sterile apex, rounded, cordate, or sagittate at the base, introrse, dehiscent throughout by a longitudinal slit, with connectives often adnate to the corolla. *Pistil*: ovary superior, composed of two carpels being barely to distinctly connate at the base and connected at the apex by the basally often cleft style; disk absent; style almost cylindrical to obconical, often widened at the apex; pistil head not coherent with the anthers (therefore pistil head still complete immediately after the corolla is shed, loose (not sticking to the anthers as in *Voacanga*) and shed with the corolla in the American *T. flavicans, T. rupicola* and *T. siphilitica*), subglobose or cylindrical and without basal ring, or more or less cylindrical, and then with a basal spreading or suberect ring of 5 or 10 acute lobes, or similar to a lampshade or conical-truncate (see fig. 59.4, p. 243), with 5 obscure angles, often slightly constricted in the middle, at the apex with 5 suborbicular or elliptic lobes and with an entire undulate or lobed ring at the base, always with a bilobed stigmoid acumen, only receptive at the base (stigma) at the sides or inside the ring beneath, when similar to a lampshade. In each cell one bipartite semiglobose placenta with several to many ovules. *Fruit* composed of two separate or less often basally united mericarps one of which sometimes remaining smaller or not developing, mature green, glaucous, orange, or yellow, sometimes dotted or warty, never spotted, subglobose to pod-like, baccate or capsular, dehiscent along an adaxial line of dehiscence or sometimes not, with a thin, thick, or even very thick wall, several- to many-seeded, with or without pulp; aril pulpy white, orange or red (in the latter case often edible), entirely or (in some American species) partly surrounding the seed. *Seeds* brown or black, nearly obliquely ellipsoid, often somewhat angular, with a deep groove to halfway its width at the hilar side and shallowly or obscurely grooved at the others, dull, rather smooth, with minute warts or with a minute honeycomb-like structure; endosperm copious, strachy, white, ruminate, enveloping the embryo; embryo mostly slightly curved, spathulate.

DISTRIBUTION: Circumtropical, 99 species.

Notes. Usually *Tabernaemontana* species ramify dichotomously and bear two inflorescences, one of which may be absent, in the forks. Exceptions are to be seen in several collections, e.g. *Leeuwenberg* 13802, *T. retusa* from Madagascar, *Harris* 9239, *T. ovalifolia* from Jamaica and *Solomon* 6149, *T. linkii* from Bolivia. In these specimens the branches form an umbel of 4 or sometimes 3 rays, each of which may be a branchlet or an inflorescence. This is the normal ramification pattern in *Rauvolfia*.

The vernacular names have rarely been noted. They refer to the shape of the fruit and are therefore used for many species, even for species of other genera of the tribe. Some species have medicinal interest, e.g. *T. sananho*. An impressive publication with 324 references on chemical compounds was written by Van Beek, T.J. & M.A.J.T. van Gessel, 1988. Alkaloids of *Tabernaemontana* species, in S.W. Pelletier (ed.), Alkaloids: Chemical and Biological Perspectives 6: 75-226. John Wiley &

Sons, New York.

The fruits of some species are edible, e.g. *T. columbiensis* and *T. undulata.*

1. KEY TO THE NEW WORLD SPECIES

As it is not easy to find useful key characters for several closely allied species, especially when the specimens to be named are incomplete, some remarks may help:

1. The branchlets are triangular in cross section only in *T. angulata, T. macrocalyx* and *T. muricata*. In all other species they are terete or elliptic in section when fresh. The branchlets of *T. sananho* are elliptic in section when fresh, and angular, not triangular, when dried.

2. The corolla lobes are upcurved in *T. distichia, T. undulata* and probably also always in *T. coriacea*. They are recurved in *T. attenuata, T. cerea, T. chocoensis, T. columbiensis, T. flavicans, T. markgrafiana, T. panamensis, T. sananho and T. siphilitica*. They are spreading and recurved later in *T. cumata, T. macrocalyx* and often also in *T. amygdalifolia.*

3. The fruits are muricate in *T. catharinensis, T. cumata, T. hystrix, T. laeta, T. lagenaria, T. linkii, T. muricata, T. salzmannii, T. solanifolia, T. vanheurckii* and mostly also in *T. heterophylla*. Some specimens of the latter species collected in Venezuela have smooth fruits.

The fruits are more or less warty in *T. apoda, T. arborea, T. cymosa* and sometimes also in *T. longipes* which has mostly smooth fruits.

1. Sepals united for at least 2 mm, 0.2-0.67 of their length, frequently white 2
   Sepals united for 0-0.5(-1) mm, mostly green 4
2. Leaves cordate or subcordate at the base, shiny or mat, but not seemingly velutinous beneath, sessile or subsessile. Northwestern South America ....**85. T. maxima**
   Leaves cuneate or rounded at the base, mostly much paler, dull and seemingly velutinous beneath; petiole 2-15 mm long 3
3. Branchlets triangular in cross section; sepals (5-)9-25 mm long. Northern South America...**83. T. macrocalyx**
   Branchlets terete; sepals 2.5-7(-8) mm long. Costa Rica to Peru and northern Brazil ....**97. T. undulata**
4. Corolla tube flask-shaped and stamens inserted 0.1-0.5 of the length of the corolla tube from the base, or narrowly cylindrical and stamens inserted 0.1-0.3(-0.4) of the length of the corolla tube from the base; inflorescence many-flowered and usually many flowers open at a time or few-flowered and lax; sepals acute or obtuse, often recurved and strap-shaped, sometimes rounded (98. *T. vanheurckii* and 80. *T. linkii*). See figs. 80, p. 323 and 98, p. 392) 5
   Corolla tube almost cylindrical and stamens inserted (0.4-)0.5-0.95 of the length of the corolla tube from the base, (from 0.4 only in 73. *T. disticha* and 97. *T. undulata*, which have upcurved corolla lobes, 81. *T. longipes* with large lax often long-pedunculate and pendulous inflorescences, and 71. *T. cuspidata* with corolla tube 6 x as long as the lobes); inflorescence few- or many-flowered and usually one or only a few flowers open at a time, lax or congested; sepals rounded or obtuse, clasping the corolla base in fresh flowers.See figs. 73 and 97...17
5. Corolla tube almost cylindrical, slightly widened only at the base; mature flower bud with a rather small mostly narrowly ovoid head; sepals sometimes recurved (75. *T. heterophylla*)...6
   Corolla tube flask-shaped; mature flower bud skittle-shaped and with a large and broad ovoid head, if not, then sepals recurved (76. *T. hystrix*) 8

6. Sepals rounded or obtuse, 1-2 x as long as wide; all leaves elliptic or nearly so; inflorescence mostly many-flowered 7
   Sepals acute or obtuse, 1.5-7 x as long as wide; often some leaves ovate, cordate or rounded at the base and sessile or subsessile; inflorescence up to 10-flowered (see also 80. *T. linkii*) ....**75. T. heterophylla**
7. Sepals ciliate, rounded; bracts 0.3-0.5 x as long as the sepals; corolla lobes pilose inside at base; fruit green or brown, often(?) turning yellow-orange, with acute or acuminate excrescences. Western South America ....**98. T. vanheurckii**
   Sepals mostly not ciliate, obtuse or rounded; bracts about as long as the sepals; corolla lobes pilose at the base for up to 0.4 of their length; fruit bright red, with blunt excrescences. South America from southern border of Suriname to Bolivia ....**80. T. linkii**
8. Leaves coriaceous or subcoriaceous, shining and more or less plane above when dried, obtuse or obscurely and obtusely acuminate at the apex; tertiary venation invisible. Brazil (Bahia, Espirito Santo) ....**93. T. salzmannii**
   Leaves membranaceous or subcoriaceous, mostly mat and with veins more or less impressed or prominent above when dried, mostly acuminate at the apex; tertiary venation usually reticulate and conspicuous 9
9. Leaves sessile or petioles up to 3 mm long, usually rounded or cordate at the base, rounded to acuminate at the apex. Eastern Brazil ....**96. T. solanifolia**
   Leaves with petioles mostly more than 4 mm long, cuneate at the base or decurrent into the petiole, usually acuminate at the apex..10
10. Sepals erect or, in dried flowers, more or less spreading, not recurved, up to twice as as long as wide...11
    Sepals recurved, at least at the apex, mostly more than twice as long as wide...14
11. Leaves mostly with conspicuous regularly arranged dots beneath; corolla pilose inside to the mouth or the base of the lobes; fruits warty, often paler-spotted. Mexico to Colombia ....**62. T. arborea**
    Leaves with or without irregularly arranged black dots; corolla lobes partly or entirely pilose inside; fruits prickly or warty, not paler-spotted 12
12. Tertiary leaf venation rather inconspicous; fruit lumpy, papillose. Cuba............... ....**61. T. apoda**
    Tertiary leaf venation rather conspicuous, reticulate; fruits prickly or lumpy ...13
13. Corolla lobes pilose inside for (0.5-)0.7-1 of their length; fruits lumpy, brown or red. Western South America from Colombia to Trinidad and south to Bolivia, also in Brazil (Acre and western Amazonas) ....**72. T. cymosa**
    Corolla lobes pilose inside for 0.1-0.35 of their length; fruits densely covered with blunt excrescences, green. Eastern Brazil..**77. T. laeta**
14. Sepals puberulous on both sides, head of mature corolla bud broadly ovoid; corolla tube 5-7 mm long. South America (French Guiana, Brazil, Peru) ....**78. T. lagenaria**
    Sepals glabrous or sometimes pubescent outside, glabrous inside; head of mature corolla bud ovoid or narrowly ovoid; corolla tube (6-)8-15 mm long 15
15. Corolla lobes pilose inside for (0.5-)0.7-1 of their length; head of mature corolla bud very large, ovoid, wider than the entire tube; leaves 2-3 x as long as wide; fruits lumpy, brown or red. Western South America from Colombia to Trinidad and south to Bolivia, also in Brazil (Acre, Roraima and western Amazonas) ....**72. T. cymosa**
    Corolla lobes pilose inside usually to the base of the lobes; leaves 2-6 x as long as wide; fruits with short acute prickles, mostly yellow 16

16. Corolla tube 0.7-1.1(-1.3) x as long as the lobes, 1.6-2.8 x as long as the calyx when sepals are recurved. Eastern Brazil to northern Argentina .... **64. T. catharinensis**
   Corolla tube 1.5-2.4 x as long as the lobes, (3-)4-8 x as long as the calyx when sepals are recurved. Southeastern Brazil .... **76. T. hystrix**
17(4). Inflorescence lax or nearly so, mostly from half as long as to much longer than the leaves; peduncle and branches usually slender, without or with a few remote pedicel scars with age (see also 82. *T. lorifera* and 92. *T. rupicola*) 18
   Inflorescence congested, mostly less than 0.3 x as long as the leaves; peduncle and branches often robust and branches densely covered with pedicel scars with age 31
18. Corolla tube 20-40 mm long; inflorescence up to 10-flowered 19
   Corolla tube 5-19 mm long; inflorescence mostly more than 12-flowered, if with less flowers then stamens exserted (59. *T. amygdalifolia*) 23
19. Corolla tube 8-13 x as long as the calyx; lobes spreading; sepals obtuse and 2-3 mm long; leaves more or less glaucous when dried, with secondary veins rather straight and up to 5 mm from each other. Northern Brazil and the three Guianas **57. T. albiflora**
   Corolla tube 0.76-7.5(-8) x as long as the calyx, lobes mostly recurved; sepals rounded, sometimes obtuse, 3-7 mm long; leaves green or brown when dried, with more or less curved secondary veins mostly at least 7 mm from each other 20
20. Tertiary leaf venation conspicuous, reticulate; corolla tube 40 mm long; follicles about 20 x as long as wide. Western South America .... **71. T. cuspidata**
   Tertiary leaf venation inconspicuous; corolla tube 20-37 mm long; follicles 2-4 x as long as wide 21
21. Sepals 2-5(-6) mm long; mature corolla bud with a very small head or not 22
   Sepals 6-12 mm long; mature corolla bud with a broadly ovoid head wider than the tube; tube 2.2-2.7 x as long as the 12-17 mm long lobes; stamens with apex 5-10 mm below mouth of corolla tube. Eastern Venezuela to Suriname .... **65. T. cerea**
22. Corolla lobes 4-6 mm long, forming a minute subglobose rounded head in the mature bud up to as wide as the part of the tube around the anthers; tube 5-7.5 x as long as the lobes; stamens with apex 1-2 mm below mouth of corolla tube. Trinidad and eastern Venezuela to French Guiana .... **63. T. attenuata**
   Corolla lobes 16-35 mm long, forming an ovoid acute or obtuse head in the mature bud wider than the tube; tube 0.9-2.7 x as long as the lobes; stamens with apex 3-5 mm below mouth of corolla tube. Venezuela, Peru and Brazil .... **74. T. flavicans**
23(18). Tertiary leaf venation inconspicuous. Jamaica.. 24
   Tertiary leaf venation mostly conspicuous and reticulate. Not in Jamaica 25
24. Inflorescence delicate, about as long as the leaves; leaves thinly papery and with flat margin when dried .... **87. T. ochroleuca**
   Inflorescence rather robust, much shorter than the leaves; leaves coriaceous and with revolute margin when dried..................................... **99. T. wullschlaegelii**
25. Leaves acutely acuminate, with tertiary venation usually not impressed above; stamens with apex 1.5-2.5 mm above mouth of corolla tube, (often included in 81. *T. longipes*) 26
   Leaves obtusely acuminate or apiculate, with tertiary venation impressed above; stamens included or with apex up to 1 mm above mouth of corolla tube 28

26. Inflorescencence often very long-pedunculate, in last branchings often rather dense; stamens inserted 0.4-0.8 of the length of the corolla tube from the base; mericarps obliquely ellipsoid or reniform, rounded or apiculate. Nicaragua to Venezuela... **81. T. longipes**
    Inflorescence entirely lax; stamens inserted 0.8-0.95 of the length of the corolla tube from the base; mericarps obliquely ellipsoid or pod-like, acuminate or caudate 27
27. Inflorescence many-flowered, 3-4 x branched; leaves large, up to 40 cm long; tertiary venation impressed above; mericarps obliquely ellipsoid, caudate. Colombia (Antioquia) and Ecuador ....**58. T. amplifolia**
    Inflorescence 2-10(-40)-flowered, 1-2 x branched; leaves smaller, rarely reaching 25 cm in length; teriary venation usually not impressed above; mericarps pod-like, acuminate. Mexico to Venezuela and Peru ....**59. T. amygdalifolia**
28. Corolla tube 12-16 mm long; stamens with apex 0.5-2 mm below mouth of corolla tube..29
    Corolla tube up to 10 mm long, if up to 12.5 mm long then stamens with apex (0-)0.5-1 mm above mouth of corolla tube...30
29. Leaves obovate or elliptic; stamens inserted 0.6-0.7 of the length of the corolla tube from the base; anthers 3-4 mm long. Puerto Rico........**88. T. oppositifolia**
    Leaves elliptic or narrowly so; stamens inserted 0.5-0.53 of the length of the corolla tube from the base; anthers 5 mm long. Jamaica...........**89. T. ovalifolia**
30. Leaves mostly elliptic or narrowly so; corolla tube 8.5-12.5 mm long; stamens with apex (0-)0.5-1 mm above mouth of corolla tube, inserted 0.7-0.8 of the length of the corolla tube, at 6.5-9.5 mm from the base; mericarps pod-like, acuminate, with lateral ridges not close to the line of dehiscence. Eastern Cuba to Tobago, but not in Jamaica and Puerto Rico ....**67. T. citrifolia**
    Leaves mostly obovate; corolla tube 6-8 mm long; stamens with apex from 0.5 mm below to 0.2 mm above mouth of corolla tube, inserted 0.5-0.7 of the length of the corolla tube, at 3.5-6 mm from the base; mericarps reniform or nearly so, rounded or apiculate. Mexico to Panama and Cuba..........................**56. T. alba**
31(17). Leaves mostly with a revolute margin, apex rounded or obtuse and margin not undulate, or apex acute or obtuse and margin undulate (see also 91. *T. panamensis*) 32
    Leaves mostly with a flat margin, apex acuminate, sometimes caudate or apiculate 34
32. Corolla tube 2-2.5 x as long as the calyx, 10 mm long; leaves obtuse or acute, undulate; fruits densely covered with blunt excrescences. Brazil (Amazonas, mainly near Manaus) ....**70. T. cumata**
    Corolla tube 4-8.5 x as long as the calyx, 10-19 mm long; leaves acute or obtuse and mostly undulate, or rounded or obtuse and not undulate; fruits smooth ..33
33. Leaves rounded or obtuse, with revolute margin; corolla tube 0.7-1.1 x as long as the lobes. Southern Venezuela and adjacent Colombia and Brazil ....**90. T. palustris**
    Leaves acute or obtuse, undulate, sinuate or entire and plane; corolla tube 1.4-1.9 x as long as the lobes. Jamaica ....**79. T. laurifolia**
34. Branchlets triangular in cross section when fresh...35
    Branchlets terete or nearly so when fresh, sometimes angular, but not triangular, when dried (see 94. *T. sananho*) 36
35. Fruit muricate; mature corolla bud forming a comparatively small broadly ovoid or subglobose head 0.12-0.16 of the bud length, 1-1.2 x as wide as the upper part of the tube, puberulous outside on the base of the lobes and sometimes at the apex of the tube. Brazil (Amazonas, near Manaus) ....**86. T. muricata**

Fruit smooth; mature corolla bud forming a broadly ovoid head 0.25-0.35 of the bud length and twice as wide as the tube, puberulous outside on the base of the lobes and on the tube from 5 mm below the mouth. Northern Brazil .... **60. T. angulata**

36. Leaves dull, seemingly velutinous and much paler beneath; corolla lobes usually upcurved in fresh flowers 37

Leaves mostly shiny or matt, not seemingly velutinous and usually not much paler beneath; corolla lobes spreading or recurved, (upcurved in 73. *T. disticha*) ..38

37. Corolla tube 2.4-5.2 x as long as the calyx; bracts 0.2-0.3 x as long as the sepals; sepals 2.5-7(-8) mm long; inflorescence erect. Costa Rica to Peru and northern Brazil .... **97. T. undulata**

Corolla tube (3.3-)5.5-10 x as long as the calyx; bracts about 0.5 x as long as the sepals; sepals 2-4 mm long; inflorescence often recurved. Ecuador, Peru and northern Brazil .... **69. T. coriacea**

38. Corolla lobes spreading or upcurved, tube not twisted; pedicels slender; peduncle mostly thin (see also 57. *T. albiflora*) 39

Corolla lobes recurved, tube twisted mainly around anthers, sometimes not (see 84. *T. markgrafiana*); pedicels rather robust; peduncle robust 41

39. Corolla lobes upcurved, tube 7-10 x as long as the calyx, 15-30 mm long; secondary veins in 5-15 pairs; dried leaves ochraceous or brownish. Colombia, Venezuela and the three Guianas .... **73. T. disticha**

Corolla lobes spreading, tube 4-6.7 x as long as the calyx, 8-14 mm long; secondary veins in 11-40 pairs; dried leaves more or less glaucous 40

40. Leaves subcordate or cuneate at the base, not decurrent into the petiole; mericarps rounded, without lateral ridges, glabrous; testa ochraceous, with long spongy papillae; style shed with the corolla. Venezuela, the three Guianas and northern Brazil .... **92. T. rupicola**

Leaves decurrent into the petiole; mericarps acuminate, with lateral ridges, papillose; testa pale brown, shortly papillose; style not shed with the corolla. Northern Brazil and the three Guianas .... **82. T. lorifera**

41. Leaves sessile or with petiole up to 1 mm long, subcordate or rounded at the base. Colombia (Choco) .... **66. T. chocoensis**

Leaves with petiole 2-22 mm long, cuneate or rounded at the base 42

42. Corolla tube 6-9 x as long as the sepals; pedicels mostly puberulous; leaves coriaceous or subcoriaceous when dried, acuminate at the apex. Panama to Peru and northwestern Brazil .... **84. T. markgrafiana**

Corolla 1.7-5.6 x as long as the sepals, if (5.7-)7-10 x as long as the sepals then leaves thinly papery when dried and caudate or acuminate at the apex and pedicels glabrous (see 68. *T. columbiensis*) 43

43. Branchlets pale-lenticellate, terete when fresh, usually greenish and not angular when dried; testa ochraceous, with long spongy papillae. South America south to Bolivia .... **95. T. siphilitica**

Branchlets not pale-lenticellate, often elliptic in cross section when fresh, usually greyish and often angular when dried; testa medium to dark brown, shortly papillose 44

44. Corolla tube 1.7-2.5 x as long as the sepals; sepals 6-12(-15) mm long; corolla yellow or orange; leaves 2.8-5 x as long as wide, up to 45 cm long. Panama, Colombia and northern Ecuador .... **91. T. panamensis**

Corolla tube 3.6-10 x as long as the sepals; sepals 2.5-6 mm long; corolla white or creamy, with a yellow throat; leaves (1.5-)2-4 x as long as wide, up to 30 cm long 45

45. Corolla tube 3.6-5 x as long as the calyx; peduncle 5-40 mm long; leaves apiculate or acuminate, papery or chartaceous when dried; mericarps obliquely subglobose, smooth when dried. Northern South America ....**94. T. sananho**
Corolla tube (5.7-)7-10 x as long as the calyx; peduncle 2-4 mm long; leaves caudate or acuminate, thinly papery when dried; mericarps obliquely ellipsoid or ovoid, minutely papillose and often wrinkled when dried. Colombia to Peru ....**68. T. columbiensis**

2. KEY TO THE SPECIES INDIGENOUS TO MEXICO AND CENTRAL AMERICA

1. Peduncle very short, 3-6 mm long, robust, (rather thin in 68. *T. columbiensis*); pedicels short; inflorescence usually few-flowered, much shorter than the leaves 2
Peduncle slender, short or long, mostly more than 8 mm long; pedicels often long; inflorescence few- or many-flowered, mostly at least half as long as the leaves 5
2. Leaves much paler, dull and seemingly velvety beneath; corolla partly pink; lobes upcurved. Costa Rica and Panama ....**97. T. undulata**
Leaves not much paler, mostly not dull and not seemingly velvety beneath; corolla yellow or white, lobes recurved..3
3. Leaves more or less regularly dotted beneath; sepals 6-12(-15) mm long; corolla tube 1.7-2.5 x as long as the calyx; mericarps subglobose, like small oranges. Panama ....**91. T. panamensis**
Leaves with scattered black dots on both sides or not; sepals 2.5-6 mm long; corolla tube 6-9 x as long as the calyx; mericarps mostly obliquely ellipsoid .4
4. Leaves coriaceous or subcoriaceous when dried; peduncle robust. Panama ....**84. T. markgrafiana**
Leaves membranaceous when dried; peduncle rather slender ....**68. T. columbiensis**
5. Stamens inserted 0.1-0.12 (up to 0.25 in South America) of the length of the corolla tube from the base; corolla tube very narrow; small leaves in unequal pairs often cordate at the base; fruits with soft prickles. Costa Rica and Panama ....**75. T. heterophylla**
Stamens inserted 0.25-0.95 of the length of the corolla tube from the base; corolla tube not very narrow; small leaves in unequal pairs never cordate at the base; fruits smooth or warty..6
6. Leaves with a regular pattern of slightly darker dots beneath; sepals often recurved at the apex; corolla tube clearly narrowed above the base; fruits light brown, warty. Mexico to Panama..**62. T. arborea**
Leaves with or without irregularly arranged or scattered black dots beneath; sepals closely clasping the base of the corolla, at least in fresh flowers; corolla tube not clearly narrowed above the base; fruits yellow, orange or green, smooth or occasionally warty (see 81. *T. longipes*)..7
7. Leaves apiculate or shortly acuminate with a usually blunt acumen, mostly with impressed reticulate venation above, with revolute margin; corolla tube 6-8 mm long; inflorescence mostly many-flowered; stamens with apex from 0.5 mm below to 0.2 mm above mouth of corolla tube; mericarps obliquely ellipsoid or reniform, rounded or apiculate at the apex. Mexico to Panama ....**56. T. alba**
Leaves acuminate with a usually acute acumen, mostly with non impressed venation and a flat margin; corolla tube (5-)7-19 mm long; inflorescence various; stamens exserted or not; mericarps various 8

8. Inflorescence mostly once branched and 2-10-flowered, lax; peduncle usually short, 1-3 mm long; stamens with apex 1.5-2.5 mm above mouth of corolla tube; mericarps pod-like, acuminate at the apex, smooth. Mexico to Panama ... . **9. T. amygdalifolia**
   Inflorescence mostly several times branched and more than 20-flowered, often congested in large branchings; peduncle often extremely long, 1-30 cm long; stamens with apex from 5 mm below to 1.5 mm above mouth of corolla tube; mericarps obliquely ellipsoid or reniform, rounded or apiculate at the apex. Nicaragua to Panama .... **81. T. longipes**

3. KEY TO THE SPECIES INDIGENOUS TO CUBA

1. Corolla tube flask-shaped; stamens inserted 0.35 of the length of the corolla tube from the base; fruits lumpy .... **61. T. apoda**
   Corolla tube nearly cylindrical; stamens inserted 0.5-0.8 of the length of the corolla tube from the base .... Go to couplet 30 of key 1, p. 219

4. KEY TO THE SPECIES INDIGENOUS TO JAMAICA

1. Leaves acuminate, with faint secondary veins and a flat margin;inflorescence delicate, about as long as the leaves. Hanover, Dolphin Head .... **87. T. ochroleuca**
   Leaves shortly and bluntly acuminate to obtuse or rounded, with conspicuous secondary veins and a revolute margin; inflorescence mostly much shorter than the leaves, with often robust peduncle 2
2. Corolla tube 5-9 mm long; tertiary leaf venation inconspicuous. Inland. .... **99. T. wullschlaegelii**
   Corolla tube 12-19 mm long; tertiary leaf venation reticulate 3
3. Inflorescence once branched into 2 racemes with robust axes covered by many bract and pedicel scars with age; leaves often sinuate and/or undulate. Coastal .... **79. T.laurifolia**
   Inflorescence 2-3 x branched, with rather slender branches, more or less paniculate; leaves entire and plane. Hanover, Dolphin Head .... **89. T. ovalifolia**

5. KEY TO THE SPECIES INDIGENOUS TO COLOMBIA

1. Sepals united for at least 2 mm, 0.2-0.67 of their length, frequently white 2
   Sepals united for 0-0.5(-1) mm, mostly green 4
2. Leaves cordate or subcordate at the base, shiny or mat, but not seemingly velutinous beneath, sessile or subsessile .... **85. T. maxima**
   Leaves cuneate or rounded at the base, mostly much paler, dull and seemingly velutinous beneath; petiole 2-15 mm 3
3. Branchlets triangular in cross section; sepals (5-)9-25 mm long.. **83. T. macrocalyx**
   Branchlets terete; sepals 2.5-7(-8) mm long................................. **97. T. undulata**
4. Corolla tube flask-shaped and stamens inserted 0.1-0.5 of the length of the corolla tube from the base, or narrowly cylindrical and stamens inserted 0.1-0.3(-0.4) of the length of the corolla tube from the base; inflorescence many-flowered and usually many flowers open at a time or few-flowered and lax; sepals acute or obtuse, often recurved and strap-shaped, sometimes rounded (98. *T. vanheurckii* and 72. *T. cymosa*). See figs. 72, p. 290 and 98, p. 392) 5
   Corolla tube almost cylindrical and stamens inserted (0.4-)0.5-0.95 of the length of the corolla tube from the base, (from 0.4 only in 73. *T. disticha* and 97. *T. undulata*, which have upcurved corolla lobes, and 1. *T. cuspidata* with corolla tube 6 x as long as the lobes); inflorescence few- or many-flowered and usually one or only a few flowers open at a time, lax or congested; sepals rounded or

obtuse, clasping the corolla base in fresh flowers. See figs. 73, p. 295 and 98, p. 392 8

5. Corolla tube almost cylindrical, slightly widened only at the base; mature flower bud with a rather small mostly narrowly ovoid head; sepals sometimes recurved (75. *T. heterophylla*) 6

Corolla tube flask-shaped; mature flower bud skittle-shaped and with a large and broad ovoid head, if not, then sepals recurved 7

6. Sepals rounded, 1-2 x as long as wide; all leaves elliptic or nearly so; inflorescence mostly many-flowered ....**98. T. vanheurckii**

Sepals acute or obtuse, 1.5-7 x as long as wide; often some leaves ovate, cordate or rounded at the base and sessile or subsessile; inflorescence up to 10-flowered **75. T. heterophylla**

7. Leaves mostly with conspicuous regularly arranged dots beneath; corolla pilose inside to the mouth or the base of the lobes; fruits warty, often paler-spotted ... .**62. T. arborea**

Leaves with or without irregularly arranged black dots; corolla lobes partly or entirely pilose inside; fruits lumpy, not paler-spotted ....**72. T. cymosa**

8(4). Inflorescence lax or nearly so, mostly from half as long to much longer than the leaves; peduncle and branches usually slender, without or with a few remote pedicel scars with age 9

Inflorescence congested, mostly less than 0.3 x as long as the leaves; peduncle and branches often robust and branches densely covered with pedicel scars with age 13

9. Corolla tube 20-40 mm long; inflorescence up to 10-flowered 10

Corolla tube 5-19 mm long; inflorescence mostly more than 12-flowered, if with less flowers then stamens exserted (59. *T. amygdalifolia*) 11

10. Tertiary leaf venation conspicuous, reticulate; corolla tube 40 mm long; follicles about 20 x as long as wide ....**71. T. cuspidata**

Tertiary leaf venation inconspicuous; corolla tube 20-35 mm long; follicles 2-4 x as long as wide ....**74. T. flavicans**

11. Inflorescencence often very long-pedunculate, in last branchings often rather dense; stamens inserted 0.4-0.8 of the length of the corolla tube from the base; mericarps obliquely ellipsoid or reniform, rounded or apiculate ....**81. T. longipes**

Inflorescence entirely lax; stamens inserted 0.8-0.95 of the length of the corolla tube from the base; mericarps obliquely ellipsoid or pod-like, acuminate or caudate 12

12. Inflorescence many-flowered, 3-4 x branched; leaves large, up to 40 cm long; tertiary venation impressed above; mericarps obliquely ellipsoid, caudate ....**58. T. amplifolia**

Inflorescence 2-10(-40)-flowered, 1-2 x branched; leaves smaller, rarely reaching 25 cm in length; teriary venation usually not impressed above; mericarps pod-like, acuminate..**59. T. amygdalifolia**

13. Leaves obtuse or rounded, with a revolute margin ....**90. T. palustris**

Leaves acuminate, sometimes caudate or apiculate, mostly with a flat margin .14

14. Leaves dull, seemingly velutinous and much paler beneath; corolla lobes upcurved in fresh flowers..**97. T. undulata**

Leaves mostly shiny or mat, not seemingly velutinous and usually not much paler beneath; corolla lobes spreading or recurved, (upcurved in 73. *T. disticha*) ..15

15. Corolla lobes spreading or upcurved, tube not twisted; pedicels slender; peduncle mostly thin...16

Corolla lobes recurved, tube twisted mainly around anthers, sometimes not (see

84. *T. markgrafiana*); pedicels rather robust; peduncle robust41 of key 1, p. 220
16. Corolla lobes upcurved, tube 7-10 x as long as the calyx, 15-30 mm long; secondary veins in 5-15 pairs; dried leaves ochraceous or brownish. ....**73. T. disticha**
Corolla lobes spreading, tube 4-6.7 x as long as the calyx, 8-14 mm long; secondary veins in 15-40 pairs; dried leaves more or less glaucous ....**92. T. rupicola**

6. KEY TO THE SPECIES INDIGENOUS TO VENEZUELA

1. Leaves rounded or obtuse at the apex, with revolute margin ....**90. T. palustris**
Leaves acuminate, caudate, or apiculate, with a mostly flat margin....................2
2. Stamens inserted 1.5-3 mm above base of corolla tube; leaves papery or membranaceous when dried 3
Stamens inserted (4-)6-23 mm above base of corolla tube; leaves various 4
3. Corolla tube very narrow, 1.7-2.7 x as long as the lobes, lobes acute, pilose inside at the base or entirely glabrous; leaves sessile or petiole 1-3(-10) mm long ....**75.T. heterophylla**
Corolla tube not very narrow, 0.6-1.3 x as long as the lobes, which are rounded and pilose inside for at least 0.6 of their length; petiole 3-20 mm long ....**72. T. cymosa**
4. Leaves sessile or with petioles up to 2 mm long, often cordate at the base 5
Leaves with petioles 3-20 mm long, not cordate at the base 6
5. Leaves 16-50 cm long, cordate or subcordate at the base; sepals 6-9 mm long ....**85. T. maxima**
Leaves 2-15 cm long, cordate to cuneate at the base; sepals 1.5-3.5 mm long ....**92. T. rupicola**
6. Inflorescence lax or nearly so, mostly from half as long to much longer than the leaves; peduncle and branches usually slender, without or with a few remote pedicel scars with age (see also 92. *T. rupicola*) 7
Inflorescence congested, mostly less than 0.3 x as long as the leaves; peduncle and branches often robust and branches densely covered with pedicel scars with age 9
7. Corolla tube 20-40 mm long; inflorescence up to 10-flowered ....Go to couplet 21 of key 1, p. 218
Corolla tube 5-19 mm long; inflorescence mostly more than 12-flowered, if with less flowers then stamens exserted (59. *T. amygdalifolia*) 8
8. Inflorescencence often very long-pedunculate, in last branchings often rather dense; stamens inserted 0.4-0.8 of the length of the corolla tube from the base; mericarps obliquely ellipsoid or reniform, rounded or apiculate **81. T. longipes**
Inflorescence not long-pedunculate, entirely lax; stamens inserted 0.8-0.9 of the length of the corolla tube from the base; mericarps obliquely ellipsoid or pod-like, acuminate ....**59. T. amygdalifolia**
9. Branchlets triangular in cross section when fresh; sepals (5-)9-25 mm long, united for 0.2-0.67 of their length...**83. T. macrocalyx**
Branchlets terete or elliptic in cross section when fresh, sometimes angular, but not triangular, when dried; sepals up to 8 mm long, mostly free or nearly so 10
10. Leaves dull, seemingly velutinous and much paler beneath; corolla lobes upcurved in fresh flowers...**97. T. undulata**
Leaves not dull and not much paler beneath; corolla lobes recurved, (upcurved only in 73. *T. disticha*) 11
11. Corolla lobes upcurved, tube not twisted, 7-10 x as long as the calyx; pedicels

slender ....**73. T. disticha**

Corolla lobes recurved, tube twisted mainly around anthers; pedicels rather robust 12

12. Branchlets mostly with pale lenticels, terete when fresh, usually not angular when dried; tertiary leaf venation usually sharp, often darker; corolla tube mostly twisted 0.5-0.6 turn to the left below the insertion of the stamens and nearly similarly to the right above this point, glabrous, or puberulous or pubescent on the part of the lobes not covered in bud; style and pistil head shed with the corolla; mericarps obliquely ovoid, ellipsoid or pod-like, smooth when fresh, but minutely tuberculate when dried; testa ochraceous, with long spongy papillae. Periodically inundated river banks ....**95. T. siphilitica**

Branchlets not pale-lenticellate, elliptic in cross section when fresh, often angular when dried; tertiary leaf venation softened, concolourous; corolla tube not or hardly twisted to the left below the insertion of the stamens, only to the right around the right around the anthers, mostly glabrous outside; style and pistil head not shed with the corolla; mericarps subglobose, faintly ridged especially near the base, smooth, also when dried; testa medium brown, with short papillae. Forest understorey, not on river banks ....**94. T. sananho**

7. KEY TO THE SPECIES INDIGENOUS TO TRINIDAD

1. Leaves dull, seemingly velutinous and much paler beneath ....**97. T. undulata**

   Leaves shiny or matt, not seemingly velutinous and not much paler beneath 2

2. Corolla tube flask-shaped and about as the long as the lobes; inflorescence many-flowered ....**72. T. cymosa**

   Corolla tube almost cylindrical and 5-7.5 x as long as lobes; inflorescence few-flowered..**63. T. attenuata**

8. KEY TO THE SPECIES INDIGENOUS TO THE GUIANAS

1. Sepals united for at least 2 mm, 0.2-0.67 of their length, frequently white ....Go to couplet 3 of key 1, p. 216

   Sepals united for 0-0.5(-1) mm, mostly green 2

2. Corolla tube flask-shaped and stamens inserted 0.1-0.5 of the length of the corolla tube from the base, or narrowly cylindrical and stamens inserted 0.1-0.3(-0.4) of the length of the corolla tube from the base; inflorescence many-flowered and usually many flowers open at a time or few-flowered and lax; sepals acute or obtuse, often recurved and strap-shaped, sometimes rounded (80. *T. linkii*, see fig. 80, p. 323) 3

   Corolla tube almost cylindrical and stamens inserted (0.4-)0.5-0.95 of the length of the corolla tube from the base, (from 0.4 only in 73. *T. disticha* and 97. *T. undulata*, which have upcurved corolla lobes); inflorescence few- or many-flowered and usually one or only a few flowers open at a time, lax or congested; sepals rounded or obtuse, clasping the corolla base in fresh flowers. See figs. 73, p. 295 and 97, p. 386 5

3. Corolla tube narrowly cylindrical; stamens inserted 0.1-0.12 of the length of the corolla tube from the base; blade of smaller leaf of a pair often cordate at the base ....**75. T. heterophylla**

   Corolla tube flask-shaped, if narrowly cylindrical then sepals rounded or obtuse and clasping the corolla base; stamens inserted 0.13-0.4 of the length of the corola tube from the base; leaf blades never cordate at the base 4

4. Sepals recurved and puberulous on both sides; corolla tube 5-7 mm long . ...**78. T. lagenaria**
   Sepals mostly erect, glabrous on both sides; corolla tube (8-)9-17 mm long.......5
5. Corolla tube almost cylindrical; lobes pilose at the base up to 0.4 of their length ....**80. T. linkii**
   Corolla tube flask-shaped; lobes pilose inside for (0.5-)0.7-1 of their length ....**72. T. cymosa**
6. Inflorescence lax or nearly so, mostly from half as long to much longer than the leaves; peduncle and branches usually slender, without or with a few remote pedicel scars with age (see also 82. *T. lorifera* and 92. *T. rupicola*) 7
   Inflorescence congested, mostly less than 0.3 x as long as the leaves; peduncle and branches often robust and branches densely covered with pedicel scars with age 9
7. Corolla tube 8-13 x as long as the calyx, lobes spreading; sepals obtuse and 2-3 mm long; leaves more or less glaucous when dried, with secondary veins rather straight and up to 5 mm from each other ....**57. T. albiflora**
   Corolla tube 0.76-7.5(-8) x as long as the calyx, lobes mostly recurved; sepals rounded, sometimes obtuse, 3-7 mm long; leaves green or brown when dried, with more or less curved secondary veins mostly at least 7 mm from each other 8
8. Sepals 2-5(-6) mm long; corolla lobes very short, 4-6 mm long, forming a minute subglobose head in the mature bud up to as wide as the part of the tube around the anthers; tube 5-7.5 x as long as the lobes; stamens with apex 1-2 mm below mouth of corolla tube ....**63. T. attenuata**
   Sepals 6-12 mm long; corolla lobes 12-17 mm long, forming a broadly ovoid head wider than the tube; tube 2.2-2.7 x as long as the lobes; stamens with apex 5-10 mm below mouth of corolla tube ....**65. T. cerea**
9. Leaves dull, seemingly velutinous and much paler beneath; corolla lobes upcurved in fresh flowers...**97. T. undulata**
   Leaves mostly shiny or mat, not seemingly velutinous and usually not much paler beneath; corolla lobes spreading or recurved, (upcurved in 73. *T. disticha*) ..10
10. Corolla lobes spreading or upcurved, tube not twisted; pedicels slender; peduncle mostly thin (see also 57. *T. albiflora*) ....Go to couplet 39 of key 1, p. 220
    Corolla lobes recurved, tube twisted mainly around anthers; pedicels rather robust; peduncle robust.................................Go to couplet 12 of key 6, p. 225

9. KEY TO THE SPECIES INDIGENOUS TO BRAZIL

1. Sepals united for at least 2 mm, 0.2-0.67 of their length, frequently white 2
   Sepals united for 0-0.5(-1) mm, mostly green 3
2. Leaves cordate or subcordate at the base, shiny or matt, but not seemingly velutinous beneath, sessile or subsessile ....**85. T. maxima**
   Leaves cuneate or rounded at the base, mostly much paler, dull and seemingly velutinous beneath; petiole 2-15 mm................Go to couplet 3 of key 1, p. 216
3. Corolla tube flask-shaped and stamens inserted 0.1-0.5 of the length of the corolla tube from the base, or narrowly cylindrical and stamens inserted 0.1-0.3(-0.4) of the length of the corolla tube from the base; inflorescence many-flowered and usually many flowers open at a time or few-flowered and lax; sepals acute or obtuse, often recurved and strap-shaped, sometimes rounded (98. *T. vanheurckii* and 80. *T. linkii*). See figs. 80, p. 323 and 98, p. 392) 4

Corolla tube almost cylindrical and stamens inserted (0.4-)0.5-0.8 of the length of the corolla tube from the base, (from 0.4 only in 73. T. disticha and 97. *T. undulata*, which have upcurved corolla lobes, 81. *T. longipes* with large lax often long-pedunculate and pendulous inflorescences, and 71. *T. cuspidata* with corolla tube 6 x as long as the lobes); inflorescence few- or many-flowered and usually one or only a few flowers open at a time, lax or congested; sepals rounded or obtuse, clasping the corolla base in fresh flowers. See figs. 73, p. 295 and 97, p. 386..8

4. Corolla tube almost cylindrical, slightly widened only at the base; mature flower bud with a rather small mostly narrowly ovoid head; sepals sometimes recurved (75. *T. heterophylla*)...Go to couplet 6 of key 1, p. 217

Corolla tube flask-shaped; mature flower bud skittle-shaped and with a large and broad ovoid head, if not, then sepals recurved (76. *T. hystrix*) 5

5. Leaves coriaceous or subcoriaceous, shining and more or less plain above when dried, obtuse or obscurely and obtusely acuminate at the apex; tertiary venation invisible. Bahia, Espirito Santo...**93. T. salzmannii**

Leaves membranaceous or subcoriaceous, mostly mat and with veins more or less impressed or prominent above when dried, mostly acuminate at the apex; tertiary venation usually reticulate and conspicuous 6

6. Leaves sessile or up to 3 mm long-petiolate, usually rounded or cordate at the base, rounded to acuminate at the apex. Eastern Brazil...........**96. T. solanifolia**

Leaves with petioles mostly more than 4 mm long, cuneate at the base or decurrent into the petiole, usually acuminate at the apex.................................7

7. Sepals erect or in dried flowers more or less spreading, not recurved, up to twice as as long as wide...Go to couplet 13 of key 1, p. 217

Sepals recurved, at least at the apex, mostly more than twice as long as wide ....Go to couplet 14 of key 1, p. 217

8(3). Inflorescence lax or nearly so, mostly from half as long to much longer than the leaves; peduncle and branches usually slender, without or with a few remote pedicel scars with age (see also 82. *T. lorifera* and 92. *T. rupicola*) 9

Inflorescence congested, mostly less than 0.3 x as long as the leaves; peduncle and branches often robust and branches densely covered with pedicel scars with age 11

9. Corolla tube 8-13 x as long as the calyx; lobes spreading; sepals obtuse and 2-3 mm long; leaves more or less glaucous when dried, with secondary veins rather straight and up to 5 mm from each other. Northern Brazil ....**57. T. albiflora**

Corolla tube 0.76-7.5(-8) x as long as the calyx, lobes mostly recurved; sepals rounded, sometimes obtuse, 3-7 mm long; leaves green or brown when dried, with more or less curved secondary veins mostly at least 7 mm from each other 10

10. Tertiary leaf venation conspicuous, reticulate; corolla tube 40 mm long; follicles about 20 x as long as wide..**71. T. cuspidata**

Tertiary leaf venation inconspicuous; corolla tube 20-37 mm long; follicles 2-4 x as long as wide..Go to couplet 22 of key 1, p. 218

11(8). Leaves mostly with a revolute margin, apex rounded or obtuse and margin not undulate, or apex acute or obtuse and margin undulate.................................12

Leaves mostly with a flat margin, apex acuminate, sometimes caudate or apiculate 13

12. Corolla tube 2-2.5 x as long as the calyx, 10 mm long; leaves obtuse or acute, undulate; fruits densely covered with blunt excrescences. Amazonas, mainly near Manaus...**70. T. cumata**

Corolla tube 4-7 x as long as the calyx, 10-19 mm long; leaves rounded or obtuse and not undulate; fruits smooth .... **90. T. palustris**

13. Branchlets triangular in cross section when freshGo to couplet 35 of key 1, p. 219
 Branchlets terete or nearly when fresh, sometimes angular, but not triangular, when dried (see 94. *T. sananho*) 14

14. Leaves dull, seemingly velutinous and much paler beneath; corolla lobes usually upcurved in fresh flowers............................Go to couplet 37 of key 1, p. 220
 Leaves mostly shiny or mat, not seemingly velutinous and usually not much paler beneath; corolla lobes spreading or recurved, (upcurved in 73. *T. disticha*) 15

15. Corolla lobes spreading or upcurved, tube not twisted; pedicels slender; peduncle mostly thin (see also 57. *T. albiflora*) ....Go to couplet 39 of key 1, p. 220
 Corolla lobes recurved, tube twisted mainly around anthers, pedicels rather robust; peduncle robust ....Go to couplet 12 of key 6, p. 225

10. KEY TO THE SPECIES INDIGENOUS TO ECUADOR

1. Sepals united for at least 2 mm, 0.2-0.67 of their length, frequently white ....Go to couplet 3 of key 1, p. 216
 Sepals united for 0-0.5(-1) mm, mostly green 2

2. Corolla tube flask-shaped and stamens inserted 0.1-0.5 of the length of the corolla tube from the base, or narrowly cyindrical and stamens inserted 0.1-0.3(-0.4) of the length of the corolla tube from the base; inflorescence many-flowered and usually many flowers open at a time or few-flowered and lax; sepals acute or obtuse, often recurved and strap-shaped, sometimes rounded (98. *T. vanheurckii*, fig.98, p. 392) 3
 Corolla tube almost cylindrical and stamens inserted (0.4-)0.5-0.95 of the length of the corolla tube from the base, (from 0.4 only in 97. *T. undulata*, which hasupcurved corolla lobes); inflorescence few- or many-flowered and usually one or only a few flowers open at a time, lax or congested; sepals rounded or obtuse, clasping the corolla base in fresh flowers. See fig. 97, p. 386 5

3. Corolla tube almost cylindrical, slightly widened only at the base; mature flower bud with a rather small mostly narrowly ovoid head; corolla lobes glabrous or only at the base pilose inside 4
 Corolla tube flask-shaped; mature flower bud skittle-shaped and with a large and broad ovoid head; corolla lobes pilose inside for at least 0.6 of their length ....**72. T. cymosa**

4. Sepals rounded or obtuse, 1-2 x as long as wide; all leaves elliptic or nearly so; inflorescence mostly many-flowered. Known from both neighbouring countries **98. T. vanheurckii**
 Sepals acute or obtuse, 1.5-7 x as long as wide; often some leaves ovate, cordate or rounded at the base and sessile or subsessile; inflorescence up to 10-flowered **75. T. heterophylla**

5. Inflorescence lax or nearly so, mostly from half as long as to much longer than the leaves; peduncle and branches usually slender, without or with a few remote pedicel scars with age ....Go to couplet 27 of key 1, p. 219
 Inflorescence congested, mostly less than 0.3 x as long as the leaves; peduncle and branches often robust and branches densely covered with pedicel scars with age 6

6. Leaves dull, seemingly velutinous and much paler beneath; corolla lobes usually upcurved in fresh flowers.............................Go to couplet 37 of key 1, p. 220

Leaves mostly shiny or matt, not seemingly velutinous and usually not much paler beneath; corolla lobes spreading or recurved..Go to couplet 42 of key 1, p. 220

11. KEY TO THE SPECIES INDIGENOUS TO PERU

1. Sepals united for at least 2 mm, 0.2-0.67 of their length, frequently white ....Go to couplet 3 of key 1, p. 216
   Sepals united for 0-0.5(-1) mm, mostly green 2
2. Corolla tube flask-shaped and stamens inserted 0.1-0.5 of the length of the corolla tube from the base, or narrowly cylindrical and stamens inserted 0.1-0.3(-0.4) of the length of the corolla tube from the base; inflorescence many-flowered and usually many flowers open at a time or few-flowered and lax; sepals acute or obtuse, often recurved and strap-shaped, sometimes rounded (98. *T. vanheurckii* and 80. *T. linkii*). See figs. 80, p. 323 and 98, p. 392) 3
   Corolla tube almost cylindrical and stamens inserted (0.4-)0.5-0.95 of the length of the corolla tube from the base, (from 0.4 only in 97. *T. undulata*, which has upcurved corolla lobes and 71. *T. cuspidata* with corolla tube 6 x as long as the lobes); inflorescence few- or many-flowered and usually one or only a few flowers open at a time, lax or congested; sepals rounded or obtuse, clasping the corolla base in fresh flowers. See figs. 71, p. 287 and 97, p. 386.....................5
3. Corolla tube almost cylindrical, slightly widened only at the base; mature flower bud with a rather small mostly narrowly ovoid head; sepals sometimes recurved (75. *T. heterophylla*)...Go to couplet 6 of key 1, p. 217
   Corolla tube flask-shaped; mature flower bud skittle-shaped and with a large and broad ovoid head...4
4. Sepals puberulous on both sides, recurved; head of mature corolla bud broadly ovoid; corolla tube 5-7 mm long ....**78. T. lagenaria**
   Sepals glabrous or sometimes pubescent outside, glabrous inside, recurved or not; head of mature corolla bud ovoid; corolla tube (8-)9-13 mm long ....**72. T. cymosa**
5. Inflorescence lax or nearly so, mostly from half as long as to much longer than the leaves; peduncle and branches usually slender, without or with a few remote pedicel scars with age 6
   Inflorescence congested, usually less than 0.3 x as long as the leaves; peduncle and branches often robust and branches densely covered with pedicel scars with age...8
6. Stamens included; corolla tube 20-40 mm long..7
   Stamens exserted; corolla tube 7-18 mm long ....**59. T. amygdalifolia**
7. Tertiary leaf venation conspicuous, reticulate; corolla tube 40 mm long; follicles about 20 x as long as wide ....**71. T. cuspidata**
   Tertiary leaf venation inconspicuous; corolla tube 20-35 mm long; follicles 2-4 x as long as wide ....**74. T. flavicans**
8. Leaves dull, seemingly velutinous and much paler beneath; corolla lobes usually upcurved in fresh flowers..............................Go to couplet 37 of key 1, p. 220
   Leaves mostly shiny or matt, not seemingly velutinous and usually not much paler beneath; corolla lobes spreading or recurved 9
9. Corolla tube 6-9 x as long as the sepals; pedicels mostly puberulous ....**84. T. markgrafiana**
   Corolla tube 1.7-5.6 x as long as the calyx; pedicels mostly glabrous 10
10. Branchlets pale-lenticellate, terete when fresh, usually not angular and greenish when dried; testa ochraceous, with long spongy papillae. ....**95. T. siphilitica**

Branchlets not pale-lenticellate, often elliptic in cross section when fresh, usually greyish and often angular when dried; testa medium to dark brown, shortly papillose ....Go to couplet 45 of key 1, p.221

12. KEY TO THE SPECIES INDIGENOUS TO BOLIVIA

1. Inflorescence congested, mostly less than 0.3 x as long as the leaves; peduncle and branches often robust and branches often densely covered with pedicel and bract scars with age 2
   Inflorescence rather lax, at least half as long as the leaves; peduncle and branches rather thin, not covered by pedicel and bract scars 3
2. Leaves dull, seemingly velutinous and much paler beneath; corolla lobes mostly upcurved in fresh flowers...**69. T. coriacea**
   Leaves mostly shiny, not seemingly velutinous and not much paler beneath; corolla lobes recurved in fresh flowers. Periodically inundated river banks ....
   ....**95. T. siphilitica**
3. Corolla tube 40 mm long; follicles about 20 x as long as wide ...**71. T. cuspidata**
   Corolla tube 6-19.5 mm long; follicles up to 3 x as long as wide........................4
4. Corolla tube almost cylindrical, slightly widened only at the base; mature flower bud with a rather small mostly narrowly ovoid head; sepals erect ....
   ....Go to couplet 6 of key 1, p. 217
   Corolla tube flask-shaped; mature flower bud skittle-shaped and with a large and broad ovoid head; sepals often recurved 5
5. Corolla lobes pilose inside for (0.5-)0.7-1 of their length, tube 2.3-5 x as long as the obtuse or acute sepals; fruits lumpy, brown or red ....**72. T. cymosa**
   Corolla lobes pilose inside only at the base, tube 1.3-2.3(-2.8) x as long as the acute sepals; fruits with short acute prickles, mostly yellow ....
   ....**64. T. catharinensis**

Species 1-55 are those distributed in the Old World (Leeuwenberg 1991, Tabernaemontana I, Kew).

ENUMERATION OF SPECIES

56. Tabernaemontana alba Mill., Gard. Dict. ed. 8. n. 2 (1768). – Type: Mexico, Veracruz, sin. loc., *Houston* anno 1730 (holotype BM; isotype CGE; phot. of holotype NY, US). Fig. 56, p. 232; map 29, p. 234

Heterotypic synonyms:

T. berteri var. *parviflora* A. DC., Prod. 8: 367 (1844), as *berterii. T. amblyocarpa* Urb., Symb. Antill. 5: 463 (1908), **syn. nov.** – Type: Cuba, Habana, *de la Sagra* 275 (holotype G-DC; isotypes FI-W,W).

T. martensii Peyr. in Linnaea 30: 31 (1859). – Type: Mexico, Veracruz, Los Baños, *Heller* 163 (holotype W; isotype NY).

T. tuxtlensis Sessé & Mociño, Fl. Mexic. ed. 1: 46 (1893) (= La Naturaleza II. 2. App. 2), not traced; ed. 2: 42 (1894), **syn. nov.** – Type: Mexico, Veracruz, Tuxtla, *Sessé* et al. (MA ?, not seen).

T. umbellata Sessé & Mociño, op. cit. 46, not traced and op. cit. 43, **syn. nov.** – Type: Mexico, Veracruz, Veracruz, *Sessé* et al. 5056 (lectotype MA, not seen; isolectotype F; phot. of lectotype WAG, designated here).

T. veracrucensis Sessé & Mociño, op. cit. 47, not traced and op. cit. 43, **syn. nov.** – Type: Mexico, Veracruz, near Veracruz, *Sessé* et al. (MA ?, not seen).

T. cymosa Sessé & Mociño, op. cit. ed. 2: 43, non Jacq. (1760), **syn. nov.** – Type: Mexico, Veracruz, Tuxtla, *Sessé* et al. 5062 (lectotype MA, not seen; isolectotype F; phot. of lectotype WAG, designated here).

T. paisavalensis Loesen. in Bull. Herb. Boiss. 2: 555 (1894). – Type: Mexico, Veracruz, Ozuluama District, near Paisavel, *Seler* 606 (holotype B†; lectotype G, designated here; isotype GH; phot. of lost holotype F, GH, MO, US).

T. chrysocarpa Blake in Contrib. Gray Herb. II. 52: 81 (1917), **syn. nov.** – Type: Belize, Manatee Lagoon, *Peck* 118 (holotype GH; isotype K).

T. amblyasta Blake in Contrib. U.S. Nat. Herb. 24: 18, pl. 6 (1922). – Type: Guatemala, Izabal, Cristina, *Blake* 7636 (holotype US; phot. WAG).

Shrub or small tree 1-15 m high. Trunk up to 60 cm in diameter; bark dark grey to light brown, longitudinally fissured. Branches pale brown, lenticellate; branchlets terete, glabrous. *Leaves* of a pair equal or unequal (larger up to 2 x as long as other and not comparatively narrower), petiolate; petiole glabrous, 5-20 mm long; (ocreae widened into intrapetiolar stipules or not); blade subcoriaceous or herbaceous when fresh, subcoriaceous or papery and often with pale venation on both sides when dried, elliptic, obovate or narrowly so, 2-4 x as long as wide, 2-23 x 1-8 cm, apex apiculate or shortly acuminate with a usually blunt acumen, cuneate at the base, entire or slightly undulate and/or sinuate, with revolute margin, glabrous on both sides, often with scattered black dots beneath, with 3-18 pairs of upcurved secondary veins; tertiary venation reticulate, usually impressed above and not very conspicuous beneath in dried leaves. *Inflorescence* very variable in size, pedunculate, 3-22 x 3-20 cm, with mostly more than 20 flowers, mostly rather lax. Peduncle slender, glabrous, 2-75 mm long; pedicels glabrous, 4-7 mm long. Bracts small, sepal-like, 0.3-1 x as long as them. *Flowers* sweet-scented, open during the day. *Sepals* pale(?) green, connate at the extreme base, erect, ovate or broadly ovate, 1-1.7(-2) x as long as wide, 2-3 x 1.5-2 mm, rounded or obtuse, glabrous outside, not ciliate, glabrous inside and with 2-7 colleters in 1 row in the middle at the base; colleters often unequal, 0.3-0.6 x 0.1-0.2 mm. *Corolla* white, creamy or yellow, thin, 10-13 mm long in the mature bud and forming a comparatively very large ovoid head 0.3-0.4 x the bud length (4-5 x 2-4 mm) with a blunt or subacute apex, glabrous outside, glabrous inside, rarely pubescent from 0.5-1 mm below the anthers almost to their apices; tube 2.4-4 x as long as the calyx, 0.6-0.9 x as long as the lobes, 6-8 mm long, almost cylindrical, 1.5-2.5 mm wide above the base, narrowed at the insertion of the stamens to 1-1.5 mm wide, widened again around the anthers to 1.5-2 mm wide, not twisted, with 5 longitudinal bilobed thickenings in the throat visible from the mouth; lobes obliquely elliptic and often falcate to dolabriform, 1.1-1.8 x as long as the tube, 1.6-2.8 x as long as wide, 7-14 x 3.5-6 mm, obtuse or rounded, slightly auriculate at the left side of the base, more or less undulate, spreading. *Stamens* with apex from 0.5 mm below to 0.2 mm above mouth of corolla tube, inserted 0.6-0.7 of the length of the corolla tube, (at 3.5-5.5 mm from the base); anthers sessile, narrowly triangular to narrowly oblong, 2.5-4 x as long as wide, 2.3-3 x 0.6-1 mm, apex acuminate, sterile for 0.2-0.3 mm, (mostly shortly) sagittate at the base, glabrous. *Pistil* glabrous, 4.5-6.5 mm long, with apex about halfway along anthers; ovary oblong or ovoid, 1.5-3 x 1-1.5 x 1-1.5 mm, of 2 separate carpels, gradually narrowed into the style; style 2-4 mm long; pistil head 1-1.2 mm high, composed of an undulate often fringed basal ring-like veil, 0.2 x 1-1.2 mm, an apically 5-lobed stipitate head 0.6-0.8 x 0.3-0.5 mm, and a bilobed stigmoid apex 0.1-0.2 x 0.1-0.2 mm. Ovules approximately 50-100 in each carpel. *Fruit* of 2 separate mericarps; mericarps orange, yellow or green, not dotted, obliquely ellipsoid or reniform, 20-40 x 15-27 x 10-25 mm, rounded or apiculate at the apex, recurved, usually with 2 ridges (one at each side of the line of dehiscence), approximately

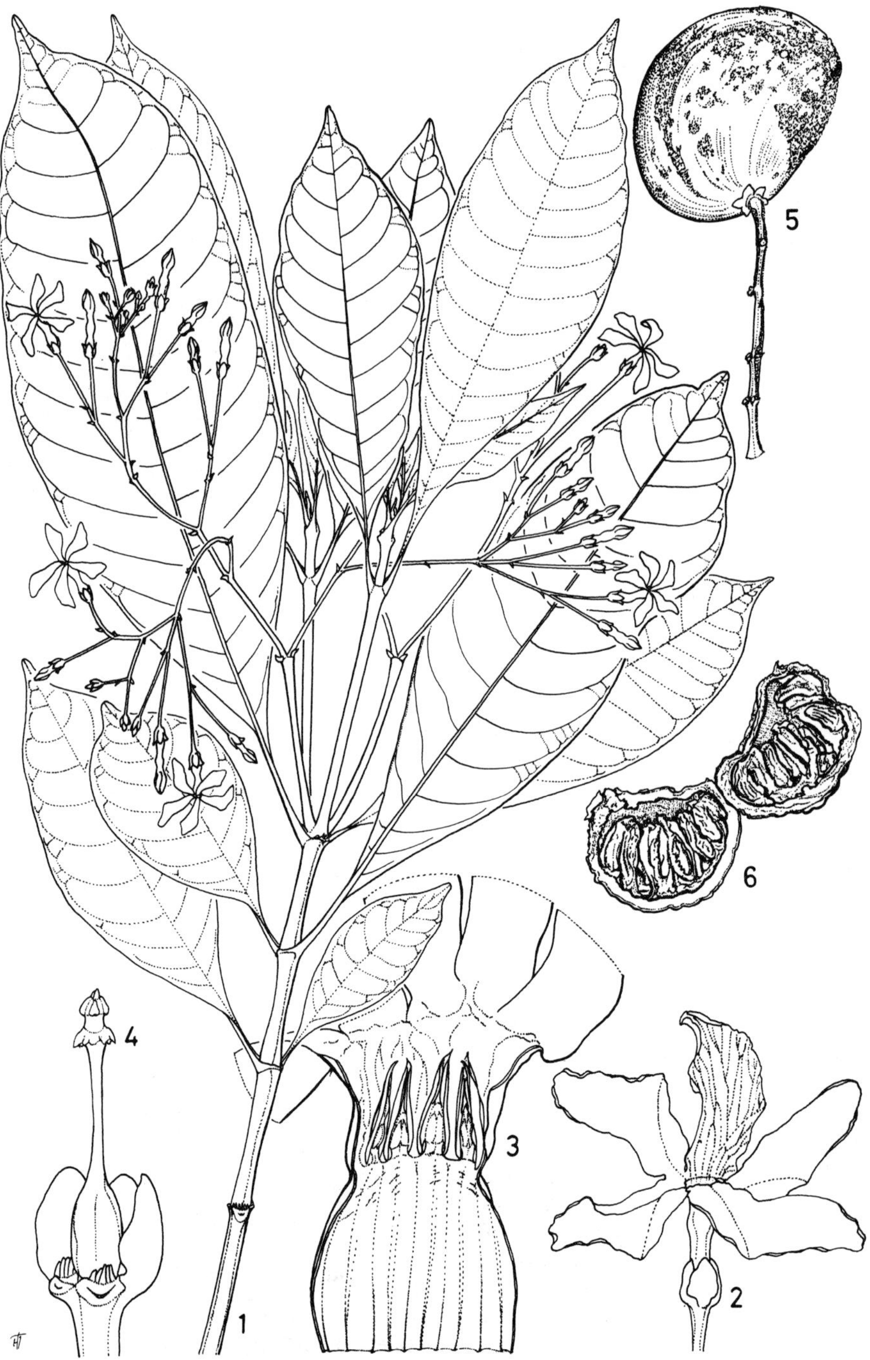

Fig. 56. *Tabernaemontana alba*. **1,** habit (x 2/3); **2,** flower (x 2); **3,** opened corolla (x 6); **4,** calyx with pistil (x 6); **5,** fruit with only one mericarp developed (x 1); **6,** open fruit with seeds (x 2/3). 1 from Croat 47888; 2-4 from Calderon 2234; 5 from Dorantes et al. 1056; 6 from A.Gentry et al. 32256.

10-50-seeded; wall 3-5 mm thick in dried fruits; aril orange or red, enveloping the seed. *Seed* dark brown, obliquely ellipsoid, 8-10 x 2.5-3 x 2-3 mm, with longitudinal grooves, papillose; embryo 6-7 mm long; cotyledons ovate, 1.4-1.7 x as long as wide, 2.5-3.5 x 1.5-2.5 mm, rounded at the apex, cordate at the base; rootlet 4 x 0.5-0.7 mm.

DISTRIBUTION: Mexico to Panama.

ECOLOGY: Moist forest or bush, often on river banks or in swamps, mostly near the coast. Alt. 0-900 m. Flowering mainly March-April and fruiting mainly June-July.

Geographical selection of the approximately 900 specimens examined:

U.S.A.: Florida: near Miami, *Small* 2273 (GH, US).

MEXICO. San Luis Potosí: Maiz, *Rivera* 101 (LL); San Dieguito, *Palmer* 137 (GH, US); Tamasopo Cañon, *Pringle* 3627 (A, GH); 32 km S of Valles, *Webster* et al. 11351 (GH, LL, MO); Tamarindo, Mun. Tampamolon, *Alcorn* 2017 (LL); Tamazunchale, *Taylor Edwards* 560 (F, LL, MO). Puebla: Fondo de la Barranca El Paraíso, *Sousa* Jan. 1962 (MEXU); Cuauhtapanaloyan, Cuetzalan, *Basurto & Patrón* 32 (MEXU); road to San Diego La Mesa, *Gómez Pompa* 15C (MEXU); Chinantla, *Galeotti* 1595 (G). Hidalgo: Cazadero, *Liebmann* 11899 (C,US). Tamaulipas: 5 km S of Ciudad Victoria, *J.S.Wilson* 12259 (LL, WIS); 88 km S of Ciudad Victoria, along Río Sabinas, *F.G.Meyer & Rogers* 2852 (BM, BR, E, G, GH, MO, U, UPS); near Gómez Fárias, *Palmer* 276 (F, GH, K, MO, NY, US); Julillo to Río Sabanas, N of Mante, *Duke* M3631 (MO); Tampico, *Berlandier* 217 (BM, G, P). Veracruz: 2 km S of Tampico, *Palmer* 419 (BM, G, GH, K, MO, NY, US); Laguna Tamiahua, *LeSueur* 354 (F, GH, LL); 19 km N of Ozuluama, *Croat 66114* (WAG); 6 km N of Huejutla, *Nee & Hansen* 18472 (F, USF, WAG); 3 km from Tepetzintla, *Chiang* 342 (F, GH); near Paisavel, Ozuluama District, *Seler* 606 (G, GH; phot. of lost B sheet F, G, GH, MO, US, type of *T. paisavalensis*); Rancho Nuevo, Mun. Cazones, *Cortés* 289 (MEXU); San Francisco, km 7 Pempoal Rumbo-Tantoyuca Road, *Lot 426* (GH); Joloapan, Mun. Papantla, *Cortés* 91 (BM, MO); 1 km E of Venustiano Carranza, *Fay & Calzada* 856 (AAU, F, GH, LL, UC, US); 20 km NE of Santa Gertrudis, Mun. Vega de Alatorre, *Castillo & Benavides* 2045 (MEXU); near Jalapa, *Pringle* 7831 (GH, MEXU, US); ibid., *Galeotti* 1577 (BR, G, P); 1 km NW of Corral Falso, Emiliano Zapata, *Marquez* et al. 875 (BM, F, MEXU, MO, NY); Mirador, *Liebmann* 12003 (C, US); Cordoba Valley, *Bourgeau* 1947 (BR, M, P, S, US, WAG) & 2222 (FI-W, G, GH, K, L, MPU, P); Los Naranjos, SE of Tierra Blanca, Bamps 5520 (BR, K, MO); 2 km NE of Tlacotalpan, Nee & Taylor 26543 (F, USF); Colonia La Palma, Catemaco, Acahual, *Martinez Calderón* 2234 (A, F, MO); Tuxtla, *Sessé* et al. 5056 (MA, not seen, F; phot. of MA sheet WAG, lectotype of *T. umbellata*) & 5062 (MA, not seen, F; phot. of MA sheet WAG, lectotype of *T. cymosa* Sessé & Mociño, non Jacq.); Biological Station Los Tuxtlas, *A. Gentry* et al. 32256 (MO); ibid., *Ibarra Manriquez* 808 (KUN, MO); E of Tebanca, *Hansen & Nee* 7632 (F, USF, WAG); Zacuapan, *Purpus* 2044 (E, F, GH, L, MO, NY, UC, US); SE of Laguna Salada, *Dorantes* et al. 1056 (F); Los Baños, 29 km from Veracruz, *Heller* 163 (NY, W, type of *T. martensii*); 10 km N of Sontecomapan, Nee 22545 (BM, F, G, GH, USF); Sontecomapan, *Galeotti* 1598 (F, G, GH, NY, P, US, W); Cerro de San Martin, *Calzada* 103 (BM, F, GH, MO); W of Coatzacoalcos, *Dwyer* 14551 (MO, WAG); Campamento Hnos., Hidalgotitlan, *Juan & Avendano* 21 (MO); 3 km from km 5 Plan de Arroyos-Alvaro Obregon Road, Hidalgotitlan, *Dorantes* et al. 2712 (BM, C, F, K, MEXU, MO, UC); sin. loc., *Houston* anno 1730 (BM, CGE; phot. of BM sheet BH, NY, US, type). Oaxaca: Omealca, *Altamirano* 436 (MEXU); Miguel Aleman, 8 km W of Nueva Patria, *Torres* 734 (MEXU); Temascal, *Torres* 2371 (MEXU); near Chiltepec, *Stevens* et al. 2462 (GH, LL, MO, NY); Jocotepec, *Liebmann* 11897 (C); La Esperanza, Mun. San Juan del Río, *Torres* et al. 9428 (MEXU, MO); Yaveo, Choapan District, *Mexia*

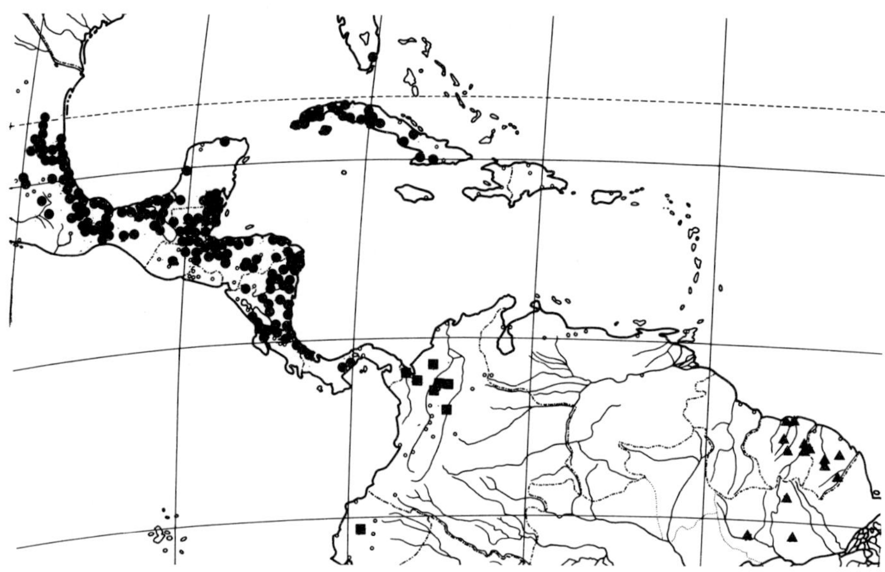

Map 29. ●. *Tabernaemontana alba*, ▲ *T. albiflora*, ■, *T. amplifolia.*

9224 (B, BH, F, G, GB, GH, K, MO, NY, S, U, UC, US); Ubero, *Ll. Williams* 9125 (F); near Mogoñé, *Alexander* 123 (NY); Juchitán, Matias Romero, *MacDougal* H334 (F, NY); S of Aserradero La Floresta, *Wendt* et al. 3159 (MEXU); 23 km N of Cardenas, towards Santa Maria Chimalapa, *Tenorio* et al. 3476 (MEXU). Chiapas: between Colonia Francisco I. Madero and Colonia A. Lopez Mateos, Mun. Cintalapa, *Breedlove* 50516 (MO); near Derna, Mun. Ocozocoautla de Espinosa, *Breedlove & Thorne* 30342 (GH, KUN, LL, MO); 17 km SE of Palenque, *Davidse* et al. 20396 (MO, WAG); km 24 Crucero Corozal-Boca Lacantum Road, Mun. Ococingo, *Martínez* 7119 (MEXU, MO); 2 km NW of El Piedron, *Stevens & Martínez* 25913 (MO); near San Quintin, *Sohns* 1599 (US). Tabasco: between Río Tonala and Sanchez Magellanes, *Barlow* 6/15 (GH, WIS); 10-40 km W of Huimanguillo, *Barlow* 30/171 (F, UC); Selva dos Montes, Mun. Villahermosa, *Cowan & Zamudio* 4668 (LL); Teapa, *Linden* 329 (F, FI, G, P); SE of Teapa, *Croat* 47888 (MO, USF, WAG); San Juan Bautista-Atasa Road, *Rovirosa* 178 (US); Santa Anna, outskirts of Macuspana, *Gilly & Hernandez* 363 (GH, MEXU); Balancan, *Matuda* 3124 (A, F, MEXU, NY). Campeche: 1 km NW of Ejido Pital, *Cabrera & de Cabrera* 8195 (MO); Isla del Carmen, *Barlow* 13/2 (GH, WIS); 12 km N of Escárcega, *Cabrera & de Cabrera* 2227 (MEXU); Esperanza, near Campeche, *Seler* 4007 (F, GH). Yucatan: San Felipe, *Lundell* 1444 (MO, NY, US). Quintana Roo: 7 km S of Ucum, along road to La Unión, *Cabrera & de Cabrera* 3667 (MEXU, MO); 8 km N of La Unión, *Davidse* et al. 20165 (MO, WAG). Sin. loc., *Sessé* et al. 5057 (MA, not seen, F; phot. of MA sheet WAG, "*T. obovata*", possibly a syntype of one of the names of Sessé & Mociño cited above as synonym).

Belize: km 3 Calcutta-Xaibe Road, *Balick* et al. 2173 (WAG); Lamani (Indian Church)-San Felipe road, *Davidse & Brant* 32724 (WAG); *Guamil, Arnason & Lambert* 17638 (MO); 15 km SW of San Ignacio, *Balick* et al. 1763 (USF); near Ridge Lagoon, *Liesner & Dwyer* 1641 (F, GH, MO, NY, US); Caves Branch, *Whitefoord* 1125 (BM, MO, NY, U); Stann Creek-Mullin's R. Road, *Gentle* 1899 (A, MO, NY); Manatee Lagoon, Peck 118 (GH, K, type of *T. chrysocarpa*); Punta Gorda, *Dwyer* 14736 (MO, WAG).

GUATEMALA. Peten: Lacandon, *Contreras* 3513 (G, LL, MO, US); km 36 Sayaxche-Arroyo Pucte Road, Ortíz 2708 (BM, F, US); La Libertad, *Lundell* 3278 (BH, BM, K, MO, WIS); Río Pasion, San Juan Acul, *Lundell* 18209 (G, US); between Cerro Ceibal and Ceibal, Steyermark 46171 (F, MO); Bajo del Hormiguero, *Contreras* 1961 (LL); Laguna Yaxja, *Harmon & Dwyer* 2741 (MO); near Poptun, *Ortíz* 957 (BM, F, MO, NY). Izabal: Modesto Mendez, *Harmon* 2557 (F, GH, MO, NY); Los Amates, *Blake* 7732 (GH, US, paratype of *T. amblyasta*); Shore of Lake Izabal, *Blake* 7853 (US, paratype of *T amblyasta*); Cristina, *Blake* 7636 (US, phot. WAG, type of *T. amblyasta*); Cienaga, *Contreras* 10862 (MO, S, US); between Puerto Berrio and Santo Tomás, *Steyermark* 42042 (A, F, MO); near Quiriguá, *Standley* 24005 (GH, NY, US). Alta Verapaz: Rubelsanto, *Lundell & Contreras* 19513 (LL); Cubilhuitz, von *Tuerckheim* 4121 (E, ECON, F, LD, U, US); Río Sebol, downstream from Carrizal, *Steyermark* 45799 (A, F, MO, NY); km 17 Sebol-Cobán Road, *Contreras* 4706 (LL). Baja Verapaz: Santa Inés, *Galusser* 11 (US).

HONDURAS. Copán: 3 km N of Santa Rosa de Copán, *Hazlett* 753 (MO); between La Florida and Espíritu Santo, *Blake* 7415 (US, paratype of *T. amblyasta*). Santa Barbara: SW of Quimistán, *Standley & Lindelie* 7354 (F); Callejones, *Molina* 5253 (F); W shore of Lago de Yojoa, near El Rincón, *Blackmore & Chorley* 3705 (BM, MO). Comayagua: Yure and Humuya Rs. confluence, *Nelson* et al. 6120 (MO); Chichipates, *Nelson* et al. 6762 (MO). Cortés: Cuyamel, *Mallery* 5036 (K, NA); Omoa, *Barkley & Hernández* 40428 (F, GH); Las Piñetas, *Molina* 10565 (F, NY); Río Chamelecón, near La Lima, *L.O.Williams & Molina* 12484 (BM, F, GH); Río Lindo, *Edwards* P-674 (A, F, MO, NY, US). Yoro: 3 km NW of Santa Rita, on road to Negrita, *Harmon & Dwyer* 3877 (AAU, MO, NY, U); 2 km N of El Progreso, *Guevara* 174 (WAG). Atlántida: near Tela, *Mitchell* 83 (A, F, GH, US); near La Ceiba, *Yuncker* et al. 8023 (BM, F, G, GH, K, MO, NY, S, US). Islas de la Bahia: near Roatan in the Island of Roatan, *Harmon & Dwyer* 3970 (MO). Colón, Trujillo, *Saunders* 517 (AAU, BM, F, LL, MO, NY, Z). Olancho: near Catacamas, *Hernández & Rendón* 5454 (MO, USF); Valle Lepaguare, *Molina* 13385 (F, NY); Puerto Siena, Potrero, *P. Wilson* 491 (F, NY, US). Gracias a Dios: near mouth of Río Platano, *A.Gentry* et al. 7477 (GH, MO); near Río Patuca, *Clewell* 4417 (M); Ahuas Bila, *Nelson & Cruz* 9481 (MO); near Mocorón, *Nelson & Vargas* 5049 (MO). El Paraíso: between El Urraco and Cifuentes, *Molina* 11418 (F).

NICARAGUA. Zelaya: near Waspam, *Bunting & Licht* 489 (F, NY); near Bilwascarma, *Davidse & Pohl* 2349 (F, K, MO, NY); S of Río Wawa, *Little* 25166 (F, MO, US); Ibo Tingni, Caño Sung Sung, *Stevens* 10630 (MO, WAG); between Rosita and Puerto Cabezas, *Stevens* 8523 (BM, MO, P, U, Z); near San Jose del Hormiguero, *Stevens* 7175 (BM, MO); S of Siuna, *Neill* 4162 (MO, WAG); Cerro Saslaya, *Neill* 1895 (MO, WAG); Prinzapolka, *Neill* 4554 (BM, MO); N of Puerto Cabezas, *Sandino* 4027 (MO, WAG); between El Empalme and Limbaika, *Stevens & Moreno* 19453 (MO); Comarca Bodega, 30 km NE of Río Blanco, *Moreno* 24074 (MO, WAG); near Macantaca Creek, *Allen* 6512 (F, GH, US); NW of Rama, *Stevens & Montiel* 17467 (MO, U, WAG); El Falso Bluff, *Sandino* 2174 (MO, WAG); Monkey Point, *Stevens* 19996 (MO, WAG). Boaco: 6 km NE of Boaco, *Moreno* 23952 (MO, WAG); 5.5 km N of Camoapa, *Moreno* 10602 (MO). Rivas: 10 km W of Rivas, *Neill* 2467 (MO, WAG); SW of Sapoá, *Sandino* 3581 (MO, WAG). Chontales: near La Libertad, *Standley 8866* (F). Río San Juan: San Carlos, *Hamblett 2069* (BM, F, MO); Río Sábalos, *Moreno* 22992 (MO, WAG); San Juan del Norte, *Araquistain* 3392 (MO, WAG).

COSTA RICA. Guanacaste: Volcán Orosí, *Wibur & Stone* 10229 (F, MO, US); Parque Nacional Santa Rosa, *Huft* et al. 2092 (MO, WAG); 10 km W of Cañas, *Raven* 20992 (F, GB, MO). Alajuela: Upala, *Herrera* 1767 (MO, WAG); near Los Chiles, *Holm &*

Iltis 701 (A, BM, F, G, P, US); Naranje, Ørsted 15141 (C). Limón: Río Toro Amarillo, *J.D.Smith* 6647 (GH, K, US); Barra del Colorado, *Stevens* 24191 (MO, WAG); near Limón, *Pittier* 4301 (G, US); between Hone Creek and Cahuita, *Gómez* et al. 20518 (WAG, WIS).

PANAMA. Bocas del Toro: Changuinola Valley, *Dunlap* 534 (F, US); Columbus Island (= Isla Colón), *Cooper* 550 (C, F, K, US); ibid., *von Wedel* 519 (GH, MO, U). Coclé: Las Minas, *Allen* 2703 (GH, MO, US); La Mesa above El Valle, *Croat* 25347 (MO).

CUBA. Pinar del Río: Bahía Honda, Bro. *Alain* 1794 (GH, US); near Guane, *Britton* et al. 9756 (K, NY); Vinales, *Killip* 13544 (US); near Pinar del Río, *Rutten-Pekelharing* 456 (U); Herradura, *Baker & Dimmock* 4820 (A, MO, NY, S, US); El Azufre, Sierra del Rosario, Bro. *Alain* 762 (GH, US); Sierra de Anafe, *P. Wilson* & Bro. *Leon* 11533 (NY). Isle of Pines (= Isla de Juventud): Siguanea Bay, *Britton & P. Wilson* 14894 (NY, US). Habana: near Habana, *Curtiss* 704 (A, BH, BM, C, E, F, FI, G, GH, HBG, K, L, M, MO, NY, P, US, WAG); ibid, *De la Sagra* 275 (FI-W, G-DC, W, type of *T. berteri* var. *parviflora*); ibid., *J.B.Rodrigues* anno 1801 (MA); between Jaimanitas and Santa Fé, *Ekman* 13670 (AAU, G, MO, NY, S); Santiago de las Vegas, *van Hermann* 839 (BM, F, K, NY, P, US); near Guines, Bro. *Leon* 2496 (NY). Matanzas: Peninsula of Zapata, *Borhidi* et al. 7 May 1970 (BP); San Juan R. Valley, *Britton* et al. 308 (F, K, NY); Agricultural Station, *Baker* 3583 (C, NSW); Peninsula of Cabo Cruz, *Ekman* 18360 (S). Cienfuegas: Horpuitas, *Deschamps* NH 12537 (BR); Cieneguita, *Combs* 18 (F, GH, K, MO, NY, P, US); near Soledad, *Howard* 4997 (F, GH, MO, NY, U, UC, US); ibid., *Jack* 4844 (A, FHO, P, US). Villaciara: Sagua, *Britton & P. Wilson* 344 (NY); Serpentine Barrens, near Santa Clara, *Howard* 5053 (GH, MO, NY, U, US); Cerro Pelo Malo, *Borhidi* et al. 400 (BP). Sancti Spiritus: Trinidad Hills, *Jack* 7104 (A); S of Sancti Spiritus, Bro. *Alain* 1539 (GH, NY, US). Camaguey, near Camaguey, *Britton* et al. 13090 (K, NY); between Queen City and Riverside, *Shafer* 1154 (F, NY, US). Granma: Yara R., *Ekman* 5633 (S). Santiago de Cuba: Manacal, near Sevilla, *Ekman* 9422 (G, NY, S). E Cuba: sin. loc., *Wright* 395 (BR, G, GH, K, MO, NY, US). Sin. loc., *Wright* 2948 p.p. (NY; other sheets seen are *T. citrifolia*).

CULT. U.S.A., Florida, Miami, Plant Introduction Station, *Gillis* 10794 (BH, MO). Mexico, Tabasco, Mun. Cardenas, Botanical Garden of CSAT, *Romero* et al. 54 (MEXU); Quintana Roo, 8 km NW of Estero Franco, Cabrera et al. 16662 (MEXU). Honduras, Atlántida, Botanical Garden Lancetilla, 10 km SW of Tela, Argueta 128 (MO). Cuba, Limones, Harvard Botanical Garden, *Singleton* 447 (ECON).

Note. Elaborate comparative studies of all specimens examined resulted in reduction of some names to synonyms of *T. alba*. The species is not very variable.

57. Tabernaemontana albiflora (Miq.) Pulle, Enum. Pl. Surinam 382 (1906). – Type: Suriname, Upper Suriname R., Victoria, *Hostmann* and/or *Kappler* 1312 (holotype U; isotypes BM, C, F, FI, FI-W, G, GENT, K, MO, P, S, W; phot. of lost B sheet F, NY). Fig. 57, p. 238; map 29, p. 234

Basionym and other homotypic synonyms:

Peschiera albiflora Miq., Stirpes Surinamenses Selectae 165, pl. 47 (1850). *Echites albiflora* (Miq.) Miers, Apoc. S. Am. 204 (1878). *Taberna albiflora* (Miq.) Mgf. in Pulle, Fl. Surinam 4, 1: 455 (1937). *Bonafousia albiflora* (Miq.) Boiteau & Allorge in Bull. Soc. Bot. France 130, Lettres 4-5: 339 (1983); in Mém. Mus. natn. Hist. Nat. IIB, 30: 100, pl. 43 (1985).

Shrub or small tree 1-4 m high. Branches pale brown, lenticellate or not, with fissured bark; branchlets terete, glabrous or sometimes puberulous at the apex. *Leaves* sessile or shortly petiolate; petiole glabrous or sometimes puberulous when young, 1-3 mm long, seemingly up to 10 mm long as blade decurrent into petiole; (ocreae not widened into intrapetiolar stipules); blade papery or subcoriaceous when dried, elliptic or narrowly elliptic, 2-4 x as long as wide, 3.5-15 x 0.8-6 cm, acuminate or caudate at the apex, cuneate at the base and decurrent into the petiole, entire, glabrous on both sides or sometimes puberulous at the base when young, dull and with or without scattered black dots beneath; with 13-23 pairs of rather straight secondary veins forming an angle of 60-80° with the costa; tertiary venation reticulate, but inconspicuous. *Inflorescence* shortly pedunculate, 3-5 x 3-5 cm, lax, 1-5-flowered. Peduncle rather thin, glabrous, 1-5 mm long; pedicels 5-15 mm long, thin, glabrous. Bracts scale-like, about half as long as the sepals; bracteoles often present on pedicels. *Flowers* fragrant. *Sepals* pale green(?), connate at the base for 0.5 mm, erect, ovate or nearly so, 1.2-2 x as long as wide, 2-3 x 1-2 mm, obtuse, sparsely puberulous or glabrous outside, ciliolate or not, glabrous inside and with 2-5 colleters in 1 row in the middle at the base; colleters 0.7-0.8 x 0-15-0.3 mm. *Corolla* white with a yellow throat, (limb ca 3 cm in diameter), 24-35 mm long in the mature bud and forming a comparatively wide ellipsoid or narrowly ovoid head 0.22-0.33 x the bud length (9-16 x 4-6 mm) with an acute or obtuse apex, glabrous outside, pubescent only on filament ridges inside among anther tails; tube 8-13 x as long as the calyx, 0.9-1.3 x as long as the lobes, 20-27 mm long, narrowly cylindrical, 2-3 mm wide above the base, slightly narrowed or not below and above the anthers to 1.5-2 mm wide, slightly widened again around the anthers to 2-2.5 mm wide, about 3 mm wide at the throat, not twisted; lobes obliquely obovate or nearly so, 0.8-1.1 x as long as the tube, 1.1-1.5 x as long as wide, 20-27 x 12-21 mm, rounded(?), not undulate, spreading. *Stamens* with apex 8-13.5 mm below mouth of corolla tube, inserted 0.4-0.5 of the length of the corolla tube, (at 9-11 mm from the base); anthers sessile, with tails 1 mm below insertion, narrowly triangular to narrowly oblong, 3.5-4 x 1 mm, apex acuminate, sterile for 0.3-0.5 mm, sagittate at the base, glabrous. *Pistil* glabrous, 10-12 mm long; ovary ovoid, 1.5-2 x 1-1.3 x 1-1.3 mm; style filiform, 7.5-9 mm long, not shed with the corolla; pistil head 1 mm high, composed of an entire flat or recurved ring 0.1-0.2 x 0.7-0.8 mm, a stipitate 5-lobed depressed globe 0.2-0.3 x 0.6 mm and a stigmoid apical part 0.1 x 0.1 mm. Ovules approximately 50 in each cell. *Fruit* of 2 separate mericarps; mericarps green(?), not dotted, pod-like, 25-40 x 6-12 x 6-10 mm, not recurved, acuminate with an often upcurved apex, with a lateral ridge at each side, minutely papillose outside, approximately 5-10-seeded; wall thin; aril not enveloping the seed. *Seed* medium brown, obliquely ellipsoid, 9-10 x 3-3.5 x 2.5-3 mm, with longitudinal grooves, dull, papillose.

DISTRIBUTION: Northern Brazil and the three Guianas.

ECOLOGY: Forest understorey, near river banks, at low altitude. Flowering mainly from August to October and fruiting probably January-February.

Geographical selection of the approximately 30 specimens examined:

SURINAME. Saramacca R. Headwaters, near Krappa Camp, *Maguire* 24888 (A, BO, F, G, K, MO, NY, P, S, U, UC, US, W); ibid., Grasi Falls, *Maguire* 24943 (A, F, G, K, MO, NY, P, U, UC, US); Suriname R., near Victoria, *Hostmann* & *Kappler* 1312 (BM, C, F, FI, FI-W, G, GENT, K, MO, P, S, U, W; phot of lost B sheet US, type); Sara Creek, S of Dam, *Florschütz* 193 (U); Lawa R. bank, opposite Stoelmans Island, *Heyde* 491 (U); Gonini R., *Versteeg* 212 (L, U); sin. loc., Hostmann 160 (BM, CGE, FI-W, G, K, P, W).

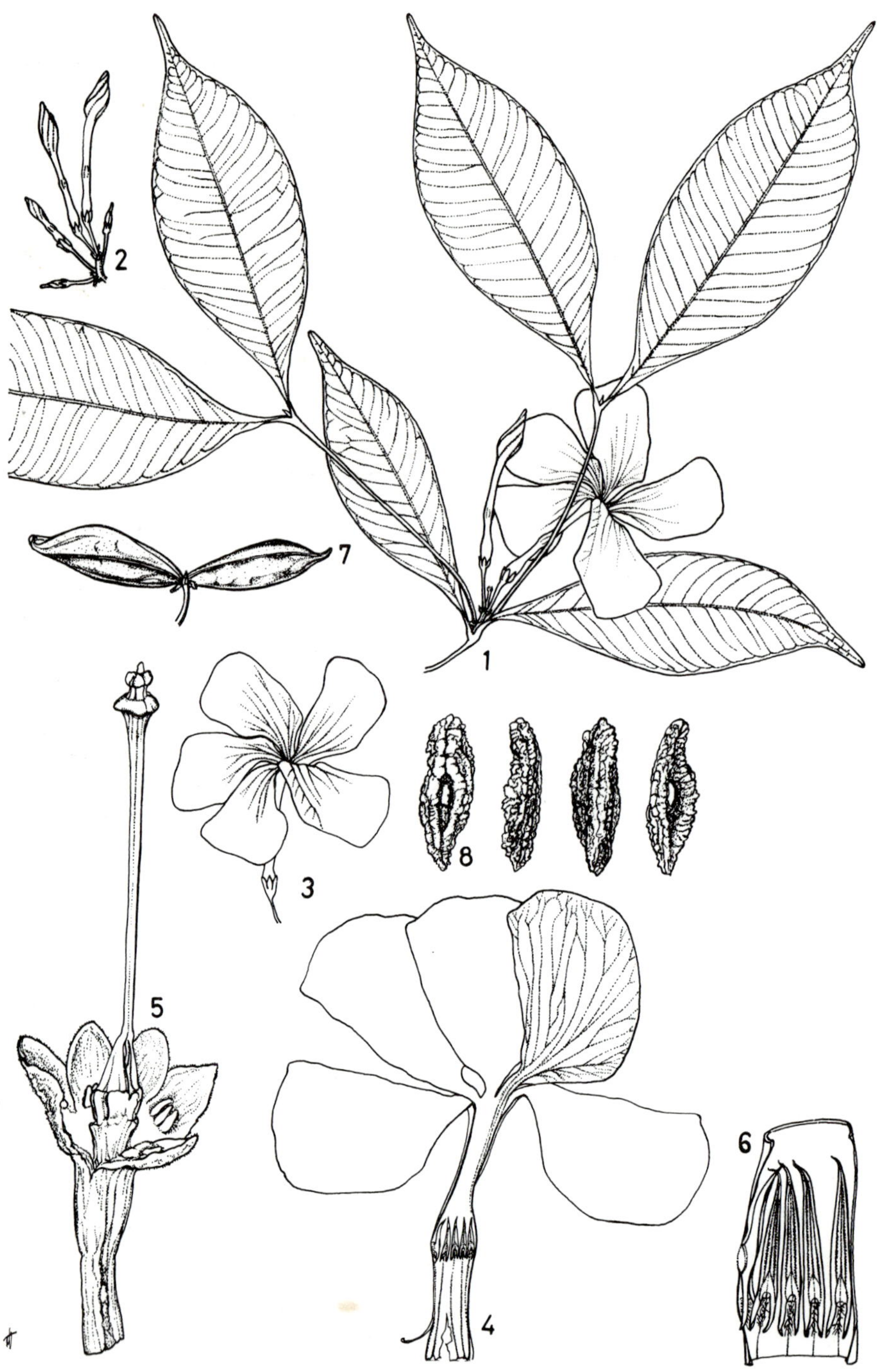

Fig. 57. *Tabernaemontana albiflora*. **1,** habit (2/3); **2,** flower buds (x 2/3); **3,** flower (x 2/3); **4,** opened corolla (x 1.6); **5,** calyx with pistil (x 6); **6,** stamens (x 6); **7,** fruit (x 2/3); **8,** seeds (x 2). 1-8 from Jacquemin 2112; 2 also from Cid et al. 1146.

FRENCH GUIANA. Maroni R., island of Abattis-Cotica Falls, *Sastre* 6458 (CAY, P); Grand Inini R. right bank, near Bicade, *de Granville* 580 (CAY, P, U); between Macaque Fall and Carbet brûlé, Petit Guaqui, *Schnell* 12011 (CAY, NY, P, U); Lower Oyapock R., 1.5 km upstream from Miriaflor, *Oldeman* B 3346 (CAY, P); Oyapock R., Maripa Fall, *Jacquemin* 1881 (CAY, P, WAG), 2112 (CAY, P, U).

BRAZIL. Pará: Mun. Oriximiná, Rio Paru do Oeste, near Raimunda, *Cid* et al. 2003 (GH, NY, US, WAG); Rio Paru do Oeste, Pancada Fall, Cid et al. 2041 (NY, US, WAG); Sete Varas airstrip on Rio Curua, *Strudwick* et al. 4021 (F, GH, K, MO, US, USF), 4444 (K, MO, US, WAG); Mun. Oriximiná, Rio Mapuera, *Cid & Ramos* 1146 (NY, US, WAG).

Note. *T. albiflora* is easily confused with *T. lorifera* and *T. rupicola*, see under *T. lorifera*.

58. Tabernaemontana amplifolia Allorge in Bull. Soc. Bot. France 130, Lettres Bot. 4-5: 344, fig. 5 (1983); Mém. Mus. natn. Hist. Nat. IIB. 30: 12, pl. 1 (1985), partly, as for collections from Colombia. – Type: Colombia, Antioquia, Planta Providencia, 35 km SW of Zaragoza, *Denslow* 2120 (holotype COL, not seen; isotypes: MO, WIS). Fig. 58, p. 240; map 29, p. 234

Small tree 3-18 m high. Trunk up to 15 cm or more in diameter. Branches pale grey-brown, lenticellate; branchlets terete, glabrous. *Leaves* petiolate; petiole glabrous, 5-50 mm long; (ocreae widened into intrapetiolar stipules); blade membranaceous or thinly papery when dried, elliptic or narrowly elliptic, 1.8-3.5 x as long as wide, 8-40 x 3.5-18 cm, acuminate or apiculate at the apex, cuneate at the base, entire, glabrous on both sides, irregularly dotted beneath, with 12-20 pairs of rather straight secondary veins forming an angle of 60-70° with the costa; tertiary venation reticulate on both sides and impressed above. *Inflorescence* pedunculate, 10-15 x 7-20 cm, many-flowered, rather lax. Peduncle robust, 25-80 mm long, glabrous; pedicels slender, 10-25 mm long, glabrous. Bracts and bracteoles sepal-like, 0.3-0.5 x as long. *Flowers* open during the day. *Sepals* pale green (?), connate at the base for 0.3 mm, erect, ovate or broadly ovate, 1.2-2.4 x as long as wide, 2.5-6 x 2-2.5 mm, rounded or obtuse, glabrous outside, not ciliate, glabrous inside and with 4-10 colleters in 1 row in the middle at the base; colleters 0.3-0.5 x 0.15 mm. *Corolla* white, 25-34 mm long in the mature bud and forming a comparatively large narrowly ovoid head 0.4-0.52 of the bud length, about 2.3 x as wide as the tube (10-16 x 5-7 mm) with an acute apex, glabrous outside, pubescent inside in a belt 4 mm wide just below the anther tails; tube 2.5-6 x as long as the calyx, 0.75-1 x as long as the lobes, 15-19 mm long, almost cylindrical, 2-3 mm wide at the base, slightly narrowed below the insertion of the stamens to 1.5 mm wide, again widened around the anthers to 2-3 mm wide, not twisted; lobes obliquely dolabriform, 1-1.3 x as long as the tube, 1.3-1.7 x as long as wide, 16-20 x 12-15 mm, obtuse, with an acute lateral lobe, neither auriculate nor undulate, spreading. *Stamens* with apex 2.5 mm above mouth of corolla tube, inserted 0.9-0.95 of the length of the corolla tube, (at 14-18 mm from the base), 1 mm below the mouth; anthers sessile, with tails 0.5 mm below insertion, narrowly triangular, 4 x 1.2 mm, apex acuminate, sterile for 0.3-0.5 mm, sagittate at the base, with tails curved towards each other. *Pistil* glabrous, 15-19 mm long; ovary ovoid, 2-3 x 1-1.2 x 1-1.2 mm; style slender, 9.8-15.8 mm long; pistil head 1.3 mm high, composed of a basal ring 0.2 x 1 mm, a stipitate 5-lobed depressed globe, 0.3 x 0.5 mm and a stigmoid apical part, 0.2 x 0.2 mm. Ovules approximately 80 in each carpel. *Fruit* of 2 separate mericarps; mericarps green, obliquely ellipsoid, 40-70 x 20-30 x 20-25 mm, recurved, caudate at the apex, tail 10-20 mm long, with a ridge at each side, smooth, often torulose when dried, approximately 10-30-seeded; wall 1-2 mm

Fig. 58. *Tabernaemontana amplifolia*. **1,** habit (x 2/3); **2,** opened corolla (x 2); **3,** calyx with pistil (x 2); **4,** sepal inside (x 6); **5,** fruit (x 2/3). 1-4 from Denslow 2120; 5 from Shepherd 313.

thick in dried fruits; aril orange or red, enveloping the seed. *Seed* dark brown, obliquely ellipsoid, 14-15 x 5-6 x 5-6 mm, with longitudinal grooves, dull, papillose; embryo 9-12.5 mm long; cotyledons ovate or oblong, 1.5-2.7 x as long as wide, 4-5.5 x 2-3 mm, rounded at the apex, cordate at the base; rootlet 5.5-6.5 x 0.7-0.8 mm.

DISTRIBUTION: Colombia (Antioquia) and Ecuador.
ECOLOGY: Wet forest, often secondary. Alt. 20-750 m. Flowering and fruiting probably throughout the year.

Specimens examined:
COLOMBIA. Antioquia: Mun. Tarazá, El Doce, *Callejas* et al. 2438 (WAG); Río Cauca, between El Doce and Barro Blanco, *Daly* et al. 5230 (K, NY, WAG); Mun. Turbo, Currulao, *Callejas* et al. 4928 (WAG); 19 km NE of Turbo, *Callejas* 4955 (MO, WAG); Turbo-Chigorodó Road, *Feddema* 2013 (S, US); Mun. Anorí, Dos Bocas-Providencia-Toma de Aljibes Road, *Callejas* et al. 4591 (MO, WAG); between Dos Bocas and Anorí, *Zarucchi* 3275 (WAG); between Providencia and Q. La Tirana, *Alverson* et al. 169 (COL, MO, NY); near Planta Providencia, *Denslow* 2120 (COL, not seen, MO, WIS, type); ibid., *Shepherd* 313 (COL, MO, WIS); ibid., *Soejarto* 3293 (MO), 4418 (F); near Segovia, *Rentería* et al. 1660 (COL, MO); 14-17 km NW of Remedios, *Callejas* et al. 5166 (K, MO, WAG); 4 km SSW of Otú, *Thomas & Castano* 5488 (WAG); Mun. San Luís, Parque Ecológico, Cañón of Río Claro, *Cogollo & Borja* 457 (MO), *Cogollo* 613 (COL, MO), 884 (MO), 910 (COL, MO), 1559 (MO), 1708 (MO), 1974 (MO); San Luís, *Rentería & Cogollo* 2644 (MO); Río Claro, road to Cairo, *Rentería & Cogollo* 2737 (MO).
ECUADOR. Cotopaxi: Quevedo-Latacunga Road, *Holm-Nielsen* et al. 2980 (F, GB, U).

Note. *Tabernaemontana amplifolia* is easily confused with the widespread *T. amygdalifolia*, which has rarely more than 10 flowers per inflorescence and smaller or more slender fruits not caudate, but acuminate.

59. Tabernaemontana amygdalifolia Jacq., Enum. Pl. Carib. 14 (1760); Select. Stirp. amer. hist. 39, t. 181. 15 (1763). – Type: Cult., Austria, Wien, Hort. Schönbrunn, Herb. Jacquin s.n. (holotype W). Fig. 59, p. 243; map 30, p. 248

Heterotypic synonyms:

Cestrum nervosum Mill., Gard. Dict. ed. 8. n. 3 (1768), non *T. nervosa* Desf. ex Poir.(1817), nec Glaziou (1910). – Type: Colombia, Bolívar, Cartagena, herb. *Miller* s.n. (holotype BM; phot. US).
T. nereifolia Vahl, Eclog. Am. 2: 21 (1807). – Type: sin. loc.,"Porto Rico", *von Rohr* 76 1/2 (holotype C-VAHL; isotype BM; phot. of holotype A).
T. jasminoides H.B.K., Nov. Gen. 3: 225 (1819). *Malouetia jasminoides* (H.B.K.) A. DC., Prod. 8: 379 (1844); Miers, Apoc. S. Am. 92 (1878), superfluous combination. *Rauvolfia laevigata* Willd. ex Roem. & Schult., Syst. Veg. 4: 805 (1819). – Type: Colombia, Bolívar, Bordones, Cumana, von *Humboldt & Bonpland* s.n. (holotype P-BO; isotype B-W 5095 as *von Humboldt* 102; phot. of holotype F, GH, US). The B-W 5095 is the holotype of the last name.
T. amygdalifolia var. *obtusiloba* A. DC., Prod. 8: 367 (1844). – Type: Colombia, Magdalena,, Santa Marta, *Bertero* anno 1822 (holotype G-DC).
T. acapulcensis Miers, Apoc. S. Am. 57 (1878). – Type: Panama, Veraguas, Tolé, *Seemann* 1221 (lectotype BM; isolectotype K, designated here).
T. occidentalis Miers, op. cit. 58. – Type: Peru, sin. loc., *Mclean* s.n. (holotype BM; isotype K).

T. dichotoma Sessé & Mociño, Fl.Mexic. ed. 1: 47 (1893), not traced; ed. 2: 43 (1894), non Roxb. ex Wall.(1829). – Type: Mexico, sin. loc., *Sessé* et al. 5061 (lectotype MA, not seen; phot. WAG, designated here).
T. deamii Donn. Sm. in Bot. Gaz. 52: 50 (1911). – Type: Guatemala, Zacapa, Montagua R., near Gualán, *Deam* 6282 (holotype US; isotype GH; phot. of holotype WAG).
T. amygdalifolia var. *glaucophylla* Allorge in Mém. Mus. natn. Hist. Nat. IIB. 30: 14 (1985). **syn. nov.** – Type: Panama, Panamá, km 12-14 El Llano-Carti Highway, *Dressler* 4356 (holotype MO).

Shrub or small tree 1-15 m high. Trunk 1-25 cm in diameter; bark light brown, fissured. Branches pale or dark brown, lenticellate; branchlets terete, glabrous. *Leaves* of a pair equal or unequal (larger up to 3 x as long as other and not comparatively narrower), petiolate; petiole glabrous, 2-25 mm long; (ocreae not or obscurely widened into intrapetiolar stipules); blade firmly membranaceous when fresh, thinly papery and often with pale venation on both sides when dried, elliptic, narrowly elliptic, or less often narrowly obovate, 2.5-4.5 x as long as wide, 3-26 x 1-9 cm, acuminate with a usually acute acumen, or in some leaves obtuse, cuneate at the base, with a mostly flat margin, entire or sometimes slightly undulate and/or sinuate, glabrous on both sides, often minutely pale-dotted above and with scattered black dots beneath, flat or sometimes with impressed reticulate tertiary venation above, with 6-12 pairs of upcurved secondary veins forming an angle of about 70-80° with the costa; tertiary venation reticulate, conspicous or not. *Inflorescence* pedunculate, 2-10 x 2-7 cm, 2-10(-40)-flowered, mostly lax. Peduncle mostly short, 1-30 mm long, glabrous; pedicels glabrous, 4-20 mm long. Bracts sepal-like, about 0.3 x as long as the sepals. *Flowers* fragrant, open during the day, sometimes present on leafless trees. *Sepals* pale green(?), connate at the extreme base, ovate or broadly ovate, 1-2 x as long as wide, 2-4 x 1.5-2.5 mm, obtuse or rounded, glabrous outside, not ciliate, glabrous inside and with 3-8 colleters in 1 row in the middle at the base; colleters often unequal, 0.3-0.7 x 0.1-0.3 mm. *Corolla* white or creamy, with yellow throat, greenish tube, and often with brown edges on lobes, thin, (12-)15-27 mm long in the mature bud and forming a comparatively large ovoid head 0.28-0.48 x the bud length (5-11 x 3-7 mm) with an acute or subacute (rarely blunt) apex, glabrous outside, entirely glabrous inside or with a villose or pilose belt 1.5-4 mm wide with recurved hairs below the anthers, or sometimes minutely pubescent from the insertion of the stamens to the base of the lobes; tube 2.5-5.5 x as long as the calyx, 0.7-1.05 x as long as the lobes, (7-)9.5-18 mm long, almost cylindrical, 2-4 mm wide above the base, narrowed above to 1.3-2 mm wide, widened again around the anthers (throat here) to 2-3 mm wide, not twisted, with a more or less lobed callose ring in the throat; lobes obliquely elliptic to almost dolabriform, 0.9-1.5 x as long as the tube, 1.5-3 x as long as wide, 9-20 x 5-9 mm, obtuse or rounded, with an acute or obtuse lateral lobe or not, slightly auriculate at the left side of the base, undulate or not, spreading and often recurved later. *Stamens* with apex 1.5-2.5 mm above mouth of corolla tube, inserted 0.8-0.9 of the length of the corolla tube, at (6-)8-16 mm from the base); anthers often blue-green, sessile, narrowly triangular, 2.5-3.3 x as long as wide, 2.5-4 x 1-1.5 mm, apex acuminate, sterile for 0.3-0.5 mm, sagittate at the base (with tails curved towards each other), glabrous. *Pistil* glabrous, (7-)9.5-15.5 mm long, with apex almost halfway along anthers and 0.5 mm above mouth of corolla tube; ovary ovoid or oblong, 1.5-3 x 1-1.5 x 1-1.5 mm, gradually narrowed into the style, of 2 separate carpels; style (4-)6-11 mm long, rather slender; pistil head 1.2-1.5 mm high, composed of an undulate basal ring like a veil, 0.2 x 1.2-1.5 mm, an apically 5-lobed stipitate head 0.8-1.1 x 0.6-1 mm, and a bilobed stigmoid apex 0.1-0.2 x 0.1-0.2 mm. Ovules

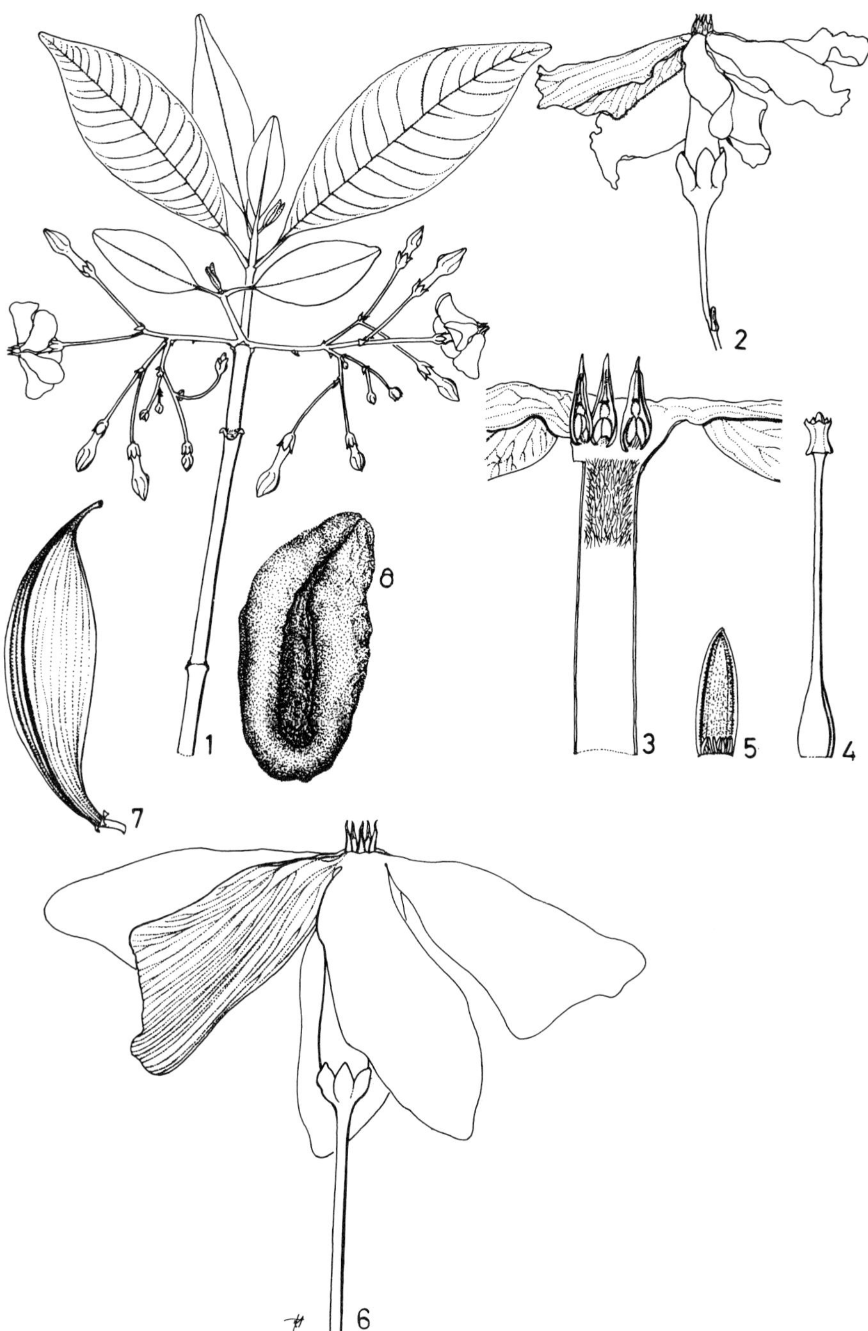

Fig. 59. *Tabernaemontana amygdalifolia*. **1,** habit (x 2/3); **2,** flower (x 1.6); **3,** opened corolla (x 4); **4,** pistil (x 4); **5,** sepal inside (x 4); **6,** flower (x 2); **7,** fruit (x 2/3); **8,** seed (x 6). 1-5 from Gaumer 555; 6 from C.L.Smith 125; 7-8 from Palmer 1535.

approximately 50-120 in each carpel. *Fruit* of 2 separate mericarps; mericarps orange, yellow or green, yellow or orange inside, not dotted, pod-like, 35-75 x 10-25 x 10-20 mm, acuminate at the apex (acumen 5-10 mm long), often slightly stipitate, not ridged, approximately 10-40-seeded; wall about 2 mm thick in dried fruits; aril orange or white, enveloping the seed. *Seed* dark brown, obliquely ellipsoid, 7 x 3-4.5 x 3-4 mm, with longitudinal grooves, papillose; embryo 5.8-6 mm long; cotyledons ovate, 2.8-3 x 2 mm, obtuse at the apex, truncate at the base; rootlet 3.5 x 0.7 mm.

DISTRIBUTION: from Mexico to Venezuela and Peru.

ECOLOGY: Forest or thickets. Alt. 0-1200 m. Flowering in northern Mexico (Sinaloa) May-July, further south flowering mainly March-June. The development of the fruits may take a whole year. In South America flowering and fruiting throughout the year.

Geographical selection of the approximately 510 specimens examined:

MEXICO. Sinaloa: Badiraguato, *Ortega* 1648 (MEXU); Culiacán, *Palmer* 1535 (GH, NY, US); km 33 Culiacán-Sanalona Road, *Vega Aviña* 748 (MEXU); San Juan, *Ortega* 4120 (US); Mazatlán, *Ortega* 7217 (BR, F, G, NSW); between Rosario and Concepción, *Rose* 3272 (US). Nayarit: Isla Ma.Magdalena, *Chiang & Flores* 1086 (MO); near San Blas, *Ferris* 5317 (A, F, US); Mazatán-Las Varas Road, *H.S. Gentry & Gilly* 10726 (LL); Acaponeta, Tepic, *Rose* 1480 (US). Jalisco: Río Cuale, Puerto Vallarta, *Carter & Chisaki* 1186 (UC); Ejido La Fortuna, Mun. La Huerta, *G.Ayala* 718 (LL); Biological Station Chamela, *Lott & Telez* 1131 (USF, WAG); ibid., *Magallanes* 4378 (MO); Tuxpan, Mexia 1039 (A, BM, F, G, GH, MO, NY, UC, US). Mexico: Nanchititla, District Temascaltepec, *Hinton* 7607 (K, LL, P, UC, US). Michoacan: Cuatro Caminos-Playa Azul Road, *Soto Nuñez & G. Silva* 4507 (MEXU); La Majada, Mun. Apatzingan, *Leavenworth & Hoogstraal* 1340 (F, GH, MO, NY). Border Michoacan and Guerrero: La Orilla, *Langlassé* 164 (F, G, GH, K, P). Guerrero: Petatlán, Montes de Oca, *Hinton* et al. 10324 (F, GH, K, LL, NY, US); 17 km NE of Zihuatanejo, *Tenorio* et al. 394 (MO); Galeana, *Hinton* 10993 (K, LL, NY, US, W); Acapulco, *Palmer* 380 (GH, US); ibid., *J.D. Rodriguez* MA 216258 (MA). Oaxaca: 8 km E of Pinotepa Nacional, *Tenorio & Torres* 205 (MEXU); Huiltepec, *Nelson* 2581 (GH, NY, P, US); W of Oaxaca, *Jürgensen* 137 (G); Chacahua National Park, *Cedillo & Torres* 1506 (LL, MEXU, MO); 8 km E of Temascal, *Janzen* 26 May 1964 (LL, UC); km 83 Tehuantepec-Ocozocantla Highway, Río Guamol, *Carlson* 1505 (F, UC); 25 km W of Puerto Angel, along road to Puerto Escondido, *Breckon* 768 (WAG, WIS); Paso Limón, Río Magdalena, Pochutla District, *Conzatti* et al. 3142 (MO, US); Salina Cruz, *Orcutt* 3290 (F, GH, K, MO, US); Yaultepec, *MacDougal* 7 May 1970 (NY). Veracruz: near Veracruz, *Galeotti* 1577 (K); Cuitláhuac, *Matuda* 1433 (A, MEXU, MO); Hueyapan de Ocampo, *Gómez-Pompa* 105 (GH, MEXU); Hidalgotitlan, *Juan* 136 (F, MEXU). Tabasco: Lázaro Cárdenas, Mun. Tacotalpa, *Cowan* et al. 2225 (LL, MEXU). Chiapas: W of Tuxtla, Gutierrez, *Chavelas* et al. 1395 (MEXU); near Cupía, SE of Tuxtla Gutierrez, *Miranda* 6052 (MEXU); San Cristobal, *Domingo* 3424 (WAG); Crucero Corozal, Palenque-Lacantum Road, *Martínez* 17812 (MEXU); NE of Bonampak, *Hoover* 228 (GH, MO, US); Tonalá, *Matuda* 17149 p.p. (K); 3 km E of Tres Picos, *Davis* 20 Mar. 1954 (LL); Acapetagua, *Matuda* 2735 (MO); Huixtla, *Matuda* 17609 (F, K, MEXU, NY). Campeche: Campeche, *Lundell* 1400 (F, GH, NY); 3 km N of Champoton, *Cabrera* 2744 (LL); 48 km N of Palenque, *Vaughan* et al. 239 (MO, Z). Yucatan: Mérida, *Schott* 432 (BM, F, MO, US, paratype of *T. acapulcensis*); 10 km N of Muna, *Butterwick* 181 (LL); Maxcanu, *Gaumer & sons* 23229 (ECON, F, G, GH, K, MO, NY, US); Dzilam, *Enriquez* 490 (MEXU, US); SE of Kancabdzonot, *Gaumer & sons* 23633 (A, BM, F, G, K, UPS, US); Izamal, *Greenman* 429 (F, GH, NY); between

Mérida and Chichen-Itza, *Utley* 11-27 (USF); 1 km SW of Yaxcabá, *Vara* & *Arias* 304 (MEXU); Chichen-Itza, *Lundell* 7312 (LL, MO); Valladolid, *Steere* 1668 (MO); Chichancanab, *Gaumer* 1844 (F, GH, MA, MO, US), 1845 (A, BM, C, F, K, S, UPS, WIS); sin. loc., *Gaumer* 555 (A, BM, BR, C, E, F, FHO, G, GH, K, NY, S, UC, US, W, WIS). Quintana Roo: Cozumel Island, *Jéllez* & *Cabrera* 1924 (BM, MEXU, MO); ibid., *Gaumer* 42 (F, GH, K, NY). Sin. loc., *Sessé* et al. 5061 (MA, not seen; phot. WAG, lectotype of *T. dichotoma* Sessé & Mociño, non Roxb. ex Wall.).

BELIZE. Stann Creek: Stann Creek, Middlesex, *Schipp* 423 (A, BM, F, G, GH, K, MO, S, UC, Z); Big Creek, *Schipp* 943 (A, BM, F, G, GH, K, MO, NY, S, UC, Z). Cayo: S of San Luis, *Dwyer* et al. 404 (A, MO). Toledo: Maya Mts., *Davidse* & *Brant* 32206 (WAG).

GUATEMALA. Peten: Lacandón, *Contreras* 3496 (F, LL); near Chinajá, *Steyermark* 45484 (F, MO); Dolores, *Contreras* 2438 (LL); Canchacan, *Lundell* 16344 (LL); Seamay, *Contreras* 6667 (F, LL). Izabal: El Estor, *Contreras* 11352 (LL, S); Between Virginia and Lago Izabal, *Steyermark* 38715 (F); Cerro San Gil, *Steyermark* 41759 (F, MO, US); Bobos District, *Record* G.17 (FHO, US). Alta Verapaz: Cubilhuitz, von *Tuerckheim* II 331 (BR, E, S, W); Trece Aguas, *Lewton* 308 (F, US). Baja Verapaz: Niño Perdido, *Lundell* & *Contreras* 20976 (LL). Chimaltenango: Volcán Fuego, *Steyermark* 52086 (F). Quezaltenango: Volcán Santa María, *Steyermark* 33506 (F). San Marcos: near Rodeo, *Standley* 68941 (F); 2 km from Finca Armenia, *Croat* 40834 (BM, MO, WAG). El Progreso: near Jícaro, *Steyermark* 43858 (F, MO, US). Zacapa: near Gualán, *Deam* 6282 (GH, US; phot. of US sheet WAG, type of *T. deamii*); Zacapa, *Deam* 158 (MO, NY, US). Chiquimula: near Chiquimula, *Standley* 71937 (F). Santa Rosa: S of Los Cerritos, *Standley* 79600 (F, MO); Sambrerito, *Heyde* & *Lux* 6344 (BM, GH, K, M, US).

EL SALVADOR. Ahuachapan: La Barra de Santiago, *De Riemer* 22 Apr. 1979 ((US). Sonsonate: near Santa Emilia, *Standley* 22131 (GH, MO, NY, US). Santa Ana: near Metapán, *Carlson* 825 (F). Chalatenango: Cantón, *Montalvo* 4643 (UC). La Unión, Laguna Maquigüe, *Standley* 20950 (GH, NY, US).

HONDURAS. Cortés: Cuyamel, *Carleton* 451 (NY, US). Atlántida: Pico Bonito, *Blackmore* & *Chorley* 4180 (BM); Coyoles, *Molina* & *Becker* 17 (F). Yoro: 8 km SW of La Florida, *Hazlett* 3114 (F, MO); between Yoro and Morazán, *Molina* 6916 (F). Comayagua: Lake Yojoa, near Pito Solo, *L.O. Williams* & *Molina* 17737 (F, US); 5 km S of La Mision, *L.O. Williams* & *Molina* 18001 (F, GH, US); near Comayagua, *Standley* & *Chacón* 5244 (F, MO). Colón: near Aguan R., *Record* & *Kuylen* H.50 (GH, US). Gracias a Dios: 3 km NW of Bulebar, *Saunders* 1133 (F, Z); Wispernini Camp, Guarunta Region, *von Hagen* 1389 (F, NY). Olancho: near Catacamas, Molina 8335 (F). El Paraíso: Valle Jamastrán, *Molina* 13770 (F, NY, US). Valle: El Coyolito, Fonseco Gulf, *Molina* 25969 (F, NY).

NICARAGUA. Nueva Segovia: 5 km from Jalapa, *Neill* 1649 (MO, WAG); 4 km NE of El Jícaro, *Araquistain* & *Moreno* 2207 (MO, U, WAG). Madriz: 20 km SE of Somoto, *Harmon* & *Fuentes* 6007 (MO). Esteli: road to Laguna de Miraflores, *Moreno* 8293 (MO, WAG). Matagalpa: Yasica R., 26 km S of Matagalpa, *Molina* 20484 (F, NY, US). León: km 37 Nueva-León Road, *Sandino* 2630 (MO, WAG); Río Sinecapa, *Stevens* 3846 (MO, U, WAG); Momotombo, *C.L. Smith* 125 (BH, BM, F, GH, MEXU, MO, NY, UC, US, WIS); NW of El Transito, *Stevens* et al. 20134 (MO, WAG). Managua: near Peninsula of Chiltepec, *Stevens* 3544 (MO); ca 15 km E of Managua, *Grijalva* 2447 (MO, WAG); near Managua, *Maxon* et al. 7547 (C, US). Masaya: Volcán Masaya National Park, *Stevens* 4317 (MO, WAG). Granada: near Granada, *Lévy* 44 (C, G, P, WAG); Casa Tejas Road, near Finca San José del Mombacho, *Moreno* 16649 (MO, WAG). Carazo: Refugio de Vida Silvestre Chococente, *Sandino* 4964 (MO, WAG). Rivas: Isla Ometepe, *Robleto* 1466 (MO,

WAG). Chontales: near Juigalpa, *Standley* 9188 (F); Santo Tomas, *Seymour* 6320 (MO). Boaco: Hacienda San Antonio, near Boaco, *Soza & Moreno* 196 (MO, WAG); 3 km N of Tecolostote, *Stevens* 22928 (MO, WAG).

COSTA RICA. Puntarenas: Monteverde, *Wibur* 14219 (GH, LL, US); near San Mateo, *Tonduz* and/or *Pittier* 4057 (BR, G, US). Alajuela: San Ramón, *Brenes* anno 1920 (NY). Heredia: Tirimbina, *Proctor* 32304 (LL, MO); Braulio Carillo National Park, *Gómez* et al. 22904 (F, MO, WAG). Limón: Llanuras de Santa Clara, Río Blanco, *Pittier* 13457 (G, US, WAG); Cerro Bitárkara, *Hammel* et al. 17676 (MO).

PANAMA. Bocas del Toro: Changuinola Valley, *Dunlap* 487 (MO, US); Water Valley, *von Wedel* 797 (GH, MO). Chiriqui: near David, *Pittier* 2825 (NY, US). Veraguas: Tolé, *Seemann* 1221 (BM, K, lectotype of *T. acapulcensis*). Coclé: 7 km N of Llano Grande on road to Coclesito, *Hammel* 2509 (MO, WAG); NW of Penónome, *Sytsma* 3897 (MO, WAG). Colón: Santa Rita Ridge, *Porter* et al. 4810 (AAU, GH, MO). Panamá: Cerro Jefe, *Dressler* 3752 (AAU, GH, MO); road from El Llano to Carti-Tupile, *Kennedy & Dressler* 2932 (F, MO, US, USF, WAG); El Llano-Carti Highway, 12-14 km N of El Llano, *Dressler* 4356 (MO, type of *T. amygdalifolia* var. *glaucophylla*); Serranía de Majé, *Huft* et al. 1697 (MO, WAG). Darién: Cerro Pirre, *Bristan* 475 (MO).

PANAMA or COLOMBIA: North coast, *Cuming* 1295 (BM, CGE, E, K, paratype of *T. acapulcensis*).

COLOMBIA. Chocó: 2 km S of Río Baudó, near Estero del Medio, *Fuchs* et al. 21943 (COL, G, K, MO, U, US). Antioquia: 12-16 km NW of Anorí, *Callejas* et al. 8772 (WAG). Valle: Río Calima, La Esperanza, *Cuatrecasas* 16742 (F, MO, US); Margarita Island, *O.O. Miller & J.R. Johnston* 90 (BM, F, GH, K, MO, NY, P, US); Río Digua, Piedra de Moler, *Cuatrecasas* 15056 (MO, US); Río Yurumanguí, El Aguacate, *Cuatrecasas* 16007 (MO). Nariño: Tumaco, *Romero-Castañeda* 5577 (AAU, MO); between El Diviso and Tumaco, *Vogel* 18 (MJG). Tolima: Honda, *Lehmann* 8512 (K). Sucre: Colosó, *Barbosa* 1474 (MO). Atlántico: between Baranoa and Polonuevo, *Dugand & Jaramillo* 2815 (COL, US); near Barranquilla, *McKee* 10459 (K, NSW, P). Bolívar: Turbaco, *Pennell* 4760 (GH, MO, NY, US); near Bordones, Cumana, von *Humboldt & Bonpland* s.n. (B-W 5095, P-BO, type of *T. jasminoides* and *Rauvolfia laevigata*); Cartagena, herb. *Miller* s.n. (BM, type of *Cestrum nervosum*); ibid., *von Humboldt* 1393 (B-W 5194); near Beltrán, Mun. Corozal, *Romero-Castañeda* 9884 (COL). Magdalena: Santa Marta, *Bertero* anno 1822 (G-DC, type of *T. amygdalifolia* var. *obtusiloba*); ibid., *Schlim* 945 (BR, F, G, K, P); ibid., near Mamatoca, *H.H. Smith* 1653 (A, BM, BR, COL, E, F, FI, G, GH, K, L, LL, MO, NY, P, S, U, UC, US, WIS); Caracolicito, Valle de Upar, *Romero-Castañeda* 6 (COL, MO); between Gaira and Santa Marta, *Fernández-Pérez* 5275 (COL, ECON); Tucurinca, *Romero-Castañeda* 395 (COL). Cesar: near Codazzi, *Haught* 2315 (COL, MO, US); Valledupar, *Hanbury-Tracy* 299 (K); Río Maracas, Becerril, *Haught* 3663 (COL, MO, US). Guajira: Serranía La Macuira, *Saravia* 3536 (COL, NY, US); Arroyo Tabaco, *Bunch* 14 Jul. 1980 (MO); Papayal, *Haught* 6470 (COL, MO, US).

VENEZUELA. Apure: between El Samán and El Mantecal, *Castroviejo & Ginés López* 198 (MA). Barinas: Caparo F.R., N of Uribante and Apure Rs., *Marcano-Berti* 2886 (MO); La Soledad, Mun. Santa Luciá, *Bernardi* 1201 (K, NY). Zulia: Maracaibo, La Ceiba, *Moritz* 357 (BM, HBG, K); Arriaga, near Maracaibo, *Pittier* 10492 (G, GH, NY, P, US). Lara: El Tocuyo, *Tamayo* 313 (US); Falcón: 7 km N of Cerro Socopo, Flora Falcón 476 (U); Golfete de Guare, *Steyermark & Manara* 110881 (COL, MO, NY, U, US); near Buruica, *Agostini* 1025 (GB, MO, NY, U, US, Z); Cerro Mampostal, *Gonzáles* 1075 (K, MO); 1 km S of Siburuá, *Wingfield* 5339 (MO, U). Carabobo: Punta Palmita, Valencia Lake, *Bunting* 3433B (NY). Aragua: 4 km E of Las Tejerias, *Morillo & Nilsson* 9651 (S, WAG); Maracay, *Vogel* 398 (HBG,

M, S, US); El Limón, *Ll. Williams* 10339 (US, paratype of *Peschiera laevifructa*); near Colonia Tovar, 40 km W of Caracas, *Fendler* 1029 (GH, GOET, K, MO), 2375 (GH, paratype of *P. laevifructa*). D.F.: Caracas: *Funck* 448 (G, P); ibid., *Vargas* 17 (FI-W, G-DC). Miranda: Río Chico, *Jahn* 1227 (GH, US). Anzoategui: near Barcelona, *Chaper* anno 1885 (P). Nueva Esparta: Salamanca, *Gines* 3686 (US); Margarita Island, *J.R. Johnston* 68 (C, F, G, GH, K, NSW, NY, S, UC, US, W, WU). Sucre: Carúpano, *Boldingh* 3983 (U); ibid., *Benitez de Rojas* 39 (MA). Monagas: ca 19 km N of Caicara, *A. Gentry & Berry* 14882 (MO, USF, WAG). Esmeraldas: 2-4 km SE of San Lorenzo, *Boom* 2640 (HBG, MO, QCA, USF, WAG); Alto Tambo, 15 km W of Lita, *Rubio & Quelal* 617 (WAG); near Zapallo, *Baslev & Steere* 3114 (AAU).

ECUADOR. Esmeraldas: 5 km N of Lita, *Acevedo & Daly* 1647 (USF); near Zapallo, *Balslev & Steere* 3114 (AAU, QCA). Pichincha: ENDESA F.R., Río Silanche, *Jaramillo* 6313 (AAU, GB, MO, QCA); near Santo Domingo, *Játiva & Epling* 524 (S, UC). Manabí: Flavio Alvaro, Chone-Santo Domingo Road, *Harling & Anderson* 18928 (GB, USF). Napo: Res. Biol. Jatun Sacha, 8 km ESE of Puerto Mishualli-Coca Road, *J.S. Miller* et. al 2480 (MO). Cotopaxi: Quevedo-Latacunga Road, *Holm-Nielsen* et al. 2980 (MO). Los Ríos: Río Palenque Science Center, km 56 Quevedo-Santo Domingo Road, *A. Gentry* 26308 (AAU). Azuay: Sanagüín Region, Río Patul, *Steyermark* 52793 (F). Guayas: 10 km W of Manglaralto, *Harling* 5010 (S); Río Congo, between Empalme and Pichincha, *Asplund* 15553 (S).

PERU. Madre de Dios: Manu National Park, Cocha Cashu Station, *Forster & Janson* 8275 (WAG). Sin. loc., *McLean* s.n. (BM, K, type of *T. occidentalis*).

SIN. LOC., erroneously Porto Rico, *von Rohr* 76 1/2 (BM, C-VAHL, phot. of C-VAHL sheet A, type of *T. nereifolia*).

CULT. Curaçao, Habaai, Bro. *Arnoldo* 1123 (U). Colombia, Magdalena, San Pedro Alejandrino Botanical Garden, *Neovis de Lopez* 586 (COL). Austria, Wien, Hort. Schönbrunn, Herb. *Jacquin* s.n. (W, type). Italy, Firenze Botanical Garden, Hort. Bot. Florent. 196 (4812) (FI). India, Calcutta, Hort. Bot. Calcutta s.n. (L, M); ibid., *Pierre* 4417 (G).

60. Tabernaemontana angulata Mart. ex Muell. Arg. in Martius, Fl. Bras. 6, 1: 72, t. 23 (1860). – Type: Brazil, Pará, Amazonas R., near Belem, *Martius* Nov. 1819 (holotype M). Fig. 60, p. 250; map. 30, p. 248

Homotypic synonyms:

Anacampta angulata (Mart. ex Muell. Arg.) Miers, Apoc. S. Am. 65 (1878). *Bonafousia angulata* (Mart. ex Muell. Arg.) Boiteau & Allorge in Bull. Soc. Bot. France 130, Lettres 4-5: 339 (1983); in Mém. Mus. natn. Hist. Nat. IIB, 30: 95, pl. 41 (1985).

Heterotypic synonym:

Bonafousia silvae Allorge in op. cit. 342, fig. 3; op. cit. 87, pl. 36, **syn. nov.** – Type: Brazil, Amapa, Rio Jari, Mt Dourado, *N.T. Silva* 1103 (holotype NY; isotype Z).

Shrub or small tree 0.50-7 m high. Trunk 1-5 cm in diameter. Branches pale brown, lenticellate, with longitudinally fissured bark; branchlets triangular in section, puberulous or glabrous. *Leaves* shortly petiolate; petiole puberulous or glabrous, 2-7 mm long; (ocreae slightly widened into intrapetiolar stipules); blade coriaceous when dried ellipic or narrowly elliptic, 1.5-4 x as long as wide, 8-36 x 3-15 cm, acuminate or apiculate at the apex, cuneate or rounded at the base, entire, glabrous above, glabrous or sparsely and minutely pubescent and with an often regular pattern of dots

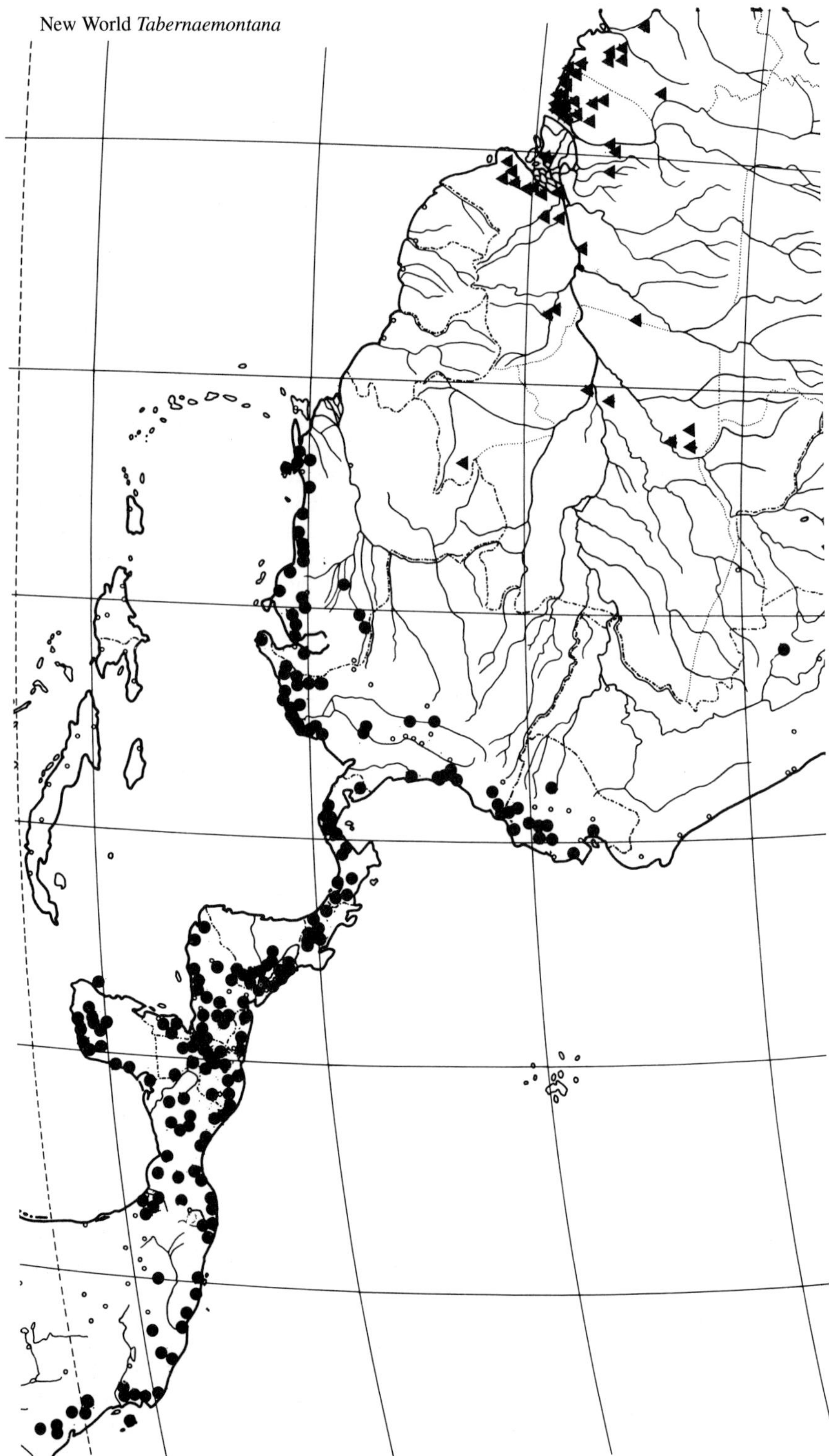

Map 30. ●. *Tabernaemontana amygdalifolia*, ▲ *T. angulata*.

beneath, with 10-17 pairs of curved or rather straight secondary veins forming an angle of 60-80° with the costa; tertiary venation reticulate, inconspicuous. *Inflorescence* shortly pedunculate, 3-8 x 2-6 cm, congested, few- or many-flowered. Peduncle puberulous or glabrous, 5-40 mm long; pedicels 1-5 mm long. Bracts sepal-like and about half as long as them; often 2 bracteoles on pedicel. *Flowers* fragrant, open during the day. *Sepals* green(?), connate at the base for 1 (-2) mm, erect, ovate or nearly so, 1.2-2 x as long as wide, 2.5-7 × 1.5-5 mm, rounded, puberulous or sometimes almost glabrous outside, mostly ciliate, glabrous inside and with 3-8 colleters in 1-2 rows in the middle at the base; colleters 0.5-1 x 0.3-0.5 mm, sometimes partly fused. *Corolla* white or limb and throat white and tube pink, 17-35 mm long in the mature bud and forming a comparatively wide broadly ovoid head 0.25-0.35 of the bud length and twice as wide as the tube with a blunt apex, puberulous outside on the tube from 5 mm below the mouth and on the base of the lobes, especially the part not covered in bud, often with an interrupted belt of pubescence 2-6 mm wide inside from 0-1 mm below the anther tails to the apex of the anthers or to halfway along them, pubescence always on filament ridges among anther tails; tube 3.3-5 x as long as the calyx, 0.9-1.5 x as long as the lobes, 13-29 mm long, almost cylindrical and 2-5 mm wide, usually slightly narrowed below the insertion of the stamens to 1.5-3 mm wide, not twisted; lobes obliquely oblong, mostly more or less falcate, 0.7-1.1 x as long as the tube, 2.5-3.6 x as long as wide, 9-25 x 3-7 mm, rounded, slightly auriculate at the left side of the base, not undulate, spreading. *Stamens* with apex 1-6 mm below mouth of corolla tube, inserted 0.55-0.75 of the length of the corolla tube, (at 9-17 mm from the base); anthers sessile, with tails 1 mm below insertion, narrowly triangular, 4-5 x as long as wide, 4.5-6 x 0.8-1.2 mm, apex acuminate, sterile for 0.4-0.7 mm, sagittate at the base, with tails usually straight. *Pistil* glabrous or occasionally ovary pubescent, 10-19 mm long; ovary ovoid, 2-4 x 1.5-2 x 1.5-2 mm, with a disk-like ring at the base 1-1.5 mm wide connecting the two separate carpels at the base; style filiform, 7-15 mm long; pistil head 1.2-1.5 mm high, composed of a basal recurved veil 0.2-0.3 x 0.9-1.2 mm, a stipitate 5-lobed depressed globe 0.3-0.4 x 0.6-0.8 mm and a stigmoid apex 0.3-0.5 x 0.1-0.2 mm. Ovules approximately 50 in each carpel. *Fruit of* 2 separate mericarps; mericarps dark brown, green when immature, obliquely ellipsoid or ovoid, not dotted, 30-50 x 15-20 x 10-20 mm, mostly recurved, acuminate or sometimes rounded at the apex, with or less often without a clear lateral ridge at each side, minutely papillose, puberulous, 5-40-seeded; wall thick; aril very small, not enveloping the seed. *Seed* pale or dark brown, obliquely and narrowly ellipsoid, 8-12 x 3-3.5 x 2-2.5 mm, with clear longitudinal grooves and a honeycomb-like structure.

DISTRIBUTION: Brazil.

ECOLOGY: Forest understorey, not inundated. Alt. low. Flowering and fruiting throughout the year with a peak in June-July.

Geographical selection of the approximately 110 specimens examined:

BRAZIL. Roraima: Serra dos Surucucus, *Prance* et al. 10045 (K, NY, S, UC, US, Z). Amazonas: Mun. Manaus, Tarumã, *Zarucchi & Morawitz* 3230 (WAG); Manaus, *Coelho* INPA 3843 (MG, MO, U); Manaus-Porto Velho Road, Castanho-Tupana Section, *M.F. Silva* et al. 189 (INPA); Mun. Humaitá, BR 230, 94 km from Humaitá, res. indig. dos Tenharim, *Cid Ferreira* 5453 (K, US, WAG); km 64 Humaitá-Jacarecanga Road, *Teixeira* et al. 1151 (F, GH, WAG); km 150 of Humaitá-Jacarecanga Road, *Teixeira* et al. 1363 (K, US, WAG). Pará: km 210 Itaituba-Humaitá Road, *Bamps* 5374 (BR, WAG); Rio Trombetas, near Cachoeira Porteira, *Campbell* et al P22359 (K, NY, U, US, Z); Mun. Oriximiná, Rio Trombetas, Minneração Santa Patricia, *Cid* et al. 1437 (CAY, GH, MG, MO, NY, US, WAG); Alter do Chão, *Ducke* 10320 (MG); Curua-una, *Staner* 84 (BR); Mun. Almeirim, Mt Dourado, *M.J.P. Pires*

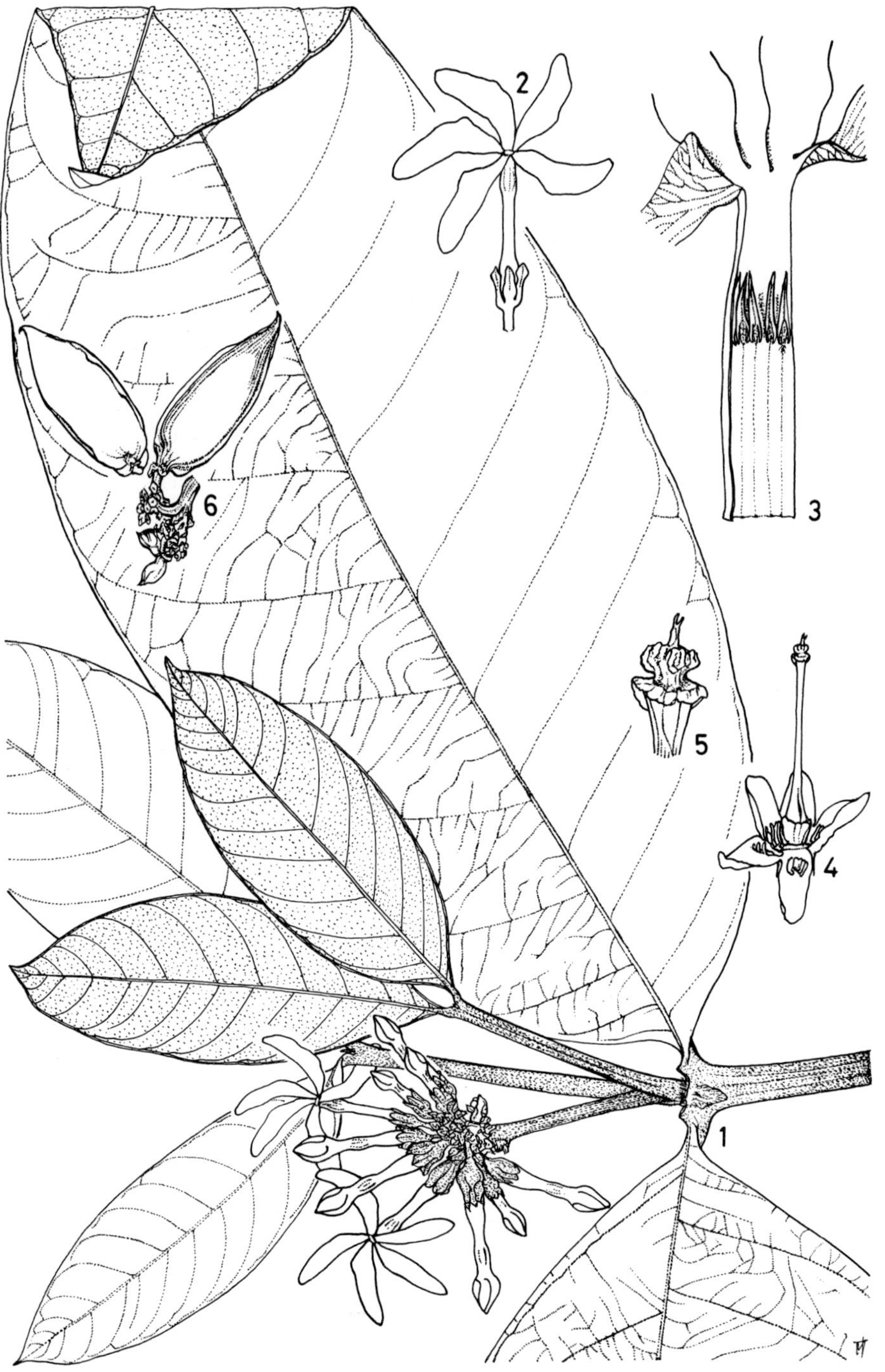

Fig. 60. *Tabernaemontana angulata*. **1,** habit (x 2/3); **2,** flower (x 1); **3,** opened corolla (x 2); **4,** calyx with pistil (x 2); **5,** pistil head (x 10); **6,** fruit (x 2/3). 1 from Austin et al. 4081; 2 from Prance et al. 1985; 3-5 from Rosa 4013; 6 from Mexia 5943.

& *Oliveira* 503 (INPA, MG, WAG); Rio Mojú, Igarapé Jambunosú, *M. Silva* 1041 (NY); Tucurui, Rio Tocantins right bank, *Lisboa* et al. 1262 (MG, NY); Bragança, *Ducke* 21583 (S, US); Tucuruí-Jatobal Road, *Rosa & Rosário* 4013 (MG, U); km 25 Reprêsa Tucuruí-Breu Branco Road, *Plowman* et al. 9556 (F, GH, NY, US, WAG); 18 km E of Tucuruí, *Daly* et al. 952 (GH, MO, US, WAG); km 309 Belém-Brasília Highway, *Maguire* et al. 56071 (NY, Z); Distr. Acará, Thomé Assú, *Mexia* 5915 (A, BH, BM, G, GB, GH, K, LL, MO, NY, S, U, UC, WIS, Z) & 5943 (A, BH, BM, G, GB, GH, K, MO, NY, S, U, UC, US, Z); Belém, *N.T. da Silva* 33 (INPA, MO); ibid., *A. Gentry & Pinheiro* 13118 (MO, NY, Z); near Belém, *Martius* Nov. 1819 (M, type); ibid. Ilha do Mosqueiro, *Killip & A.C. Smith* 30388 (MO, NY, US); 5 km from Vigia, *Duarte* 9777 (M, Z); Castanhal, *Mexia* 5911 (A, BH, BM, G, GB, GH, K, MO, NY, S, U, UC, US, WIS, Z); Rio Capim, *J. Huber* 704 (MG); Belém-Bragança Railway, Igarapé-assú, *Kuhlmann* RB 21845 (U); Mun, São Francisco do Pará, *N.C. Bastos* et al. 157 (MG, NY, WAG); near Marco, *Baker* 404 (MG); Paragominas, *Rosa* 2431 (MG, NY); Road BR 22, Capanema to Maranhão km 64, near Piritoro, *Prance & Pennington* 1985 (K, NY, S, US, Z); Tenoné, E.F. Bragança, *M. Bastos* 16 (F, GH, MO, NY, RB); 10 km S of Gurupi, Belém-Brasília Highway, *Prance & N.T. Silva* 58961 (GH, K, NY, P, S, U, US, Z); 13 km N of Bragança, *Davidse* et al. 18036 (NY, US, WAG); 30 km S of Gurupi, *Prance & N.T. Silva* 58966 (GH, K, NY, S, U, US, Z). Amapa: Rio Jari, Serra da Arumanduba, *Egler & Irwin* 45931 (K, NY, US, Z); Ariramba, *Rosa* et al. 4354 (MG); Mun. Mazagão, morro do Felipe V, *M.J.P. Pires* et al. 694 (INPA, MG, WAG); Macapá-Monte Dourado Road, *Rabelo* et al. 3067 (K, WAG); Rio Araguari, *J.M. Pires* et al. 50934 (NY, US, Z); Matapi, *Ribeiro* 1604 (NY); Igarapé do Lago, Macapá, *Rabelo* 805 (MG); Rio Pedreira, *J.M. Pires & Cavalcante* 52210 (K, NY, S, US, Z); Monte Dourado, *N.T. Silva* 1103 (NY, Z, type of *Bonafousia silvae*), 1483 (NY, Z). Maranhão: Mun. Santa Luzia, road to Rio Mutum, Rio Pindaré left bank, *M.F.F. Silva* et al. 1016 (MG); Mineirinho, Rio Pindaré, *Jangoux & Bahia* 920 (MG, Z); Colônia Betel, Rio Alto Turiaçu, *Jangoux & Bahia* 779 (MG, Z); Nova Esperança, *Jangoux & Bahia* 103 (MG); Turiaçu, km 6 of BR 106, Maracaçumé-Santa Helena, *Rosa & Vilar* 2836 (MG, NY); Mun. Monção, Urutawi, *Balée* 3555 (NY); Bom Passar, 24 km from Peritoró, *Jangoux & Bahia* 1056 (Z).

61. Tabernaemontana apoda C. Wright in Sauvalle, Ann. Cien. Habana 7: 102 (1870), n.v.; in Sauvalle, Fl. Cubana 118 (1873); León & Alain, Fl. de Cuba 4: 173 (1957). – Type: Cuba, sin. loc., *C. Wright* 3625 (holotype K; isotypes GH, NY, US, phot. of US sheet in WAG). Fig. 61, p. 252; map 31, p. 256

Homotypic synonym:

Peschiera apoda (C.Wright) Mgf. in Notizbl. Bot. Gart. Berlin 4: 171(1936).

Shrub or small tree 2-9 m high. Bark grey. Branches pale grey-brown, not lenticellate; branchlets terete, glabrous. *Leaves* petiolate; petiole glabrous, 6-15 mm long; (ocreae slightly widened into intrapetiolar stipules); blade subcoriaceous when dried, elliptic, 1.8-2.4 x as long as wide, 8-18.5 x 4-8.5 cm, acuminate at the apex, cuneate at the base, entire, glabrous on both sides, with scattered black dots beneath, with 8-17 pairs of upcurved secondary veins forming an angle of 60-80° with the costa; tertiary venation more or less reticulate, rather inconspicuous. *Inflorescence* pedunculate, 4-6 x 4-9 cm, many-flowered, rather lax. Peduncle very short, 2-3 mm long, glabrous; pedicels 5-10 mm long, rather slender, glabrous. Bracts sepal-like, 0.7-1 x as long as the sepals. *Flowers: Sepals* pale green(?), connate at the base for 0.3 mm, erect, ovate, 2 x as long as wide, 3 x 1.5 mm, acute, glabrous outside, cilio-

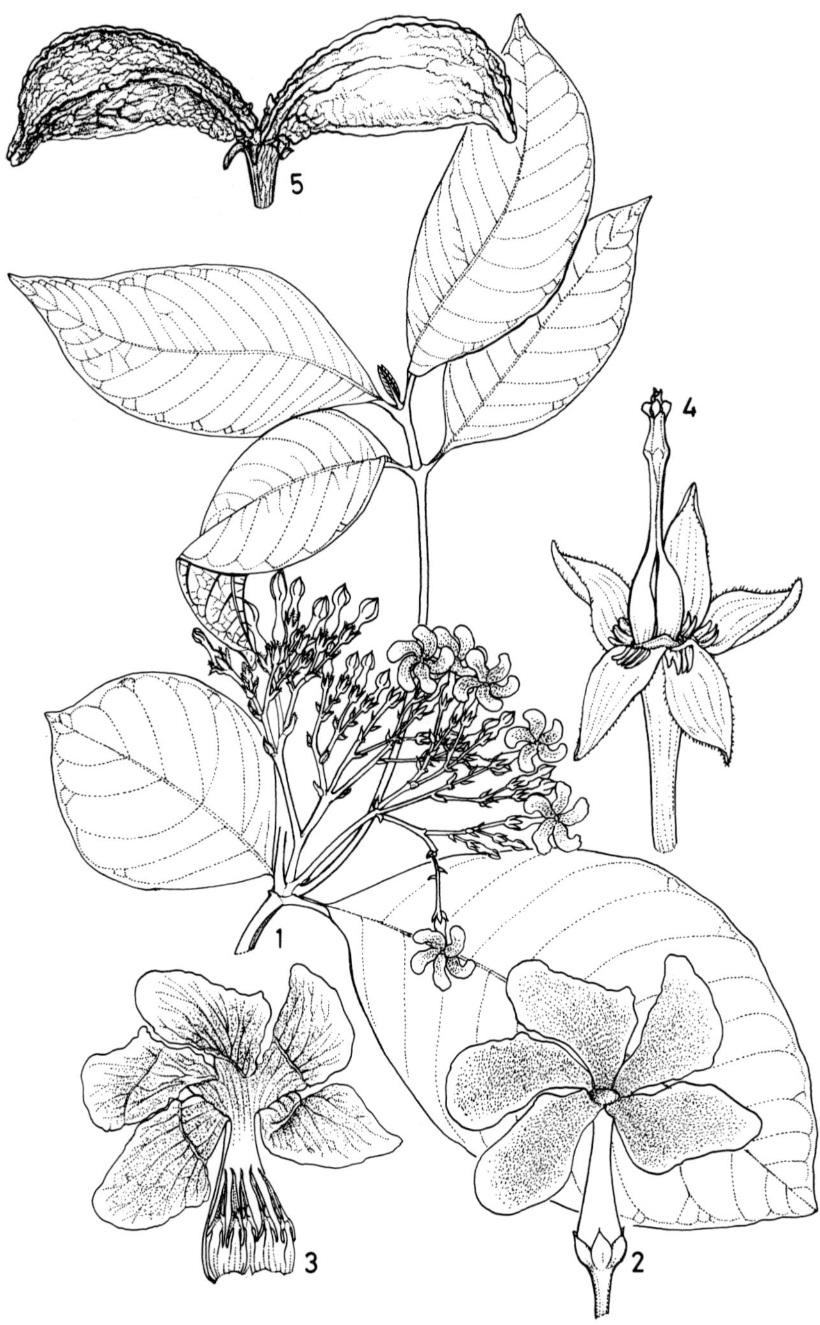

Fig. 61. *Tabernaemontana apoda*. **1,** habit (x 2/3); **2,** flower (x 2); **3,** opened corolla (x 2); **4,** calyx with pistil (x 8); **5,** fruit (x 1). 1-4 from Wright 3625; 5 from Bijhouwer 516.

late, glabrous inside and with 4 colleters in 1 row in the middle at the base; colleters 0.5 x 0.2 mm. *Corolla* white(?), 15-18 mm long in the mature bud and forming a comparatively large ovoid head 0.37-0.4 x the bud length (6-7 x 4-5 mm) with a blunt apex, glabrous outside, with a belt of pubescence inside from 1 mm above the apices of the anthers to the basal 4 mm of the lobes of which upper part less hairy; tube 3-3.3 x as long as the calyx, 0.9 x as long as the lobes, 9-10 mm long, flask-shaped, 3 mm wide above the base which is around the stamens, narrowed above to 1.5 mm wide, again widened at the throat to 2 mm wide, not twisted; lobes obliquely obovate or nearly so, 1.1 x as long as the tube, 1.8-2 x as long as wide, 10-11 x 5-6 mm, rounded, spreading. *Stamens* with apex 2.5 mm below mouth of corolla tube, inserted 0.35 of the length of the corolla tube (at 3-3.5 mm from the base); anthers sessile, with tails 1 mm below insertion, narrowly triangular, 5 x 1.5 mm, apex acuminate, sterile for 0.5 mm, sagittate at the base, glabrous. *Pistil* glabrous, 4-4.8 mm long; ovary ovoid, 1.5 x 1 x 1 mm, without basal thickening; style 1-1.9 mm long; pistil head 1.4 mm high, composed of a basal ring 0.3 x 0.8 mm, a broadly stipitate 5-lobed depressed globe 0.3 x 0.6 mm, and a stigmoid apical part 0.2 x 0.2 mm. Ovules approximately 100 in each carpel. *Fruit* of 2 separate mericarps; mericarps obliquely ellipsoid or obovoid, 35-45 x 15-25 x 14-20 mm, mostly recurved, lumpy, papillose, apiculate or acuminate at the apex, not ridged, approximately 10-30-seeded; wall rather thick, 2-3 mm thick in dried fruits; aril surrounding the seed. *Seed* dark brown, obliquely ellipsoid, 8-9 x 4-5 x 3 mm, with longitudinal grooves, dull, papillose; embryo 7.5 mm long; cotyledons ovate, 3.5 x 2.5 mm, obtuse at the apex, cordate at the base; rootlet 5 x 0.8 mm.

DISTRIBUTION: Endemic to Cuba.

ECOLOGY: Thicket. Flowering season unknown. The only flowering specimen ever collected, the type, has no date on the label.

Specimens examined:

CUBA. Cienfuegos: Cassia Hill, Soledad, *Bijhouwer* 516 (WAG); Limones, *Jack* 5361 (A, F, P, US) & 5365 (A, K, NY, P, SING, US); Vilches Hill, *Jack* 5680 (A, F, K, NY, S, US). Sancti Spíritus: near Sancti Spíritus, *Shafer* 12096 (NY). Sin. loc., *Wright* 3625 (GH, K, NY, US; phot. of US sheet WAG, type).

62. Tabernaemontana arborea Rose ex Donn. Sm. in Bot. Gaz. 18: 206 (1893). – Type: Guatemala, Department Quezaltenango, Río Ocosito, *J.D. Smith* 2766 (holotype US; isotypes F, G, GH, K, M, MO, P, UPS, WU; phot. of holotype WAG).
Fig. 62, p. 254; map 31, p. 256

Homotypic syononym:

Peschiera arborea (Rose ex Donn. Sm.) Mgf. in Notizbl. Bot. Gart. Berlin 14: 173 (1938).

Heterotypic synonym:

T. schippii Standl. in Publ. Field Mus. Nat. Hist. Chicago, Bot. Ser.8: 34 (1930). *P. schippii* (Standl.) Mgf. in Notizbl. Bot. Gart. Berlin 14: 173 (1938). – Type: Belize, Stann Creek Railway, Big Creek, *Schipp* 168 (holotype F; isotypes BM, G, GH, K, MO, UC,US, Z; phot. of US sheet WAG).

Tree 4-25 m high. Trunk 15-60 cm in diameter; bark dark greyish to light brown, scaly, 1.3 cm thick. Branches pale or dark grey-brown, lenticellate; branchlets terete, glabrous. *Leaves* of a pair equal or unequal (larger up to 3 x as long as the other and not comparatively narrower), petiolate; petiole 3-10 mm long; (ocreae not widened

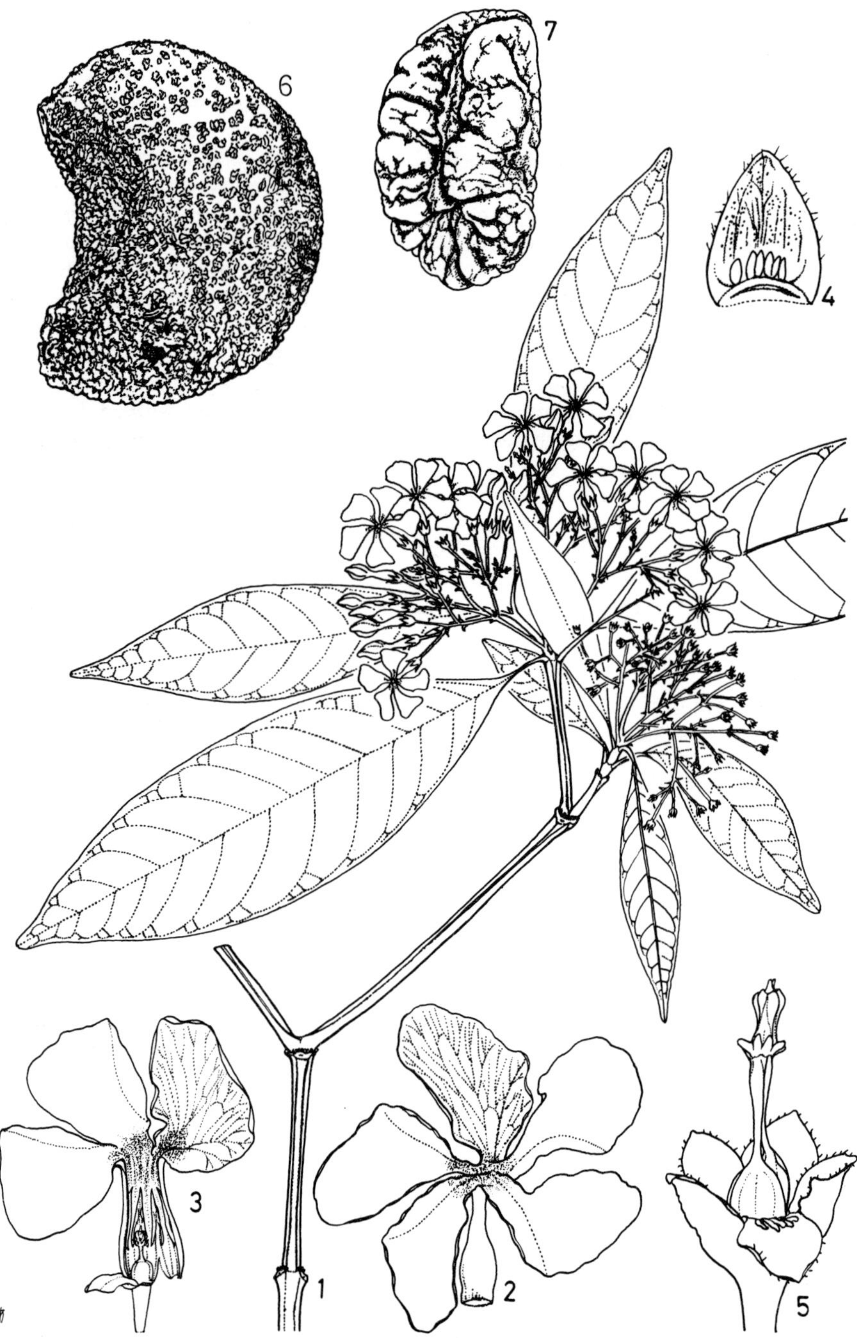

Fig. 62. *Tabernaemontana arborea*. **1,** habit (x 2/3); **2,** corolla (x 2); **3,** opened flower (x 2); **4,** sepal inside (x 6); **5,** calyx with pistil (x 4); **6,** one mericarp outside (x 2/3); **7,** seed (x 12). 1 from Contreras 1389; 2-4 from Proctor 32235; 5 from Contreras 4551; 6-7 from Standley 23765.

into intrapetiolar stipules); blade thinly papery and sometimes with paler venation on both sides when dried, elliptic, obovate or narrowly so, 1.8-4 x as long as wide, 3-19 x 1-7.2 cm, acuminate, apiculate or sometimes obtuse at the apex (with a blunt acumen), cuneate at the base, with flat or slightly revolute margin, entire, glabrous on both sides, mostly with many clear and sometimes with few scattered black dots beneath, with 7-12 pairs of upcurved secondary veins forming an angle of 60-80° with the costa; tertiary venation not very conspicuous. *Inflorescence* very shortly pedunculate, 3-8 x 5-10 cm, many-flowered, rather dense. Peduncle glabrous, 1-3 mm long; pedicels glabrous, 5-20 mm long. Bracts sepal-like, 0.3-1 x as long as them. *Flowers* fragrant, open during the day, mostly many at a time. *Sepals* pale green (?), connate at the exteme base, ovate or oblong, 1-2.3 x as long as wide, 1.7-3.5 x 1-2 mm, obtuse, with a mostly recurved apex, glabrous outside, sometimes ciliate, glabrous inside and with 3-8 colleters in 1 row in the middle at the base; colleters 0.3-0.6 x 0.1-0.2 mm. *Corolla* white, yellow in the throat, with tube often creamy, or entirely yellow, thin, 9-17 mm long in the mature bud, skittle-shaped and forming a comparatively very large ovoid head 0.3-0.5 of the bud length (3-7 x 1.5-4 mm) with an acute, acuminate or sometimes blunt apex, glabrous outside, pilose inside from the insertion of the stamens to the mouth or the base of the lobes; tube 2-4 x as long as the calyx (when sepals straightened out, if not 2.4-4.5 x as long), 0.6-1 x as long as the lobes, 7-10 mm long, flask-shaped, 2-3 mm wide around the anthers (= above the base), narrowed above to 1-2 mm wide, twisted 0-0.1 turn just below the mouth, without thickenings; lobes dolabriform to obliquely elliptic, 1-1.8 x as long as the tube, 1.5-2.5 x as long as wide, 7-15 x 3.5-9 mm, obtuse, neither auriculate nor undulate, spreading. *Stamens* with apex 1-4 mm below mouth of corolla tube, inserted 0.25-0.4 of the length of the corolla tube (at 2-3 mm from the base); anthers sessile, narrowly triangular, 2.7-3.5 x as long as wide, 2.5-4.5 x 1-1.5 mm, apex acuminate, sterile for 0.2-0.3 mm, sagittate at the base (tails straight), glabrous. *Pistil* glabrous, 3.5-4.5 mm long, with apex about halfway along anthers; ovary ovoid or subglobose, mostly laterally compressed, 1.2-1.5 x 1-1.2 x 0.8-1.2 mm, of 2 separate carpels, rather abruptly narrowed into the short style; style 1-1.8 mm long; pistil head 1.2-1.5 mm high, composed of an upcurved 10-lobed basal ring 0.2 x 0.8-1.5 mm, a stipitate apically 5-lobed head 0.7-1.1 x 0.4-0.8 mm, and a stigmoid apex 0.1-0.3 x 0.1-0.2 mm. Ovules approximately 70 in each carpel. *Fruit* of 2 separate or occasionally fused mericarps; mericarps light brown, warty and often paler-spotted, obliquely ellipsoid or reniform, 45-75 x 35-45 x 25-35 mm, rounded or minutely and bluntly apiculate at the apex, approximately 20-40-seeded; wall 3-7 mm thick in dried fruits; aril orange. *Seed* dark brown, obliquely ellipsoid, 10-12 x 5-5.5 x 4-5 mm, with longitudinal grooves, with honeycomb-like structure; embryo 7-7.5 mm long; cotyledons broadly ovate, 1.1-1.2 x as long as wide, 3.5-4 x 3-3.5 mm, rounded at the apex, cordate at the base; rootlet 4 x 0.8-0.9 mm.

DISTRIBUTION: Mexico to Colombia.

ECOLOGY: Forest. Alt. 0-800 m. Flowering almost exclusively April-May. No clear-cut fruiting season deduced, but the fruit may need an entire year to develop.

Geographical selection of the approximately 100 specimens examined:

MEXICO. Veracruz: 10 km E of Coatzacoalcos, *Croat* 40048 (MO); S of Poblado Balzapote, San Andres Tuxtla, *Villegas* H.114 (F, MEXU, MO); Biological Station Los Tuxtlas, *Ibarra* 1495 (MEXU, MO); Colonia 1 de Mayo, *Ortega* 1440 (F). Oaxaca: Ubero, *Ll. Williams* 9111 (F, MO, S, US, WIS); Plan de Aguila, Chiltepec, *Martínez Calderón* 1389 (L, MEXU, MO). Tabasco: San Juan Bautista-Atasa Road, *Rovirosa* 148 (K); Villahermosa, Selva Dos Montes, *Cowan* & *Zamudio* 4668 (LL, MO). Chiapas: 12 km E of Pico do Oro, *E. Martínez* 18437 (MEXU); Cerrito,

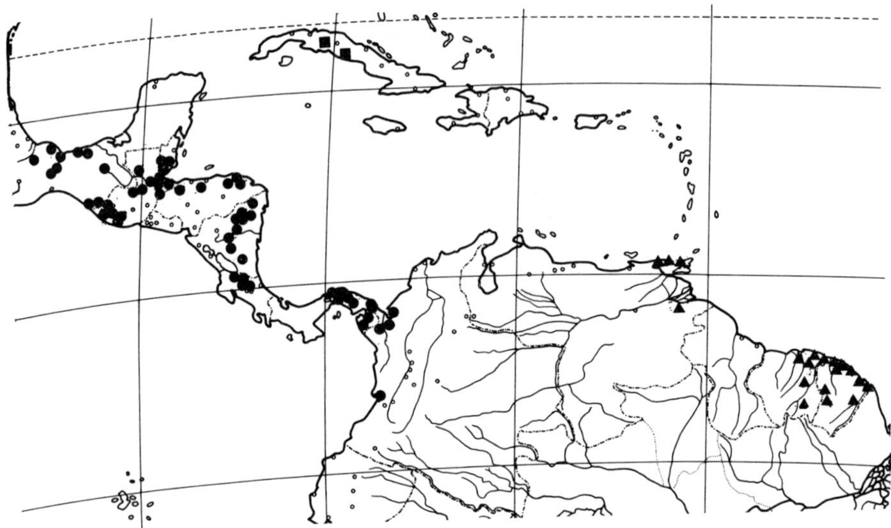

Map 31. ■. *Tabernaemontana apoda*, ● *T. arborea*, ▲, *T. attenuata*.

Acapetahua, *Matuda* 17716 (F, K, MEXU); Cantón El Tesoro, 8 km from Tapachula, *Ventura* & *Lopez* 3456 (MO).

BELIZE. Stann Creek: Stann Creek Railway, *Gentle* 2688 (A, F, K, LL, NA, US); km 35 Stann Creek Valley, *Schipp* 975 (A, BM, F, G, GH, K, MO, S, UC, Z); Big Creek, *Schipp* 168 (BM, F, G, GH, K, MO, UC, US, Z; phot. of US sheet WAG, type of *T. schippii*); Cockscomb Basin, Jaguar Res., *Balick* et al. 2674 (WAG). Cayo: near San Ignacio, *Balick* et al. 2239 (WAG). Toledo: San Pedro Columbia, *Proctor* 36053 (BM, IJ, MO); near Colombia For. Stat., N of San Antonio, *A. Gentry* 8142 (MO).

GUATEMALA. Peten: La Libertad, *Aguilar* 326 (A, K, MO, WIS); Chinchila, km 8 San Luis-Sebal Road, *Contreras* 6432 (F, LL). Izabal: near Lago Izabal, *G.C. Jones* et al. 3076 (F, U, US); near Quiriguá, *Standley* 23765 (GH, MO, US); near Puerto Barrios, *Standley* 73125 (F). Alta Verapaz: Cubilhuitz, *von Türckheim* 7938 (F, GH, K, M, MO, US); Sebol, *Contreras* 4551 (F, G, LL, MO, US). San Marcos: Cafetal above Rodeo, *Steyermark* 37934 (F, MO). Quezaltenango: Río Ocosito, *J.D. Smith* 2766 (F, G, GH, K, M, MO, P, UPS, US, WU; phot. of US sheet WAG, type). Suchitepéquez: Finca La Argentina, *Pinkus* 1170 (US); Retalulea, *Bernoulli* & *Cario* 1832 (K). Solala: sin. loc., *L. Rodriguez* 21181 (P).

HONDURAS. Cortés: San Pedro Sula, *Thieme* 5345 (BM, F, GH, K, US); near La Lima, *Standley* & *Chacón* 7246 (F, GH, MO). Atlántida: Lancetilla Valley, *Hottle* 70 (F). Colón: Rio Esperanza, Puerto Sierra, P.Wilson 596 (NY). Gracias a Dios: 3 km NW of Bulebar, *Saunders* 1237 (Z); Río Platano, *A. Gentry* et al. 7516 (GH, MO).

NICARAGUA. Zelaya: Bonanza-Siempreviva Road, *Stevens* 7996 (MO); 10 km NE of Siuna, *Neill* 3728 (MO, WAG); Río Wawa, *Little* 25075 (US); road to Mina Nueva America, *Stevens* 8450 (MO, U, WAG); Río Danlí, *Ortiz* 1252 (MO); Kurinwacito, *Moreno* 23853 (MO); 1 km N of El Zapote, *Nee* & *Vega* 27930 (MO). Chontales: 4 km NW of Santo Domingo, *Grijalva* et al. 3844 (MO, WAG). Matagalpa: 11 km SW of Río Blanco, Matiguás, *Moreno* 23964 (MO, WAG).

COSTA RICA. Alajuela: Upala, Río Palo Quemado, *Herrera* 1831 (MO, WAG); Colonia Puntarenas, *Herrera* 1338 (MO, WAG); between Florencia and Ciudad Quesada, *Poveda* & *Castro* 3944 (MO). Heredia: Tirimbina, *Proctor* 32235 (LL, MO).

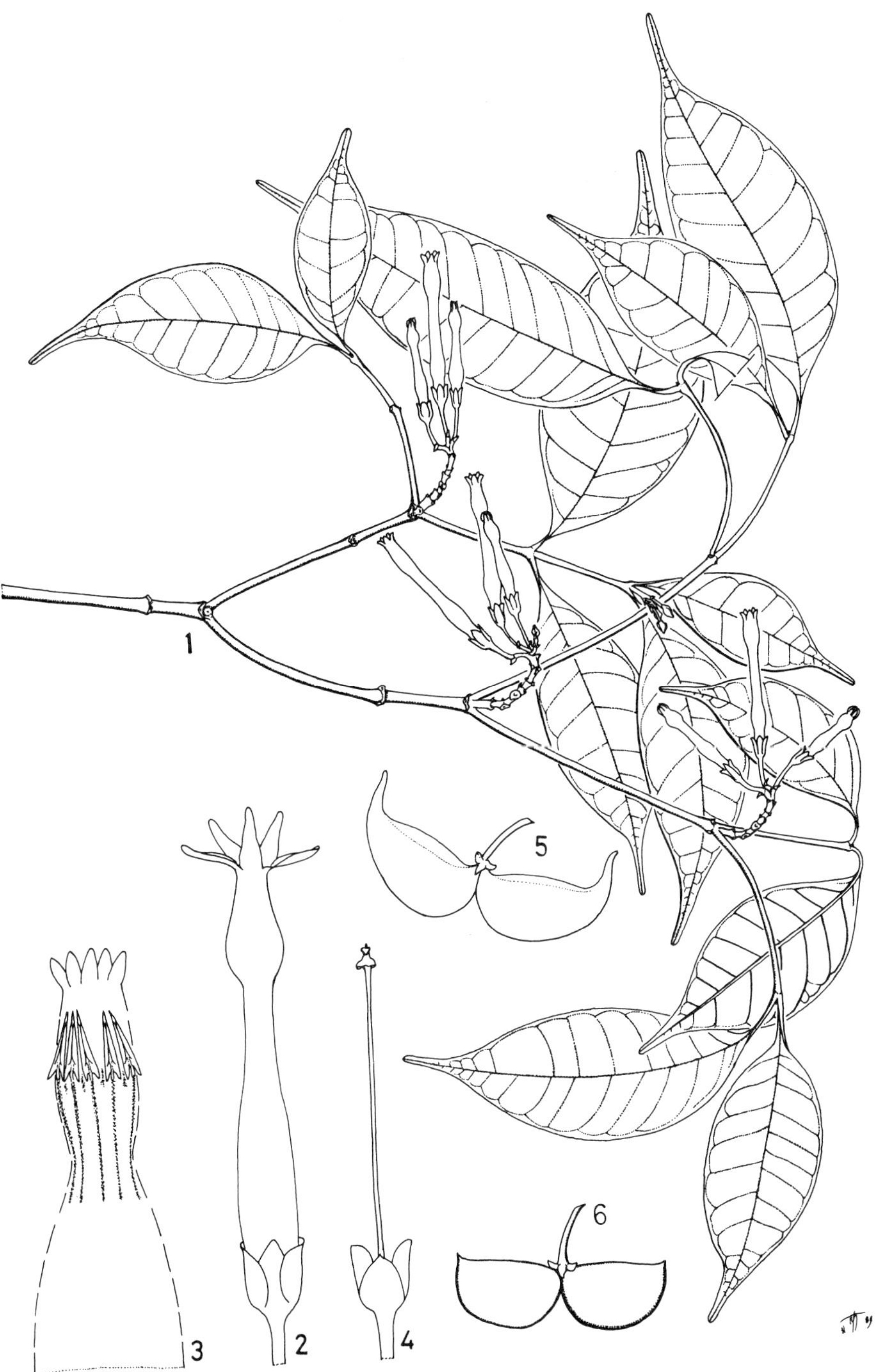

Fig. 63. *Tabernaemontana attenuata*. **1,** habit (x 2/3); **2,** flower (x 2); **3,** opened corolla (x 2); **4,** calyx with pistil (x 2); **5-6,** fruits (x 2/3). 1-4 from Cremers 6017; 5 from Broadway 2709; 6 from Moretti 1132.

PANAMA. Canal Zone: Barro Colorado Island, *Croat* 8559 (F, MO). Panamá: Campana Hill, *Alston* 8924 (BM, F, P); Río Bayano, just above mouth of Río Espave, *A. Gentry* 3785 (BM, MO); between Charco Rico and the headwater of Río Ipeti Grande, *Churchill & de Nevers* 4329 (MO, WAG). San Blas: Cerro Brewster, *de Nevers* et al. 5503 (WAG); Llano-Carti Road, S of Cuna Indian Res., *McCook* et al. 1000 (US); Perme, *Cooper* III 283 (F, GH, MO, UC, US). Darién: near Marraganti, *R.S. Williams* 648 (US); El Cupe R., 20 km SSW of Boca de Cupe, *McPherson* 14991 (WAG); Río Manene to mouth of Río Cuasi, *Kirkbridge & Bristan* 1392 (MO).

COLOMBIA. Choco: below first rapid on Río Truando, *Duke* 12233 (MO, US). Antioquia: Mun. Chigorodó, Bohios, *Rentería & Cogollo* 3795 (MO). Cordoba: Quimarí, *von Sneidern* 5747 (A, AAU, C, COL, LL, S, UC, WAG). Valle: Bajo Colima area, ca 15 km N of Buenaventura, *A. Gentry* et. al 53698 (MO, WAG).

Note. *Tabernaemontana arborea* is easily confused with *T. cymosa.* They differ from each other as follows:

Corolla lobes glabrous inside or pilose only at the base; leaves mostly with many clear dots and sometimes also with black dots beneath. Central America and Colombia .... **T. arborea**

Corolla lobes pilose inside for 0.5-1 of their length; leaves sometimes with some black dots beneath. Western South America .... **T. cymosa**

63. Tabernaemontana attenuata (Miers) Urb. in Fedde, Repert. 13: 471 (1915). – Type: Suriname, sin. loc., *Hostmann* 1314 (lectotype BM; isolectotypes C, CGE, G, GH, K, MO, P, S, U, W, designated here). Fig. 63, p. 257; map 31 p. 256

Basionym and homotypic synonym:

Bonafousia attenuata Miers, Apoc. S. Am. 51 (1878). *Anartia attenuata* (Miers) Mgf. in Notizbl. Bot. Gart. Berlin 14:165 (1938).

Shrub or small tree, 1.50-10 m high. Trunk 2-9 cm in diameter; bark dark brown, finely and longitudinally fissured, lenticellate, thin. Branches pale or dark brown, lenticellate; branchlets terete when fresh, glabrous. *Leaves* of a pair equal or unequal (larger up to twice as long as the other and similarly shaped), petiolate; petiole glabrous, 3-10 mm long; (ocreae not widened into intrapetiolar stipules); blade subcoriaceous to membranaceous when dried, elliptic or narrowly elliptic, 2-4(-6) x as long as wide, 4-14.5 x 0.7-5 cm, acuminate to caudate at the apex, with an obtuse acumen, cuneate at the base or decurrent into the petiole, entire or sometimes undulate, glabrous and with scattered black dots on both sides, with 5-10 pairs of upcurved secondary veins forming an angle of 60-80° with the costa; tertiary venation inconspicuous. *Inflorescence* shortly pedunculate, 3-7 x 2-6 cm, 1-7-flowered, rather lax. Peduncle rather robust, 3-5 mm long, seemingly longer in older inflorescences, especially when they bear both flowers and fruits and then covered with bracts; pedicels glabrous, 5-8 mm long. Bracts 0.2-0.5 x as long as the sepals, triangular, obtuse. *Flowers* open during the day. *Sepals* green, subequal, connate at the base for about 0.5 mm, erect, ovate, 1-2 x as long as wide, 3-6 x 2.5-4 mm, rounded, entire, glabrous outside, not ciliate, glabrous inside and with 6-12 colleters in 1-2 rows on each sepal at the base forming a continuum in the calyx; colleters 0.3-0.7 x 0.1-0.2 mm. *Corolla* creamy or pale yellow, sometimes (?) with a greenish tube and a (darker) yellow throat, 22-32 mm long in the mature bud, forming a comparatively small subglobose head narrower than the part of the tube around the anthers 0.11-0.13 of the bud length (3-4 x 3-4 mm) with a rounded apex, glabrous outside, not ciliate, with a pubescent belt 7-8 mm wide inside from the insertion of the stamens downwards, with a callose ring in the throat; tube 4.7-8 x as long as the calyx, 5-7.5 x as long as the lobes, 23-32 mm long, almost cylindrical, widened above the base and 2.5-5 mm wide, narrowed below the insertion of the stamens to 1.5-2.5 mm wide, widened again around the anthers to 3-5 mm wide, not twisted; lobes obliquely strap-shaped, 0.13-0.2 x as long as the tube, 2-3 x as long as wide, 4-6 x 1.5-2 mm, curved, rounded, undulate, recurved. *Stamens* with apex 1-2 mm below the mouth of corolla tube, inserted 0.7-0.8 of the length of the corolla tube (at 18-23 mm from the base); anthers sessile, narrowly triangular, 3-4 x as long as wide, 4-5.2 x 1-1.5 mm, apex acuminate, sterile for 0.3 mm, sagittate at the base, glabrous. *Pistil* glabrous, 20-25 mm long, with apex 0.2-0.4 way along the anthers; ovary ovoid, 3-5 x 2-3 x 2-3 mm, acuminate, of two carpels basally connate for 0.5-1 mm, with a disk-like thickening at the base; style filiform, 15-20 mm long, persistent; pistil head composed of an undulate ring 0.2-0.5 x 1-1.5 mm, an obovoid distally 5-lobed central part 0.6-0.8 x 0.6-1 mm and a stigmoid apex about 0.2 x 0.2 mm. Approximately 100 ovules in each carpel. *Fruit* of 2 separate mericarps; mericarps green (?), obliquely ellipsoid, 22-40 x 13-23 x 10(?)-15(?) mm, acuminate or apiculate at the mostly recurved apex, slightly recurved or not, not ridged, smooth, about 20-40-seeded; wall about 2 mm thick; aril orange-red. *Seed* medium or dark brown, obliquely ellipsoid, 7-12 x 3-5 x 3-4 mm, with deep longitudinal grooves, papillose.

DISTRIBUTION: Northern South America.

ECOLOGY: Forest understorey, not on river banks, mostly not far from the coast. Alt. 0-500 m. The knowledge of the seasonal aspects is very incomplete. The species is found flowering in Venezuela April-September, in Trinidad April-August, and in the Guianas November-December. No clear-cut fruiting season.

Most of the specimens examined:

VENEZUELA. Sucre: Peninsula de Paria, Cerro Patao, *Steyermark & Agostini* 91135 (MO, US); Cerro de Río Arriba, *Steyermark & Rabe* 96272 (NY, US). Delta Amacuro: ENE of El Palmar, *Marcano-Berti* 222 (K, MO, NY, U, US).

TRINIDAD. Saint Andrews: Matura F.R., *Philcox* et al. 7329 (K, NY, P). Saint George: Maracas Bay Road, *R.S. Williams* et al. Herb. Trin. 11239 (K); Arima-Blanchisseuse Road, *Webster* et al. 9690 (S, U); Blanchisseuse Road, *Broadway* 5640 (BM, BR, E, F, K, MO, UC).

SURINAME. Zanderij I, *Stahel* 303 (A, B, CAY, K, MO, NY, S, U, UC, WAG, WIS, Z); near Moengo, *Lindeman* 6065 (A, U); Mapane Creek area, *Elburg* LBB 9884 (MO, NY, U, Z); 8 km ESE of Brownsberg village, *van Donselaar* 1900 (U); sin. loc., *Hostmann* 1314 (BM, C, CGE, G, GH, K, MO, P, S, U, W, lectotype).

FRENCH GUIANA. Montagne Bellevue de l'Inini, *de Granville* et al. 8000 (CAY, MO, P, U); Saint Laurent de Maroni, *Bèna* Serv. For. 7055 (CAY, NY, P, U); Mana Road, Serv. For. 7253 (CAY, NY, P); Acarouany (= Karouany), *Sagot* 391 (BM, BR, K, P, U), 993 (BM, K, NY, P, U, W); 13 km SW of Sinnemary, piste Saint Elie, *Leeuwenberg* 11653 (CAY, MO, P, USF, WAG); ibid., *Moretti* 1132 (CAY, P, WAG); Montsinery, *Jacquemin* 2413 (CAY, P, U); ibid., *Cremers* 6017 (CAY, WAG); Cayenne, *Martin* s.n. (G-DC, K, P, paratype of *T. oblongifolia*); Montagne de Kaw, *de Granville* 2942 (CAY, P, WAG); ca 2 km W of Ouanary, *de Granville* 6751 (CAY); Camopi R., Mt Alikene, *Garnier* 34 (CAY); Tampok R., Koumakou Fall, *Moretti* 626 (CAY, P, WAG); sin. loc., *Martin* s.n. in *Rudge* s.n. (BM).

Notes. Flowers and a fruit together in a single inflorescence in *Berti* 222 (MO). *Tabernaemontana attenuata* is closely allied to *T. cerea* (q.v.) and *T. flavicans*.

64. Tabernaemontana catharinensis A. DC., Prod. 8: 365 (1844). – Type: Brazil, Santa Catarina, Santa Catarina Island, *Gaudichaud* 186 (lectotype G-DC; isolectotypes F, FI-W, G, P; phot. of lost B sheet GH, MO, US, designated here).
Fig. 64, p. 261; map 32, p. 264

Homotypic synonym:

Peschiera catharinensis (A. DC.) Miers, Apoc. S. Am. 41 (1878).

Heterotypic synonyms:

T. affinis Muell. Arg. in Martius, Fl. Bras. 6, 1: 83 (1960), partly, at least for lectotype, **syn. nov.** *P. affinis* (Muell. Arg.) Miers, op. cit. 40; Allorge in Mém. Mus. natn. Hist. Nat. II B. Bot. 30: 142 (1985), partly, excl. syn. *T. bracteolaris*. – Type: Brazil, Rio de Janeiro, Corrego, *Pohl* 1842 (lectotype W; isolectotypes BR, F, GH, K; phot. of NY sheet P, designated by Allorge, 1985).

T. affinis var. *lanceolata* Muell. Arg. in op. cit. 84. – Type: Brazil, sin. loc., *Sellow* 1642 (121) (syntype B†; lectotype K, designated here).

T. australis Muell. Arg., l.c., **syn. nov.** *P australis* (Muell. Arg.) Miers, op. cit. 46; Ezcura in Darwiniana 23: 441, fig. 21: (1981). – Type: Brazil, Rio Grande do Sul, *Isabelle* s.n. (holotype P; isotype G; phot. of holotype GH, US).

T. hilariana Muell. Arg. in op. cit. 85. *P. hilaraina* (Muell. Arg.) Miers, op. cit. 41. *P. australis* var. *hilariana* (Muell. Arg.) Allorge in op. cit. 148. – Type: Brazil, São Paulo, Oreas Mts, *Saint-Hilaire* 1069 (holotype P; phot. GH, US).

T. acuminata Muell. Arg. in Linnaea 30: 406 (1860). *P. acuminata* (Muell. Arg.) Miers, op. cit. 43. – Type: Bolivia, sin. loc., *Cuming* 121 (holotype W).

P. albidiflora Miers, op cit. 39. – Type: Brazil, Rio de Janeiro, Itagohy, *Bowie & Cunningham* 16 Sept. 1815 (holotype BM).

T. hybrida Hand.-Maz., Asclep. und Apoc. 9, t.33, fig. 1 (1910), preprint of Denkschr. Akad. Wiss. Math.-Nat. Kl. 79, 2, which never was issued. – Type: Brazil, São Paulo, near Itapetininga, *Wacket* anno 1902 (holotype WU).

T. salicifolia Hand.-Maz., op. cit. 9, t. 33, fig. 2, not Wall. (1829). – Type: Brazil, São Paulo, between Fazenda Bela Vista and Salto Grande de Paranapanema, *Wettstein & Schiffner* July 1901 (holotype WU).

Shrub or small tree 1-9 m high. Trunk 2-35 cm in diameter; bark pale grey or grey-brown. Branches pale or dark brown, lenticellate; branchlets terete, glabrous or pubescent. *Leaves* of a pair equal or unequal (larger up to twice as long as other and not comparatively narrower), petiolate; petiole glabrous or pubescent, 1-12 mm long; (ocreae not widened into intrapetiolar stipules); blade thinly papery and often with paler venation on both sides when dried, elliptic or narrowly elliptic, sometimes narrowly obovate, 2-6 x as long as wide, 3-20 x 0.7-6 cm, acuminate, acute or sometimes obtuse at the apex, cuneate at the base or decurrent into the petiole, entire, glabrous or pubescent on both sides, or pubescent only beneath, often with scattered black dots beneath, with 8-21 pairs of upcurved secondary veins forming an angle of 50-80° with the costa; tertiary venation reticulate, conspicuous, often impressed above in dried leaves. *Inflorescence* mostly shortly pedunculate, 2-6 x 2-6 cm, 5-30-flowered, rather dense. Peduncle short or obsolete, glabrous or pubescent, 0-10 mm long; pedicels glabrous or pubescent, 2-15 mm long. Bracts sepal-like or very narrowly elliptic, 0.7-1.5 x as long as the sepals. *Flowers* fragrant, open during the day, usually many at a time. *Sepals* green, connate at the extreme base, narrowly ovate or almost oblong, unequal, 2-4 x as long as wide, 2.5-6.5 x 1-2.2 mm, acute, with recurved apex, glabrous on both sides and not ciliate, or pubescent on both sides and ciliate, with 5-12 colleters in 1-2 rows (at the same level) in the middle at the base or forming a con-

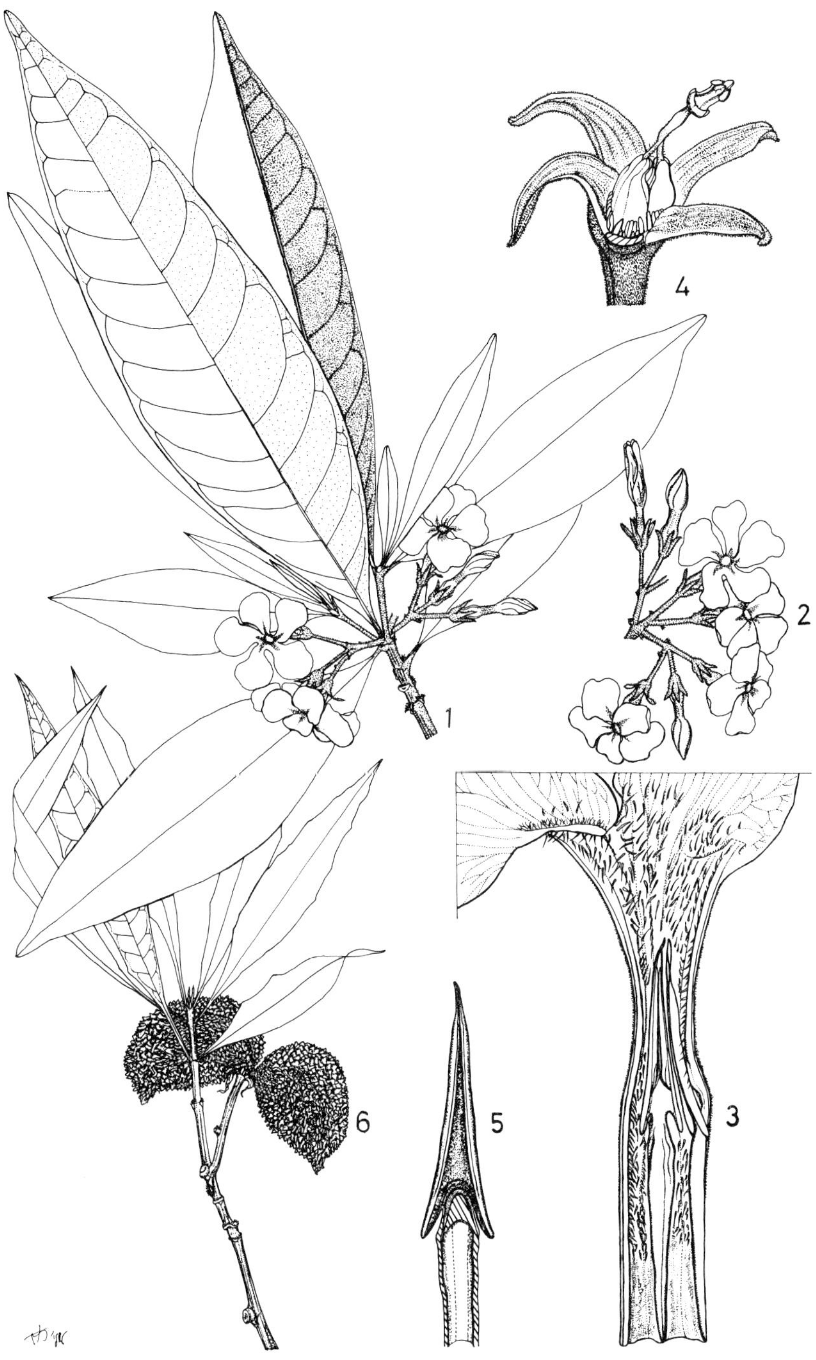

Fig. 64. *Tabernaemontana catharinensis*. **1,** habit (x 2/3); **2,** inflorescence (x 2/3); **3,** opened corolla (x 6); **4,** calyx with pistil (x 6); **5,** stamen (x 8); **6,** fruiting branch (x 2/3). 1-5 from Lindeman & de Haas 2729; 6 from Simonis et al. 93.

tinuous ring of 50-60 colleters in the flower; colleters often unequal, 0.2-0.6 x 0.1-0.3 mm. *Corolla* white, sometimes greenish at the base, 9-19 mm long in the mature bud, skittle-shaped and forming a comparatively large narrowly ovoid head 0.4-0.5 of the bud length (4-9 x 2-4 mm) with an acute or obtuse apex, glabrous outside or sometimes pubescent on the tube and less often also on the lobes, pilose-pubescent inside from 1.5-2 mm above the base (= 0.5-1.5 mm below the insertion of the stamens) or less often from the insertion of the stamens to the base of the lobes; tube 1.3-2.3 x as long as the calyx (when sepals straightened out, if not 1.6-2.8 x), 0.7-1.1 x as long as the lobes, 6-14 mm long, widened above the base (around the anther bases) and 2.5-3.5 mm wide, above them narrowed to 1.3-2.2 mm wide, widened again at the throat or not, not twisted, without thickenings; lobes dolabriform, (0.7-)0.9-1.7 x as long as the tube, 1-1.6 x as long as wide, 7-13 x 4-11 mm, obtuse at both corners or sometimes acute at one of them, neither auriculate nor undulate, spreading. *Stamens* with apex 0.5-6 mm below mouth of corolla tube, inserted 0.23-0.46 of the length of the corolla tube (at 2.5-3.5 mm from the base); anthers sessile, narrowly triangular, 3-4 x as long as wide, 3-5.5 x 1-1.5 mm, apex acuminate, sterile for 0.3-1 mm, sagittate at the base (tails straight), glabrous. *Pistil* glabrous, 3.7-5.8 mm long, with apex about halfway along anthers; ovary ovoid or subglobose, often laterally compressed, 1.5-2.8 x 1-1.8 x 1-1.5 mm, of 2 separate carpels rather abruptly narrowed into the short style; style 0.9-2.2 mm long; pistil head 1.3-1.5 mm high, composed of firstly an almost cylindrical basal part 1.1-1.3 x 0.5-0.7 mm, narrowed in the middle, with 5 acute lobes at the base, 5 longitudinal grooves and with 5 rounded lobes at the apex all together 0.5-0.8 mm wide, and secondly a stigmoid apical part 0.2-0.3 x 0.1-0.2 mm. Ovules approximately 50-80 in each carpel. *Fruit* of separate mericarps; mericarps green, yellow, orange, or red, often orange or dark red inside, with short acute, often wart-like prickles, obliquely ellipsoid, 15-35(-50) x 10-25(-30) x 10-20 mm, rounded or acute (or occasionally apiculate with recurved acumina) at the apex, recurved, not ridged, 8- to ca. 20-seeded; wall 1-3 mm thick in dried fruits; aril red, enveloping the seed. *Seed* dark brown or black, obliquely ellipsoid, 6.5-8 x 3.5-5 x 3-3.5 mm, with narrow longitudinal grooves, papillose; embryo 4.5-5 mm long; cotyledons broadly ovate, 1.1-1.2 x as long as wide, 1.8-2.5 x 1.5-2.2 mm, rounded at the apex, cordate at the base; rootlet 2.5-3.2 x 0.5-0.8 mm.

DISTRIBUTION: Eastern and southern Brazil, Bolivia, Uruguay, Paraguay and northern Argentina.

ECOLOGY: Forests, gallery forests, thickets, or open places. Alt. 0-600 m. Flowering mainly October-November and fruiting mainly May-June.

Geographical selection of the approximately 540 specimens examined:

BRAZIL. Maranhão: 2 km NW of Presidente Dutra, *Daly* et al. 603 (WAG); 40 km E of Barra do Corda toward Presidente Dutra, *Daly* et. al 555 (WAG). Ceará: Fortaleza, *Ducke* 2362 (Z); 8 km W of Sobral, *Cutler* 8085 (MO, US); 3 km W of Eusebio, *Drouet* 2671 (A, MO, S, US); Serra da Ibiapaba, S.Felix, *Dahlgren* 935 (MO, P, US); between Guaraciaba do Norte and Ipu, *Jangoux* et al. 1828 (K, WAG); Pta do Tabagé, *Allemao* 973 (P). Pernambuco: near Mana, *Ule* 4292 (HBG). Bahia: between Igaporã and Caetité, *Pereira* 10444 = *Duarte* 9533 (Z); Agua Preta, *Bondar* 2387 (US). Minas Gerais: 40 km N of Itaobim, *Magelhães* 16000 (MO); Itaobim, *Pereira* 10441 = *Duarte* 9231 (Z); Itinga, *Duarte* 8563 = *Pereira* 9333 (Z); Mun. Diamantina, Senador Mourão, *Hatschbach* 40875 (MU, Z); Lagoa Santa, *Segadas-Vianna & Lorêdo* 1060 (US). Mato Grosso do Sul: Rio Dourados, *Hatschbach* 45938 (C, G, US, Z); Amambaí, *Hatschbach & Kummrow* 48417 (G, WAG, Z). Rio de Janeiro: between Macaé and Barra, *Trinta* 1025 = *Fromm* 2101 (M, Z); Nova Friburgo, *Curran* 670 (GH, MO, US); Petropolis, *Pohl* 1842 (BR, F, GH, K, NY, W;

phot. of NY sheet P, lectotype of *T. affinis*); Corrego, *Pohl* 1844 (BR, K, M, W, paratype of *T. affinis*); near Cachoeira de Macacú, *E. Santos* et al. 2034 (HBG, K, M, U, Z); Iguaçu National Park, *Duarte & Pereira* 1717 (MO); Itaguaí, *Bowie & Cunningham* s.n. (BM, type of *Peschiera albidiflora*). São Paulo: 9 km W of Santa Cruz do Rio Pardo, *Archer* 4201 (A, MO, NA, UC, US); Parque Estudual das Fontes do Ipiranga, *Melo* 276 (WAG); Mun. Rincão-Frente Unica, *Pozetti* 2 (Z); Mun. Boa Esperançá do Sul-Pedra Branca, *Pozetti* 1 (Z); Amparo, *Hoehne* 20608 (K); Mogi-Guaçu, *Custodio Filho* 404 (WAG, WIS); Jaguariuna, *Roman* 1(BR); Vitoriana-Santa Maria do Araquá Road, *Amaral* et al. 51 (SP); Mun. Garça, Fazenda São Francisco, *Capelossa* 16 (F); between Fazenda Bella Vista and Salto Grande de Paranapanema, *Wettstein & Schiffner* July 1901 (WU, type of *T. salicifolia* Hand.-Maz. not Wall.); Iepê, *Hatschbach & Kummrow* 37296 (C, HBG, M, NY, UC, WAG, Z); near Mirante do Paranapanema, *Tsugaro & Guinoza* B-2101 (A); Teodoro Sampaio, *Dias & Sério* 6134 (F); near Caçapava, *Pabst* 7138 (K, M, U, Z); São José dos Campos, *Loefgren* 376 (S); Jundiaí, *Hoehne* 28333 (BM, US); Campinas, *Campos Novaes* 378 (US, WU); km 223 São Paulo-Botucatu Road, *Hallard* 6 (BR, WAG); near Itapetininga, *Wacket* anno 1902 (WU, type of *T. hybrida*); Santos, *Mosèn* in Glaziou 3436 (P); Ilha Porchat, Herb. Univ. Brasilia 15 Nov. 1932 (US); Juquiá, *Kuhlmann* 3096 (K, US); Apiaí, *Puiggari* 3125 (P); Oreas, *Saint-Hilaire* 1069 (P; phot. GH, US, type of *T. hilariana*); sin. loc., *Lindberg* 646 (BR, S, paratype of *T. affinis* var. *lanceolata*); *Saint-Hilaire* D 729 (P). Paraná: Mun. Alvorado do Sul, Rio Vermelho, *Hatschbach* 25312 (C, UC, Z); Lobato, *Hatschbach & J.M. Silva* 50665 (C); Mun. Pôrto Rico, Ilha do Mutum, *Hatschbach* 22174 (C, UC, Z); S. Izabel do Ivai, *Hatschbach* 17088 (L, Z); Rio Jaguariaiva, *Hatschbach* 35426 (BH, HBG, UC, Z); Joaquim Távora, *Hatschbach* 39293 (BH, UC, WAG); 4 km N of Venceslau Braz., *Lindeman & de Haas* 3129 (F, K, U); between Bosque do Lapar and Londrina, *Kuniyoshi* 4001 (Z); Paranavaí, *Hatschbach & Guimãraes* 22164 (MO, S, UC); S of Rio Ivaí, ca. 30 km E of Cianorte, *Lindeman & de Haas* 761 (P, U, USF, W); Mun., Icaraima, Paredão das Araras, *Hatschbach* 19062 (C, UC); Jacareí, *Dusén* 15527 (A, BM, G, K, LD, NY, S, US); Mun. Adrianopolis, Barra Grande, *Hatschbach* 11698 (Z); Itararé, *Dusén* 10587 (BM, G, GH, K, S, US); Mun. Cerro Azul, Rio Turvo, *Hatschbach* 40327 (AAU, BH, C, MO, NY, WAG); Jaguariaíva, *Hoehne* SP 23420 (US); Mun. Tibagi, Jaguatirica, *Hatschbach* 2865 (US, Z); SE of Tibagi, *Reiss* 165 (GH, WIS); 7 km E of Candido d'Abreu, *Lindeman & de Haas* 2729 (C, K, P, U, US, USF, W, WIS); Mun. Campo Mourão, road to Barboza Ferras, *Hatschbach* 12970 (P, UC, US, Z); Mun. Guaira, Guaira-Ocoi Road, *Butttura* 187 (MU); Mun. Guaraqueçaba, Rio de Cedro, *Hatschbach* 18683 (C); Mun. Paranaguá, Caiobá, *Hatschbach* 13066 (K, US, Z); Mun. Morretes, Prainhas, *Hatschbach* 15288 (MO, P, UC, US, WIS, Z); Mun. Guaratuba, Morro do Cristo, *Hatschbach* 43254 (HBG, US, Z); Mun. Rio Branco do Sul, Curiola, *Hatschbach* 17591 (L, US, Z); 1 km N of Ponte de Amizade, *Lindeman & de Haas* 5460 (BR, F, K, U, US). Santa Catarina: Ilha de Santa Catarina, *Gaudichaud* 186 (F, FI-W, G, G-DC, P; phot. of lost B sheet GH, MO, US, lectotype); ibid., *d'Urville* anno 1826 (G-DC, paratype); ibid., *Reitz & Klein* 939 (M, UC); Pilões Palhoça, *Reitz & Klein* 3902 (G, GH, M, MO, U, US, WIS); Brusque, *Reitz* 3152a (M, MO, UC, US); Ibirama, *Klein* 672 (M, S, UC); Mun. Uruguai, Villa Rica For., *L.B. Smith & Reitz* 12944 (C, MA, P, US, WAG, WIS, Z); Mun. Itapiranga, Rio Uruguai, *L.B. Smith & Klein* 13136 (MA, P, US, WAG, WIS, Z); Lauro Müller, *Reitz & Klein* 7486 (Z); Jacinto Machado, *Reitz & Klein* 9270 (Z). Rio Grande do Sul: Terra de Areia, *Wasum* et al. 6335 (MA, US); Lagoa Pinguela, *Weberling* 6373 (ULM, WAG); Parque do Turvo, Rio Uruguai, *Lindeman* et al. ICN 8883 (U); between Mato Grande and Giruá, *Hagelund* 3409 D (Z); Granja Piratini, Santo Angelo, *Hagelund* 7865 D (Z); 6 km N of Itaquí, *Lindeman* et al. ICN 21082 (U); Rio Ibicuy, *Rosengurtt*

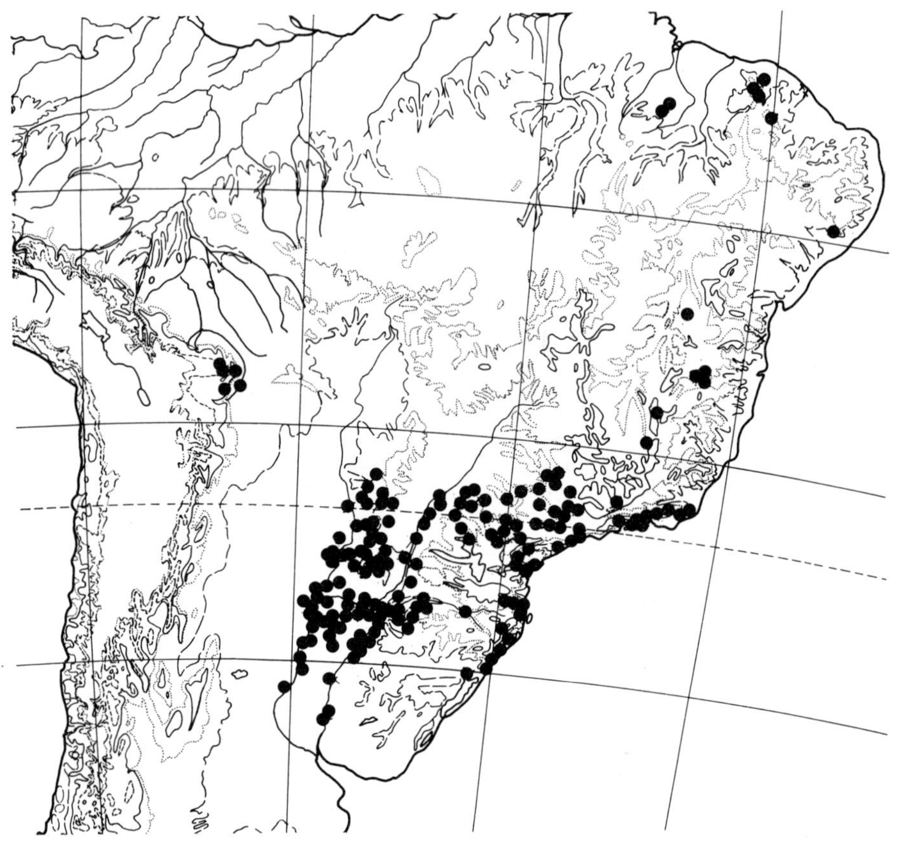

Map 32. *Tabernaemontana catharinensis.*

el al. 9408 (K, NA); Porto Alegre, *Lindman* A 649 (S, UPS); sin. loc., *Isabelle* anno 1835 (G, P; phot. of P sheet GH, US, type of *T. australis*). Brazil, sin. loc. *Sellow* 1642 (121) (K, lectotype of *T. affinis* var. *lanceolata*).

BOLIVIA. Santa Cruz: km 12 Santa Cruz-Cotoca Road, *Nee* 37939 (LL, USF, WAG); Santa Cruz, *Brooke* 5803 (BM); 20 km from Santa Cruz, *Quevedo* 201 (WAG); Bañado del Piray, *Steinbach* 6630 (A, G, K); Amboro National Park, Río Saguayo, *Nee* 41003 (WAG); R Seco, 100 km S of Santa Cruz de la Sierra, *Krapovickas* & *Schinini* 32473 (F, U, Z); Montes Piray, *Steinbach* 27289 (F). Bolivia, sin. loc., Cuming 121 (W, type of *T. acuminata*).

PARAGUAY. Presidente Hayas: Cerrito, Cercanías del Río Verde, *Zardini* et al. 2623 (MO); Cruz Indigena, *Sparre* & *Vervoorst* 918 (BR, L). Concepción: Estancia San Rafael, *Eliceche* 127 (MA); between Río Apa and Río Aquidaban, *Fiebrig* 4939 (G); 28 km W of Concepción, *Schinini* 9061 (Z). Amambay: Cerro Corá National Park, *Fernández Casas* & *Molero* 6080 (G, MA, MO); ibid., *W. Hahn* 1780 (G, GH, K, MO, WAG); ibid., *Simonis* et al. 93 (G, MO, U, WAG); Bella Vista, *Mereles* 614 (G); between Bella Vista and Fuerte San Carlos, *L. Bernardi* 19381 (BM, F, G, U). San Pedro: Primavera, *Woolston* 696 (C, K, P, S); Estancia Carumbé, District Lima, *Pedersen* 9436 (C, K, L, U, Z). Canendiyu: near Igatimi, *Hassler* 4720 (A, BM, G, K, MO, MPU, P, S, UC, W). Alto Parana: Pto Stroessner, km 12 Sapiranguy, *Stutz de*

Ortega 2452 (C, G). Caaguazú: Arroyo Yuquyry, *Zardini & Velázquez* 23764 (WAG); near R. Capibary, *Hassler* 5929 (G); Yhú, *Sparre & Vervoorst* 2008 (P, W); Arroyo Yakare'i, *Zardini & Aguayo* 10698 (WAG). Guaíra: Villa-Rica, *Balansa* 1358a (BM, G, K, LD, P); ibid., *Jörgensen* 3443 (A, C, K, MA, MO, S, US); 8 km S of Melgarejo, *Zardini* 11289 (WAG). Cordillera: Eusebio Ayala-Parque Nacional Vapor Cué Road, *Zardini & Velázquez* 21943 (WAG); Meseta Ybytú Silla, *Zardini* 7360 (G, MO); Cordillera de Altos, *Fiebrig* 129 (A, BM, E, G, GH, GOET, K, L, M, W); ibid., *Hassler* 2059 (A, BM, G, K, MO, MPU, P, S, UC, W); near Lago Ypacaray, *Hassler* 12260 (A, BM, C, E, G, GH, K, L, MO, S, UC, US, Z); San Bernardino, *Sparre & Vervoorst* 876 (L, P, W). Central: Estero del Ypoá, Lomas Valentinas, *Zardini & Guerrero* 32127 (WAG); Paraguay R., near San Antonio, *Zardini & Velázquez* 19013 (WAG); Asunçión, *Balansa* 1358 (BM, G, GOET, K, LD, P, S); Limpio, Paso Correo, *Zardini* et al. 2721 (G, MO, WAG). Paraguari: Cerro Chololó, *Schinini* 13376 (C, G, MO, U, Z); Cerro Palacios, *Ortiz* 711 (G, MA, MO); Parque Barrientos, Ybicuí, *L. Bernardi* 18133 (BM, F, G, K, MO, U, Z); Pirareta, *Pedersen* 9278 (A, C, K, L, P, Z). Neembucú: Humaitá, *T. Meyer* 16197 (W). Misiones-Neembucú border: Estero Cambá, *L. Bernardi* 18385 (BM, F, G, MO). Misiones: Santiago, *Pedersen* 5183 (A, C, E, K, L, P, S, UC, US, Z). Itapuá: Villa Encarnación, *Shrottlhy* 33 (G); El Tirol, 19.5 km NNE of Encarnación, *M.S. Forster* 83-6 (US).

Uruguay. Artigas: Isla Correntina, *Herter* 1304 (Z). Río Negro: Arroyo Negro, Dumas, *A.L. Cabrera* 3268 (S). Paysandú: Isla Caridad, *Schroeder* 17301 (GB).

Argentina. Jujuyi: Sierra Santa Barbara, *Venturi* 9634 (BM, MO). Formosa: Costa Alegre, N of Primavera, *Morel* 7022 (L); Dep. Pilcomayo, Isla Yabay Guazú, *Morel* 8066 (B, S); Puerto Porteño, *Morel* 2100 (C, LL); 2 km from Laguna Blanca, Morel 6342 (C, W). Chaco: Margarita Belén, *Aguilar* 1110 (C, K); Colonia Benítez, *Schulz* 441 (C, MO, S); Las Palmas, *Jörgensen* 2087 (GH, MO, US). Santa Fé: Reconquista, *Job* 931 (GH); between Cayastá and Helvecia, *Ragonese* 2776 (MO). Entre Rios: Isla del Puerto, Dep. Uruguay, *T. Meyer* 10460 (UC); Federación, *Jurado* 27/1579 (MO). Corrientes: Guayquiraró, *Quarín* et al. 2107 (Z); Estancia San Augustín, 12 km NE of C.Pellegrini, *Krapovickas* et al. 29451 (WIS); Paso de los Libres, *Schinini* 7608 (G); near Goya, *Curran* 228 (BM, US); Laguna Iberá, Paso Picada, *Tressens* et al. 3586 (GH, K); Arroyo Villanueva, *Rodrigo* 635 (F); Tres Ceros, *Renvoize* et al. 2983 (K, MO, U, WAG); La Cruz, *Parodi* 12416 (GH); Río Aguapey, *Lourteig* et al. 2848 (P); Bella Vista, Río Paraná, *Tressens & Pire* 65 (LL, WIS); Dep. Santo Tomé, Estancia Garruchos, *Pedersen* 9251 (C, K, L, P, S, UC, US, Z); Dep. Saladas, San Lorenzo, *Schwarz* 9460 (P); Pago de los Deseos, *Schwarz* 8947 (LD, UPS); Río Santa Lucia, *Schwarz* 85 (A, U, US); Dep. Mburucuyá, Estancia Santa Teresa, *Pedersen* 7477 (A, C, E, K, L, P, S, UC, US, Z); Dep. Concepción, Carambola, *Pedersen* 11869 (A, C, L, Z); 12 km NE of San Miguel, *Schinini & González* 9349 (AAU, Z); 5 km E of Governador Virasoro, *Schinini & Carnevali* 10570 (MO, Z); Dep. San Cosme, Costa Toledo, *Krapovickas & Cristóbal* 11603 (UC); Dep. Empedrado, Estancia La Yela, *Pedersen* 8698 (A, C, E, K, L, P, S, UC, US, Z); E of San Luis del Palmar, *Renvoize* et al. 3601 (K, MO, P, U, WAG); Dep. General Paz, Arroyo Santa Isabel, *Schwarz* 8263 (L, P, W); Berón Astrada, *Schwarz* 8331 (LD, NSW); Itá-Ibaté, *Huidobro* 2046 (S, UC, W); Dep. Ituzaingó, Ruta 39, *Krapovickas* et al. 26401 (AAU, G, Z); Arroyo Garapé and Río Paraná, 45 km E of Ituzaingó, *Tressens* et al. 366 (AAU, G, Z); Estancia Santa Tecla, *Schwarz* 8050 (L, P, W); 28 km SE of Concepción, *Schinini* et al. 13053 (G, WIS); near San Cosme, *Pedersen* 5512 (A, C, K, L, P, S, US, Z). Misiones: El Tigre, *Schwarz* 3647 (S); Dep. Apostoles, San José, *Schwarz* 3547 (BM); 8 km E of Leandro N.Alem, *Maruñak* 151 (C, MO, P, WIS); Dep. Candelaria, San Juán, *Schwindt* 930 (L, NSW); Posadas, *Ekman* 1588 (LD, MO, S); Dep. San Ignacio, Arroyo Nancanguozú, *Schwarz* 5058 (L, P, W); Puerto Rico,

Montes 3349 (K, US); La Gruta, *Chodat* 192 (G); Eldorado, T. Meyer 6716 (U); Puerto Segundo, *Montes* 9519 (BH, C, NSW).

CULT. U.S.A., Florida, Fairchild Tropical Garden, Stanley Kiem Coll. 794 from Jaguariava, Paraná, Brazil, *Gillis* 9222 (BH, IBSC, MO, USF). Brazil, D.F., Biological Station UNB, from seeds collected in Minas Gerais, *Heringer* 17178 (K, MO, US, Z), 17956 (K, MO, US, WAG); Minas Gerais, Ituiutaba, *Macedo* 4925 (S, US); São Paulo, Botanical Garden, *Sendulsky* 906 (WAG). Paraguay, Central, Asunción, Botanical Garden, *Little* 40091 (G, GH, MO). Italy, Firenze, from Brazil, Botanical Garden, *Corradi* 1 Sept. 1957 (FI). Indonesia, Bogor, Botanical Garden XVI I B 1, *Bisset* s.n. (Z), of same plant, *Koerniasih* 37 (BH, BO, L).

Notes. *Tabernaemontana catharinensis* is variable in the shape of the leaves, the indumentum, and the length of the corolla and consequently it has been described under several names. Often the more narrowly leaved specimens were named *T. australis. T. affinis* was based on specimens of *T. catharinensis* (*Pohl* 1842, the lectotype, 1844 and *Gaudichaud* 97) and *T. hystrix* (*Clausse*n 325, *Pohl* 5107, *Saint Hilaire* 32 and *Weddel* 876).

T. catharinensis has corolla lobes hairy inside only at the base in almost all collections. However, a few specimens, all known from Maranhão and Ceará in NE Brazil, have corolla lobes hairy inside all over. These could be considered as a distinct forma.

65. Tabernaemontana cerea (Woods.) Leeuwenberg in Novon 1: 105 (1991). – Type: Guyana, NW portion of Kanuku Mts, Mt Iramaikpang, *A.C. Smith* 3606 (holotype MO; isotypes A, B, G, K, MG, NY, P, S, U, US, W).

Fig. 65, p. 267; phot. 1; map 33, p. 271

Basionym and homotypic synonym:

Stemmadenia cerea Woods. in Bull. Torr Bot. Cl. 75: 557 (1948). *Anartia cerea* (Woods.) Allorge in Mém. Mus. natn. Hist. Nat. IIB.30: 58, pl. 22 (1985).

Shrub or small tree 1-6 m high. Trunk 1-8 cm in diamater; bark black. Branches pale or dark brown, lenticellate; branchlets terete when fresh (?), glabrous. *Leaves* of a pair equal or unequal (larger up to twice as long as other and similarly shaped), petiolate; petiole glabrous, 3-20 mm long; (ocreae not widened into intrapetiolar stipules); blade subcoriaceous when dried, elliptic or narrowly elliptic, 1.8-4 x as long as wide, 4.5-15 x 1.5-6.5 cm, acuminate to caudate at the apex, with an obtuse acumen, cuneate at the base or decurrent into the petiole, entire, glabrous and with scattered black dots on both sides, with 4-8 pairs of upcurved secondary veins forming an angle of 60-80° with the costa; tertiary venation inconspicuous. *Inflorescence* shortly pedunculate, 5-7 x 2-5 cm, 1-5-flowered. Peduncle robust, glabrous, 6-20 mm long, covered with bracts; pedicels glabrous, 7-15 mm long. Bracts small, 0.1-0.2 x as long as the sepals, ovate or triangular, obtuse or rounded. *Flowers* reflexed or pendulous. *Sepals* greenish-white, subequal, connate at the base for 0.5-1 mm, erect, ovate, 1.3-2.5 x as long as wide, 6-12 x 3-6 mm, rounded, entire, glabrous outside, not ciliate, glabrous inside and with 20-25 colleters in 2-3 rows at the base; colleters (about 100-125) forming a continuum in the calyx, 0.5-0.8 x 0.2 mm. *Corolla* white or yellow or with creamy tube and yellow limb, 35-37 mm long in the mature bud and forming a comparatively rather wide broadly ovoid head 0.14-0.2 of the bud length (5-7 x 5-8 mm) with an obtuse or almost rounded apex, glabrous outside, not ciliate, with a pubescent belt 6-7 mm wide inside from the insertion of the stamens downwards and a callose ring in the throat; tube 3-4.3 x as long as the calyx, 2.2-2.7 x as long as the lobes, 30-37 mm long, almost cylindrical, slightly widened above the base to 3-5 mm wide, mostly slightly narrowed below the insertion of the stamens to 2-5 mm wide,

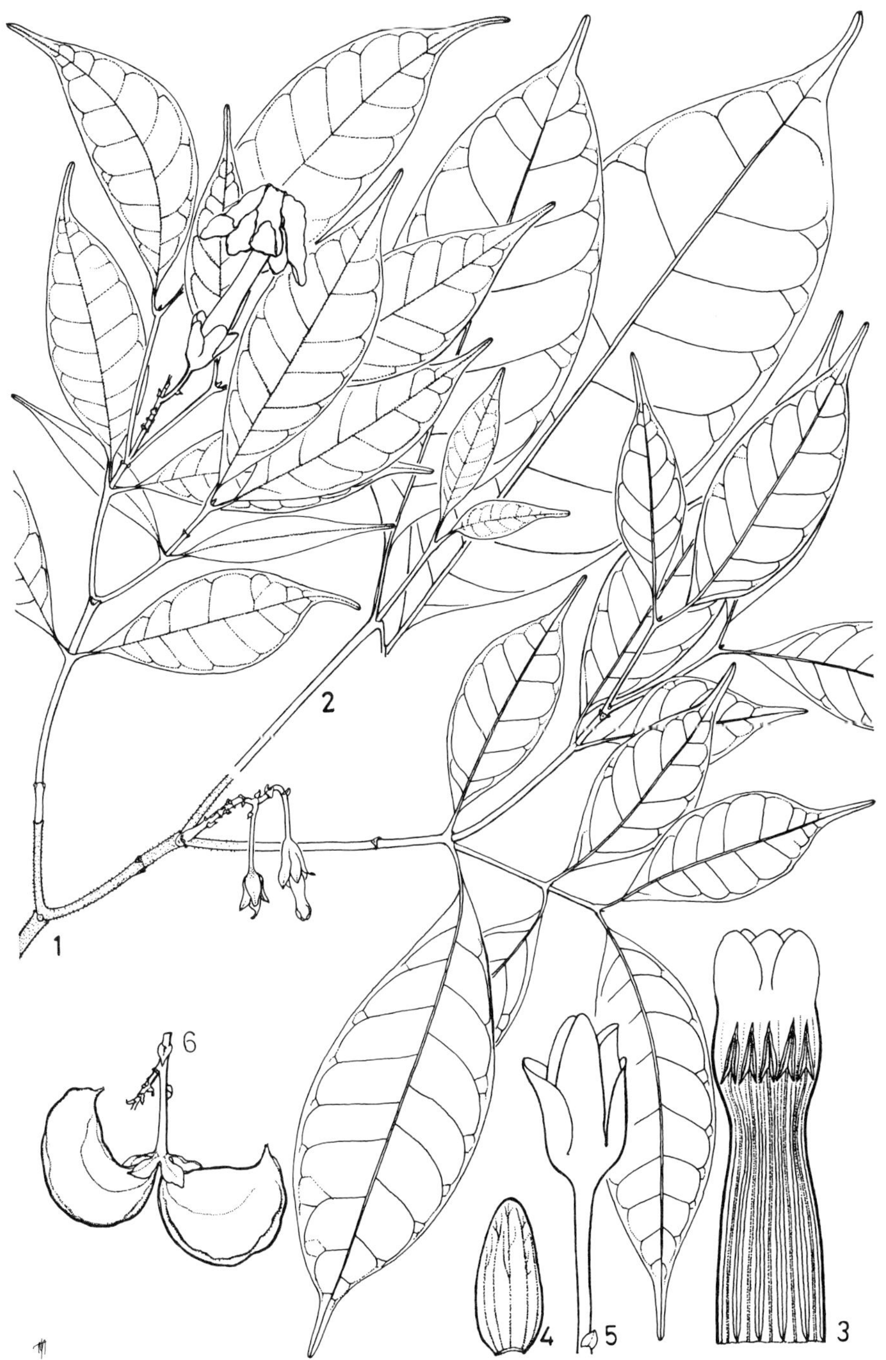

Fig. 65. *Tabernaemontana cerea.* **1-2,** habit (x 2/3); **3,** opened flower bud (x 2); **4,** sepal (x 2); **5,** calyx (x 2); **6,** fruit (x 2/3). 1 from A.C.Smith 3606; 2-5 from Maguire et al. 23536; 6 from A.L. Bernardi 6587.

widened again around the anthers to 3.5-7 mm wide, not twisted; lobes obliquely obovate or elliptic, 0.4-0.48 x as long as the tube, 1.9-3 x as long as wide, 12-17 x 4-9 mm, rounded, undulate, recurved. *Stamens* with apex 5-10 mm below mouth of corolla tube, inserted 0.57-0.7 of the length of the corolla tube (at 20-21 mm from the base); anthers sessile, narrowly triangular, 2.5-4 x as long as wide, 4.5-5.2 x 1.2-1.5 mm, apex acuminate, sterile for 0.3 mm, sagittate at the base, glabrous. *Pistil* glabrous, 22-27 mm long, with apex 0.3-1 way along anthers (0-3.5 mm below apex of anthers), without abscission layer at the apex of the ovary; ovary ovoid, 2.5 x 2 x 2 mm, acuminate, with a disk-like thickening at the base; style filiform, 18-23 mm long, persistent; pistil head composed of an undulate ring 0.3 x 1-1.5 mm, a broadly obovoid apically 5-lobed central part 0.7-0.8 x 0.7-1 mm and a stigmoid apex 0.2 x 0.2 mm. Ovules approximately 100 in each carpel. *Fruit* of two separate mericarps; mericarps green, obliquely ellipsoid, 25-70 x 17-25 x 15-25 mm, apiculate at the apex, smooth, without ridges (?), often slightly torulose when dried, about 20-seeded; wall about 2 mm thick in dried fruits; aril red, enveloping the seed. *Seed* dark brown, obliquely ellipsoid, 10-11 x 3-4 x 2-3 mm, with longitudinal grooves, papillose.

DISTRIBUTION: Northern South America (Eastern Venezuela, Guyana, Suriname).
ECOLOGY: Forest understorey. Alt. 40-1200 m. Flowering and fruiting throughout the year.

Specimens examined:
VENEZUELA. Bolívar: near Guayaraca, *Steyermark* 94160 (MO, NY, US); Kavanayén, *Lasser* 1852 (NY, VEN); Río Icabarú, near Campo Diamantifero Uaiparú, *A.L. Bernardi* 6587 (NY); 29 km ENE of Icabarú, *Steyermark* et al. 117654 (MO). Bolívar-Delta Amacuro Border: Río Grande o Toro, *Breteler* 3847 (HBG, K, L, M, MO, NY, P, S, U, UC, US, WAG, Z). Delta Amacuro: ENE of El Palmar, *Marcano-Berti* 88 (BR, GH, K, NY, P, S, U, US); Santa Elena, Mata Cutia, *Rosa & Nascimento* 3315 (MG).
GUYANA. Demerara County, *Abraham* 311 (K); Kakaralli For., *Maguire & Fanshawe* 33417 (K, NY); Puruni, Marabukea Hill, For. Dept. 7743 (K, NY); Fillipui, Mapauyek, For. Dept. 7880 (K, NY); West Demerara, Mabura Hill, *ter Steege & de Jager* 422 (U, WAG); km 115 Bartica-Potaro Road, *Fanshawe* F.D.5572 (G, K, P, S); Upper Mazaruni R. Region, Kamarang trail, *Boom* et al. 8419 (WAG); ibid., Karowtipu Mt, *Boom & Gopaul* 7437 (K, US, WAG, Z); Cuyuni-Mazaruni Region, Eping R., *McDowell & Stobey* 3998 (US, WAG); km 123 Bartica-Potaro Road, Morabukea For., *Tutin* 212 (BM, K, RB, U, US), 228 (BM, K), 301 (BM, CGE, K, U, US); Manabadin, Christianburg, *Anderson* 445 (K); Tumatumari, *Gleason* 160 (GH, K, NY, US), 285 (GH, NY); Mahdia R., Potaro R., For. Dept. 3729 (K, NY); Waraputa Compartment, ca 25 km S of Mabura, *Polak & Maas* 507 (U); Potaro R. Gorge, Tukeit, *Maguire & Fanshawe* 23536 (A, F, K, MO, NY, P, US); Itabru Fall, Berbice R., For. Dept. 2671 (K); NW portion of Kanuku Mts, Mt Iramaikpang, *A.C. Smith* 3606 (A, B, G, K, MG, MO, NY, P, S, U, US, W, type); Wabuwak, Kanuku Mts, For. Dept. 5796 (K, NY).
SURINAME. Wilhelmina Gebergte, 1.5 km SE of Julianatop, *Irwin* et al. 55079 (C, F, MO, NY, US).

Note. *Tabernaemontana cerea* is similar to *T. flavicans* and *T. attenuata* in the leaves, corolla tubes and fruits. These species differ as described in couplets 21-22 of key 1, p. 218.

66. Tabernaemontana chocoensis (A. Gentry) Leeuwenberg in Agric. Univ. Wageningen Pap. 87, 5: 4 (1988). – Type: Colombia, Choco, Upper Río Truandó, La Teresita, *A. Gentry* 9406 (holotype COL, not seen; isotypes MO, Z).

Fig. 66, p. 270; map 33, p. 271

Basionym:

Bonafousia chocoensis A. Gentry in Ann. Miss. Bot. Gard. 68: 116 (1981); Allorge in Mém. Mus. natn. Hist. Nat. IIB. 30: 109, pl. 48 (1985).

Shrub 1 m high or more. Trunk 2.5-5 cm in diameter; bark pale brown, deeply and irregularly corrugated. Branches pale brown, lenticellate; branchlets terete, glabrous. *Leaves* sessile or with petiole up to 1 mm long; petiole glabrous; (ocreae obscure); blade thinly papery when dried, elliptic or narrowly elliptic, 1.9-2.7 x as long as wide, 7-21 x 3.5-10 cm, caudate at the apex, acumen 10-30 mm long, rounded to subcordate at the base, entire, glabrous on both sides, with scattered black dots beneath or not, with 8-14 pairs of rather straight secondary veins forming an angle of approximately 60° with the costa; tertiary venation reticulate. *Inflorescence* pedunculate, 2 x 2 cm, 2-flowered. Peduncle short, glabrous, 5 mm long; pedicels glabrous, 3 mm long. Bracts sepal-like and similarly sized. *Flowers* open during the day. *Sepals* white (?), free nearly so, erect, broadly ovate, 2 x 2 mm, obtuse, glabrous outside, not ciliate, glabrous inside and with 3-4 colleters in 1 row in the middle at the base. *Corolla* light yellow or white with a yellow throat, 16 mm long in the mature bud and forming a narrowly ovoid head 0.4 of the bud length (6 x 3 mm) with a blunt apex, glabrous outside, with a pilose belt 3 mm wide inside, 2 mm below and 1 mm above insertion of stamens; tube 6 x as long as the calyx, 0.9 x as long as the lobes, 14 mm long, cylindrical, 3 mm wide above the base, narrowed to 1.7 mm below and above the anthers; around them 2 mm wide, twisted 0.5 turn around the anthers; lobes obliquely obovate, 1.07 x as long as the tube, 15 x 8 mm, rounded, neither auriculate, nor undulate, recurved. *Stamens* with apex 2 mm below mouth of corolla tube, inserted 0.65 of the length of the corolla tube, (at 9 mm from the base); anthers sessile, with tails 0.5 mm below insertion, narrowly triangular, 3.5 x 1 mm, twisted with corolla, apex acuminate, probably sterile for 0.3 mm, sagittate at the base with straight tails. *Pistil* glabrous, 10 mm long, with apex halfway along anthers; ovary ovoid, about 2 x 1.5 x 1.5 mm; style filiform; pistil head about 1.2 mm high, composed of a basal ring, a stipitate 5-lobed depressed globe and a stigmoid apex. *Fruit* of 2 separate mericarps; mericarps green (?), obliquely ellipsoid, 25-30 x 15-20 x 15-20 mm, smooth, subacute or shortly acuminate at the apex, with faint lateral ridges, approximately 10-20-seeded; wall about 1 mm thick in dried fruits; aril probably only at the hilar side. *Seed* dark brown, obliquely ellipsoid, 7-9 x 4 x 3 mm, with longitudinal grooves, dull, papillose.

Distribution: Colombia (Chocó).
Ecology: Swamp forest. Alt. 100-200 m.

Specimens examined:

Colombia. Chocó: trail from Camp Teresita to Río Salada (fr. June) *Duke* 12230 (ECON, MO, NY, P, paratype); near Helipad, Río Truandó, at junction with Quebrada Buche (fl. fr. Apr.) *Duke* 15743 (MO, paratype); Upper Río Truandó, La Teresita (bud Jan.) *A. Gentry* 9406 (COL, not seen, MO; phot. of MO sheet P, type); Río Truandó, between the confluence of Ríos Chintadó and Saladó (fr. Oct.) *Romero-Castañeda* 6110 (COL).

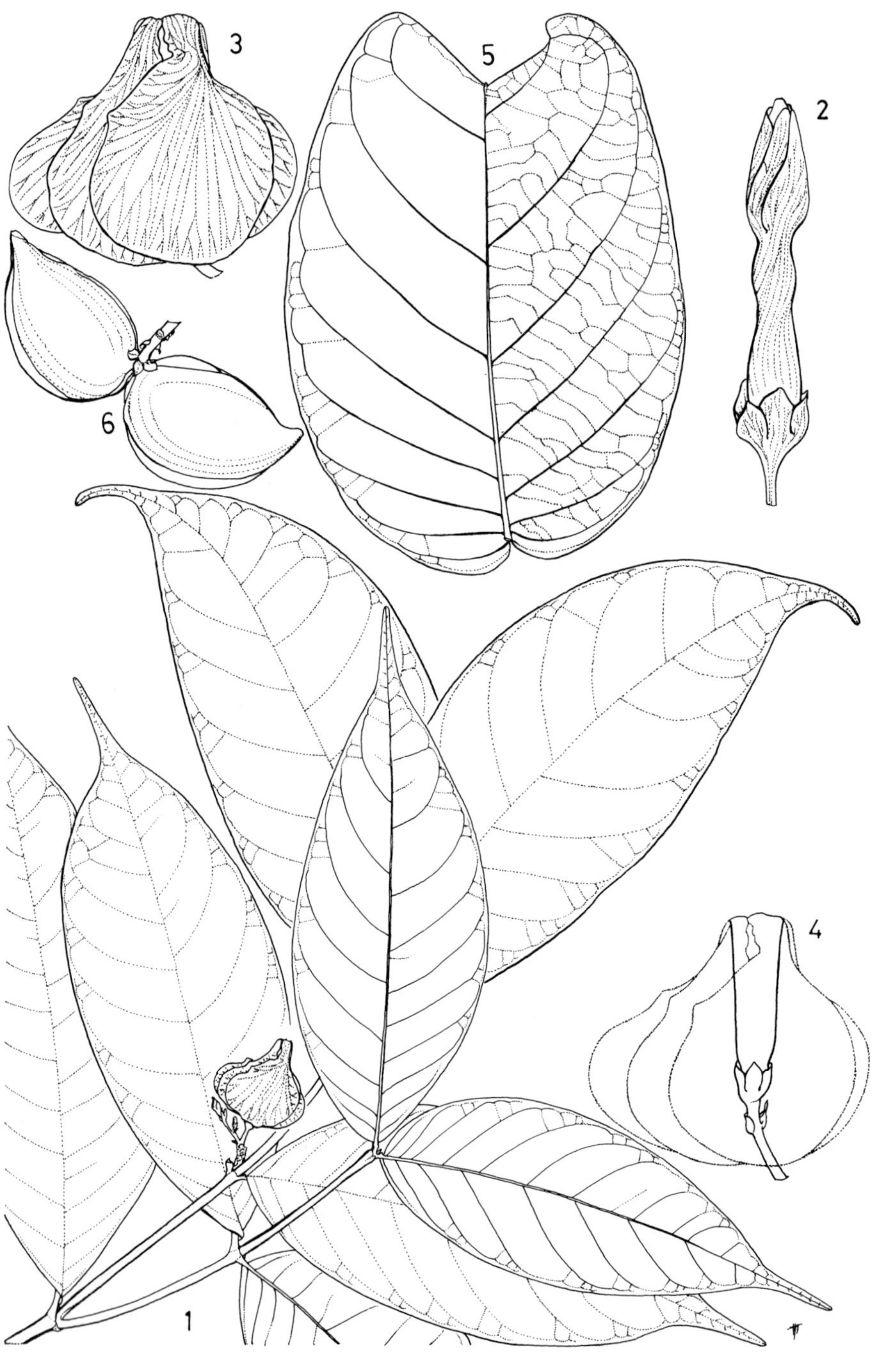

Fig. 66. *Tabernaemontana chocoensis*. **1,** habit (x 2/3); **2,** flower bud (x 2); **3,** flower (x 2); **4,** flower with corolla lobes as if transparent to show tube and calyx (x 2); **5,** leaf with damaged apex (x 2/3); **6,** fruit (x 1.2). 1, 3 and 6 from Duke 15743; 2 and 4-5 from A.Gentry 9406.

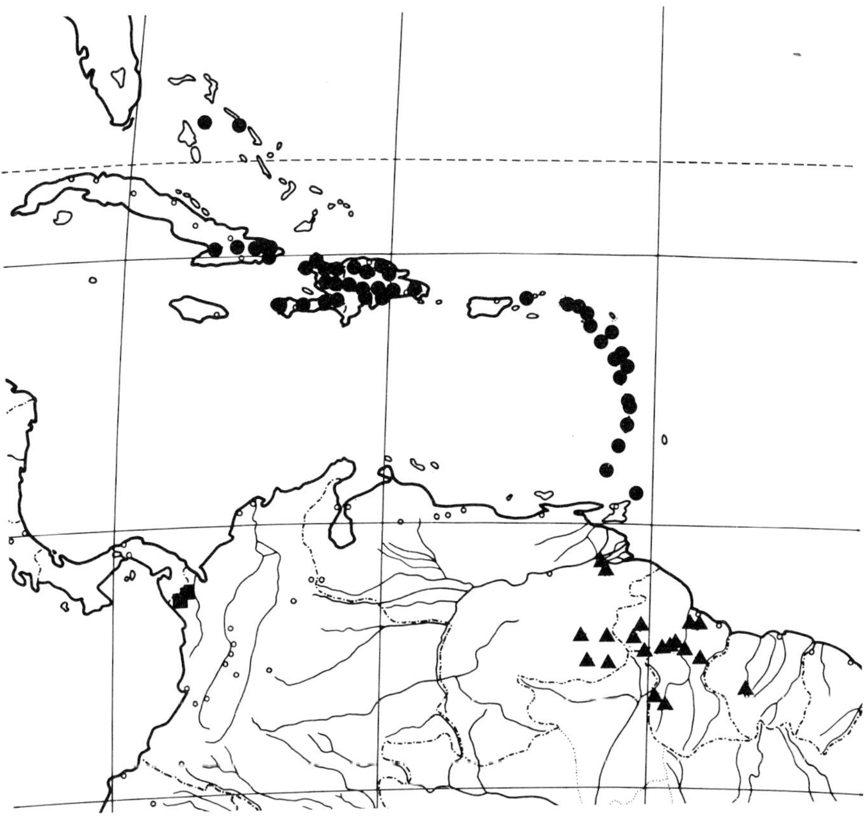

Map 33.▲. *Tabernaemontana cerea*, ■. *T. chocoensis*, ●, *T. citrifolia*.

67. Tabernaemontana citrifolia L., Sp. Pl. 210 (1753). – Type: Plumier, Gen. 18 (1703) (lectotype, designated by Leeuwenberg in *C.E. Jarvis* et al., A list of Generic Names and their Types 92 (1993)).

Fig. 67, p. 273; phot. 2, 3; map 33, p. 271

Heterotypic synonyms:

T. lanceolata Miers, Apoc. S. Am. 268, t.7 (1878). *T. citrifolia* var. *lanceolata* (Miers) Allorge in Mém. Mus. natn. Hist. Nat.IIB, 30: 18 (1985). – Type: Eastern Cuba, sin. loc., *Wright* 2948 p.p. (holotype BM; isotypes G, GH, GOET, K, MO, P). The NY sheet of this number belongs to *T. alba*.

Shrub or small tree 1-8 m high. Branches pale brown, lenticellate; branchlets terete, glabrous. *Leaves* of a pair equal or unequal (larger up to twice as long as other and not comparatively narrower), petiolate; petiole glabrous, 3-23 mm long; (ocreae widened into intrapetiolar stipules or not); blade subcoriaceous or coriaceous and often with pale venation when dried, elliptic or narrowly elliptic, 2-6 x as long as wide, 2-21 x 1-8.5 cm acuminate or apiculate at the apex with a usually blunt acumen, cuneate at the base, entire, with a revolute margin, glabrous on both sides, with scattered black dots beneath or not, with 8-17 pairs of rather straight or upcurved secondary veins forming an angle of 60-80° with the costa; tertiary venation reticulate, often impressed above in dried leaves, conspicuous beneath. *Inflorescence* pedunculate, 5-10 x 5-10 cm, 3-20(-30)-flowered, lax. Peduncle slender, glabrous, 3-25 mm long; pedicels glabrous, 5-12 mm long. Bracts sepal-like about 0.3 x as long as them. *Flowers* fragrant, open during the day. *Sepals* pale green (?), connate at the extreme base, erect, ovate or broadly ovate, 1-2 x as long as wide, 2-3.5 x 1.5-2 mm, obtuse or rounded, glabrous outside, not ciliate, glabrous inside and with 1-6 colleters in 1 row in the middle at the base; colleters mostly unequal, 0.2-0.6 x 0.1-0.2 mm. *Corolla* white, cream or yellow, or limb white and tube yellow, 11-16 mm long in the mature bud and forming a comparatively wide ovoid head 0.3-0.44 of the bud length (3-7 x 2.5-5 mm) with a blunt apex, glabrous outside, glabrous inside or sometimes with minute glandular hairs around the anthers; tube 3-5 x as long as the calyx, 0.7-1.1 x as long as the lobes, 8.5-12.5 mm long, almost cylindrical, 2-3.5 mm wide above the base, narrowed at the insertion of the stamens to 1.2-2 mm wide, widened again around the anthers to 2-3 mm wide, not twisted, with 5 longitudinally bilobed thickenings in the throat, visible in the mouth; lobes obliquely elliptic to dolabriform, 0.9-1.4 x as long as the tube, 2-3 x as long as wide, 8-14 x 3.5-6.5 mm obtuse or rounded, (not or obscurely auriculate at the left side of the base), spreading. *Stamens* with apex (0-)0.5-1 mm above mouth of corolla tube, inserted 0.7-0.8 of the length of the corolla tube (at 6.5-9.5 mm from the base); anthers sessile, narrowly triangular, 2.5-3 x as long as wide, 2.5-3 x 1-1.2 mm, apex acuminate, sterile for 0.3-0.4 mm, (mostly shortly) sagittate at the base, glabrous. *Pistil* glabrous, 8-11 mm long, with apex about halfway along anthers; ovary ovoid or oblong, 2-4 x 1-1.5 x 1-1.5 mm, of 2 separate carpels, gradually narrowed into the style; style 5-7 mm long; pistil head 1-1.2 mm high, composed of an undulate basal ring-like veil 0.2 x 1-1.5 mm, an apically 5-lobed stipitate depressed globe 0.3-0.5 x 0.6-0.8 mm, and a stigmoid apex 0.1-0.2 x 0.1-0.2 mm. Ovules approximately 50-70 in each carpel. *Fruit* of 2 separate mericarps; mericarps green, bright red or yellow inside, not dotted, obliquely pod-like, 25-50 x 7-15(-20) x 7-12(-20) mm, acuminate at the apex, recurved, with one ridge at each side clearly separated from the line of dehiscence, approximately 10-50-seeded; wall 1 mm thick in dried fruits; aril orange, enveloping the seed. *Seed* dark brown, obliquely ellipsoid, 6.5-11 x 3-4 x 3-3.5 mm, with longitudinal grooves, papillose; embryo 4.5-6 mm long; cotyledons ovate, 1.2-2 x as long as wide, 1.8-3 x 1.5-2 mm, rounded at the apex, (sub)cordate at the base; rootlet 3-3.5 x 0.5-0.6 mm.

DISTRIBUTION: Eastern Cuba, Hispaniola, Bermudas and Lesser Antilles South to Tobago.

ECOLOGY: Forest or bush. Alt. 0-900 m. Flowering throughout the year with a peak in April.

Geographical selection of the approximately 280 specimens examined:

BAHAMA ISLANDS. Eleuthera: Tarpum Bay, *Correl* 48939 (MO, NY, US). Nassau: Hotel Victoria, New Providence, *Fairchild* 2595 (MO, US).

CUBA. Granma: near Bayamo, *Shafer* 12384 (NY, US). Santiago de Cuba: Jimbambay R., *Ekman* 9557 (B, MO, S); Cabañas R., Moa, Bro. *Alain* & Bro. *Clemente* 940 (GH, US). Guantánamo: Cerro de Cananova, Bro. *Clemente* 6236 (GH, US); Lower Valley of Río Miel, *Shafer* 4348 (F, NY, US); near Imías, Fre. *Leon* 12168 (GH, NY, US). E Cuba: Toscano (not localized), *Wright* 2948 p.p. (BM, G, GH, GOET, K, MO, P, type of *T. lanceolata*).

HAITI. near Jean Rabel, *Leonard* 12596 (NY, US); Port de Paix, *Ekman* 3635 (C, LL, S, US); Bayeux, *Nash* 76 (K, NY); Gonaïves, *Buch* 233 (IJ); near Dondon, *Leonard* 8710 (F, US); Lower Valley of Tiburon R., *Christ* 2281 (Z); near Les Cayes, *Ekman* 4 (A, S); Salagnac, *Sastre* & *Polynice* 7100 (K, P); Mariani, *Ekman* 9962 (G, K, LL, NY, S, US); between Mariani and Gressier, *Proctor* 10978 (IJ, US); near

Fig. 67. *Tabernaemontana citrifolia.* **1,** habit (x 2/3); **2,** leaf (x 2/3); **3,** flower (x 4); **4,** opened corolla (x 4); **5,** sepals inside (x 6); **6,** pistil (x 6); **7,** fruit (x 2/3). 1 and 3-6 from P.Beard 1019; 2 and 7 from Ekman 3583.

Mirebalais, *Bailey* 107 (BH, US); Manneville Spring, Thomaseau, *Holdridge* 1082 (LL, NY, US); near Etang Saumatre, *Leonard* 3545 (GH, NY, US); near Parédon, *Proctor* 10969 (IJ, US).

SANTO DOMINGO. Guazumal, *Jiménez* 4535 (US); Guayubín, *Abbot* 913 (US); near San Juan de la Maguana, *Howard* 8794 (BM, C, FI, GH, MO, NY, P, S); near Barahona, Father *Miguel Fuertas* 1127 (BM, BP, E, F, FI, G, GH, HBG, K, L, NY, P, U, US, W, WAG, Z); ibid., *von Tuerckheim* 2818 (A, BM, BP, BR, E, F, G, GH, HBG, K, L, MO, P, S, U, US, W, Z); Puerto Plata, *Eggers* 1511 (A, BM, BR, C, G, GOET, HBG, K, L, LE, M, NY, P, US, WU); Mao R., La Hoya, *Valeur* 270 (A, F, G, GH, MO, NY, S, US); Río Yaque, near Santiago, *Howard* 9699 (BM, FI, GH, MO, NY); NW of Azua, *Rose* et al. 3926 (NY, US); Sabana Roble, Prov. Peravia, *Maas & Zanoni* 6392 (U, WAG); Santo Domingo, *von Tuerckheim* 2606 (BR, G, M, NY); Samaná Peninsula, E of Guazuma, *Zanoni* et al. 29262 (USF).

SAINT THOMAS. sin. loc., *Finlay* 110 (P).

SAINT MARTIN. Mt Paradis, *Boldingh* 3213 B (NY, U).

SABA. Booby Hill, Bro. *Arnoldo* 3285 (A, U, WAG).

SAINT EUSTATIUS. Slope of the Quill, near Glassbottle, *Boldingh* 1207 B (U, WAG).

Saint Kitts. Lower portion of Wingfield Ravine, Proctor 19644 (A, BM, IJ).

NEVIS. near Charlestown, *Simmonds* Apr. 1957 (K).

ANTIGUA. Wallings, *J.S. Beard* 282 (A, K, MO, NY, U, UC).

MONTSERRAT. Woodlands Estate, *Proctor* 18820 (A, BM, IJ).

GUADELOUPE. near Basse Terre, Father *Düss* 2617 (F, GH, MO, NY, US); Gosier, Les Grandes Fonds, *Jéremie* 300 (A, P); Baie du Moule, *Sastre & Jéremie* 1870 (A, MO, P, U). Marie Galante. 2 km SW of St. Louis, *Proctor* 20224 (BM, IJ); Saragot, *Barrier* 3187 (A, WAG).

DOMINICA. St. Joseph, Layou R., *Wasshausen & Ayensu* 305 (IJ, US); Layou R. valley, Clark Hall Estate, *Ernst* 1440 (GH, P, US); between Prince Rupert Bay and Douglas Bay, *Hodge* 675 (GH, NY, US).

MARTINIQUE. sin. loc., *Sieber* 75 (BH, BM, FI-W, G, K, M, MEL, MO, NY, P, W); near Fort de France, *L. Hahn* 1457 (BM, BR, G, MPU, P); Mont Parnasse, *L. Hahn* 122 (BM, FI, G, GH, P, UC, W); Belleville, N of Précheur, *Larsen* 35357 (AAU).

SAINT LUCIA. The Morne, Castrie, *P. Beard* 1019 (F, GH, MO, S, US); near the Sulphur Springs at Sufriere, *Howard* 11566 (A, BM, MO, S, US); Castries R., *Proctor* 17541 (A, BM, IJ).

SAINT VINCENT. Kingstown, *P. Beard* 1400 (F, GH, MO, NY, S, UC, US); near Georgetown, *Proctor* 26144 (GH, IJ, USF); near Fort Charlotte, *Morton* 4845 (GH, US).

GRENADA. Balthazar, *J.S. Beard* 215 (A, K, MO, NY, U); Morne Délice, *P. Beard* 1291 (F, GH, K, MO, NY, S, UC, US).

TOBAGO. Pigeon Hill, *Cowan & Forster* 1453 (G, K, MO, US); ibid., *Sandwith* 1802 (K, NY, S, U).

SIN. Loc. LINN 304.2.(LINN, phot. F).

CULT. U.S.A., Florida, Miami, *Popenoe* 327 (BH). Hawai'i, Oahu, *Teraoka* 186 (US). Bahamas, *Brace* 494 (F); *Robinson* 51 (K).

Notes. In Cuba, Hispaniola and in the Bahamas *T. citrifolia* has mostly narrower leaves than elsewhere. In Cuba and Hispaniola they are 2.4-6 x as long as wide, in the Bahamas 3.4-6 x and in the other islands 2-4 x. One of the narrow leaved specimens, *Wright* 2948 p.p., is the type of the synonym *T. lanceolata*. As none of the other characters corroborates the distinction of the latter as a separate taxon, it is reduced to a synonym here.

68. Tabernaemontana columbiensis (Allorge) Leeuwenberg in Agric. Univ. Wageningen Pap. 87, 5: 5 (1988), (*as colombiensis*). – Type: Colombia, Nariño, Esprielle, Tumaco, *Castañeda* 2799 (holotype MO; isotype COL).

Fig. 68, p. 276; map 34, p. 277

Basionym:

Bonafousia columbiensis Allorge, Ann. Miss. Bot. Gard. 68: 679, fig. 2 (1981); Mém. Mus. natn. Hist. Nat. IIB, 30: 111, pl. 49 (1985), partly, excl. *Romero Castañeda* 2920 which belongs to *T. panamensis*.

Shrub or tree 1-13 m high. Trunk up to at least 15 cm in diameter; bark pale grey-brown, longitudinally fissured, corky; wood brownish-white. Branches pale grey-brown, lenticellate; branchlets terete, glabrous. *Leaves* petiolate; petiole glabrous, 2-12 mm long (ocreae widened into intrapetiolar stipules); blade thinly papery when dried, elliptic or narrowly elliptic, 2.2-4 x as long as wide, 6-36 x 2.5-16 cm, acuminate or caudate at the apex, cuneate at the base, entire, glabrous on both sides, with or without scattered black dots beneath, with 7-15 pairs of upcurved secondary veins forming an angle of 50-60° with the costa; tertiary venation reticulate. *Inflorescence* shortly pedunculate, 3-4 x 3-4 cm, 2-10-flowered, congested. Peduncle robust, glabrous, 2-4 mm long; pedicels 3-5 mm long, glabrous. Bracts scale-like, 0.3 x as long as the sepals. *Flowers* open during the day. *Sepals* white, connate at the base for 0.3-0.5 mm, erect, ovate or circular, 1-1.5 x as long as wide, 2.5-3.5 x 2-3 mm, rounded, glabrous outside, ciliate or not, glabrous inside and with 3-6 colleters in one row in the middle at the base; colleters 0.3-0.5 x 0.1-0.2 mm. *Corolla* white or cream, orange or yellow in the throat or on the base of the lobes, (17-)25-33 mm long in the mature bud and forming a comparatively small subglobose head 0.14-0.2 of the bud length (3-6 x 3-6 mm) with a rounded or blunt apex, glabrous outside, with a pilose-pubescent belt 4-7 mm wide inside just below the anthers; tube (5.7-)7-10 mm x as long as the calyx, 1.6-2.5 x as long as the lobes, (17-)23-30 mm long, almost cylindrical, 2-4 mm wide at the base, widened above to 3-6 mm wide, again narrowed around the stamens and from there almost perfectly cylindrical and 3-5 mm wide, twisted 0-0.1 turn to the left in basal half and 0.3-0.6 turn to the right from just below the insertion of the stamens or less often not twisted at all; lobes obliquely oblong, 0.57-0.65 x as long as the tube, 2.6-3.3 x as long as wide, 10-18 x 3-7 mm, rounded, not auriculate, undulate, recurved. *Stamens* with apex 3-5.5 mm below mouth of corolla tube, inserted 0.55-0.6 of the length of the corolla tube, (at (11-)15-21 mm from the base); anthers sessile, with tails 1 mm below insertion, narrowly triangular, (3-)4.5-5.5 x (1-)1.5 mm, apex acuminate, sterile for 0.2-0.5 mm, sagittate at the base. *Pistil* glabrous, (13-)16-24 mm long; ovary ovoid, 2-3 x 1.5-2.5 x 1.5-2.5 mm; disk-like thickening 0.8-1 mm high; style filiform, (10-)13-20.5 mm long; pistil head about 1 mm high, composed of a basal ring 0.5 x 1.2 mm, a stipitate 5-lobed depressed globe about 0.5 x 0.3 mm and a cleft stigmoid apex 0.2 x 0.2 mm. Ovules approximately 30 in each carpel. *Fruit* of 2 separate mericarps; mericarps yellow, obliquely ellipsoid or ovoid, not dotted, 25-55 x 10-27 x 10-25 mm, recurved or not, acute or acuminate (seen from the sides) or rounded (seen from above) at the apex, with a lateral ridge at each side and a raised line along the line of dehiscence, minutely papillose or smooth, often wrinkled when dried and almost mature, approximately 10-30-seeded; wall about 2 mm thick in dried fruits; aril white, enveloping the seed. *Seed* black or dark brown, obliquely ellipsoid or polyhedral, 7-12 x 3-5 x 3-4 mm, with longitudinal grooves, papillose; embryo 7-8 mm long; cotyledons 1.4-2.1 x as long as wide, 3-3.5 x 1.8-2.5 mm, rounded at the apex, subcordate or rounded at the base; rootlet 4-4.5 x 0.8-0.9 mm.

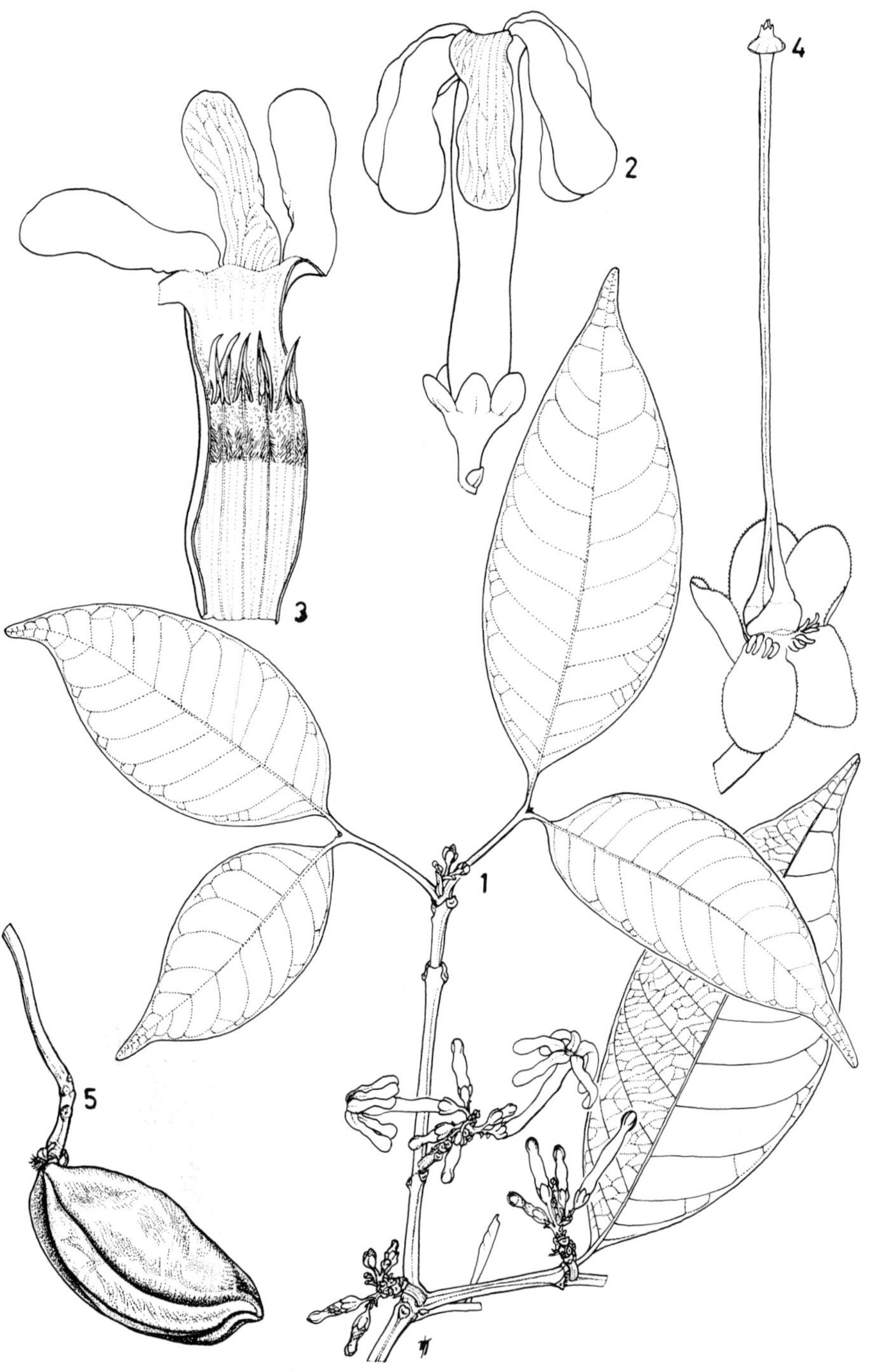

Fig. 68. *Tabernaemontana columbiensis*. **1,** habit (x 2/3); **2,** flower (x 2); **3,** opened corolla (x 2); **4,** calyx with pistil (x 4); **5,** fruit (x 2/3). 1-3 from Cazalet et al. 5166; 4 from Harling 124; 5 from Cogollo 979.

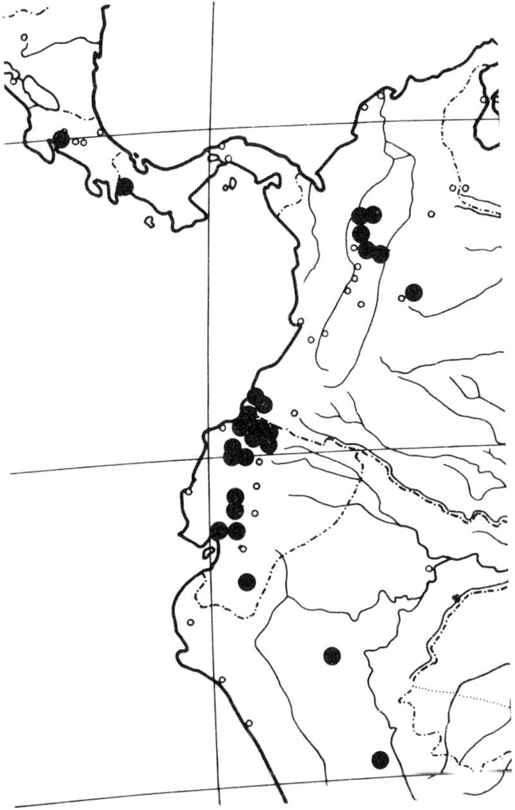

Map 34. *Tabernaemontana columbiensis.*

DISTRIBUTION: Costa Rica to Peru.

ECOLOGY: Wet forest understorey. Alt. 0-900 m. Probably flowering and fruiting throughout the year.

USES: Fruits edible (teste *Rimbach* 45).

Geographical selection of the approximately 70 specimens examined:

COSTA RICA. Puntarenas: Fila de Cal, between Las Cruces and Nelly, *Gómez* 19612 (MO).

PANAMA. Chiriqui: San Bartolo Limite, 20 km W of Puerto Armuelles, *Busey* 547 (MO).

COLOMBIA. Antioquia: Mt Chimborazo, *Spruce* July 1860 (K, paratype); Mun. Zaragoza, Providencia, near Mazo, *Goudot* 1 (K); San Luís-Independencia Road, *Romero-Castañeda* 1553 (COL); Mun. San Luís, Río Claro, *Cogollo* 979 (MO); ibid., road to Cairo, *Rentería & Cogollo* 2728 (MO); near La Josefina, Quebrada La Salada, *Hoyos & Hernández* 792 (MO). Nariño: Tumaco, *Romero-Castañeda* 2799 (MO, type); S of Tumaco, *Romero-Castañeda* 5402 (AAU, COL).

ECUADOR. Esmeraldas: San Lorenzo, *Játiva & Epling* 726 (S, UC, US); Río Zapallo Grande, tributary of Río Cayapa, *Barford* 41044 (AAU, QCA, WAG); ibid., *Kvist & Asanza* 40739 (QCA, WAG); Río San Miguel, *Harling* 4650 (S). Carchi: km 7 Lita-Alto Tambo Road, *van der Werff* et al. 12072 (WAG); 25 km NW of El Chical, Parroquia Maldonado, *Rubio* et al. 981 (WAG). Imbabura: 2 km from Lita, *D'Arcy*

14851 (MO). Pichincha: km 12 Patricia Pilar-24 de Mayo Road, *Dodson* et al. 8705 (MO); near Santo Domingo, *Játiva & Epling* 527 (US, paratype); ibid., *Penninton* 86 (K, paratype); 20 km W of Santo Domingo, *Cazalet & Pennington* 5166 (B, K, UC, US, paratype); Finlandia, 16 km E of Santo Domingo, *A. Gentry* et al. 12159 (AAU, MO, Z); ENDESA For. Res., Río Silanche, *Jaramillo* 7463 (AAU, GB, MO, QCA); ibid., 6 km WNW of P Vincente Maldonado, *Harling & Anderson* 23263 (GB, QCA, USF). Los Rios: Hacienda Clementina Samana, *Harling* 124 (S); Río Palenque Biol. Stat., km 56 Quevedo-Santo Domingo Road, *Dodson* 5025 (MO), 5457 (MO, US, paratype), 6611 (MO); ibid., *Dodson & A. Gentry* 10043 (MO, USF). Guyas: Tafalla, Guyaquil, *Ruiz & Pavón* anno 1800 (F; phot. of MA sheet WAG); Teresita, 3 km W of Bucay, *Hitchcock* 20460 (US). Morona-Santiago: Los Encuentros, *Sparre* 19010 (S). Foot of W Cordillera, *Rimbach* 45 (A, F, paratype).

PERU. San Martín: Lamas, *Knapp & Mallet* 6996 (MO, WAG). Huanuco: region of Pucallpa, *Morawetz* 31-31888 (WAG, Z).

Notes. *Tabernaemontana columbiensis* resembles *T. sananho* strikingly as far as the foliage is concerned, although its leaves are usually narrower. The two species differ mainly as follows:

Corolla lobes 0.57-0.65 x as long as the tube; tube (17-)23-30 mm long; mericarps obliquely ellipsoid or ovoid, with a lateral ridge at each side, minutely papillose or smooth, often wrinkled when dried...................................... **T. columbiensis**

Corolla lobes 0.8-1.3 x as long as the tube; tube 14-25 mm long; mericarps obliquely subglobose, with 2 faint ridges, especially at the base, smooth, not wrinkled when dried... **T. sananho**

69. Tabernaemontana coriacea Link ex Roem. & Schult., Syst. 4: 431 (1819). – Type: Brazil, sin. loc., herb. *Link* s.n. (holotype B†; excellent photographs of it G, GH, MO, NY, US). Fig. 69. p. 280; map 35, p. 282

Homotypic synonyms:

Anacampta coriacea (Link ex Roem. & Schult.) Mgf. in Notizbl. Bot. Gart. Berlin 14: 162 (1938). *Bonafousia coriacea* (Link ex Roem. & Schult.) Boiteau & Allorge in Bull. Soc. Bot. France 130, Lettres Bot. 4-5: 340 (1983); Mém. Mus. natn. Hist. Nat. II B, 30: 83 (1985).

Heterotypic synonyms:

T. rubrostriolata Mart. ex Muell. Arg. in Martius, Fl. Bras. 6, 1: 71 (1860). *Anacampta congesta* Miers, Apoc. S. Am. 65 (1878). – Type: Brazil, Pará, near Santarém, *Spruce* 232 (holotype G; isotypes AWH, BM, CGE, FI-W, GH, K, M, NY, P, TCD, W; phot. of G sheet GH, MO, NY, US; phot. of W sheet GH, MO, US; the BM sheet holotype of *Anacampta congesta*).

T. submollis Mart. ex Muell. Arg. in op. cit. 70, **syn. nov.** *Anacampta submollis* (Mart. ex Muell. Arg.) Miers, op. cit. 67. *Bonafousia submollis* (Mart. ex Muell. Arg.) Boiteau & Allorge, l.c. and in op. cit. 77. – Type: Brazil, Amazonas, mouth or Rio Negro, *Spruce* 1666 (holotype G; isotypes AWH, BM, BP, BR, C, CGE, F, FI, GH, GOET, K, LD, MPU, NY, P, TCD, W).

T. riedelii Muell. Arg. in op. cit. 72. *Taberna riedelii* (Muell. Arg.) Miers, op. cit. 64. *Anacampta riedelii* (Muell. Arg.) Mgf. in op. cit. 163. – Type: Brazil, Amazonas, Rio Madeira, near Borba, *Riedel* s.n. (cited as 1378 by Mueller) (holotype B†; phot. G, GH, MO, NY, US; lectotype P, designated here; isotype G).

T. acutissima Muell. Arg. in op. cit. 73. *A. acutissima* (Muell. Arg.) Miers, op. cit. 66.

– Type: Brazil, Amazonas, Amazonas R., between Coari and Rio Negro, *Poeppig* Mar. 1832 (holotype W; isotype F).

T. brachyantha Woods. in Ann. Miss. Bot. Gard. 47: 76 (1960), non Stapf (1894). *Bonafousia brachyantha* Boiteau & Allorge in Bull. Soc. Fr. 130, Lettres Bot. 4-5: 340 (1983); Mém. Mus. natn. Hist. Nat. IIB. 30: 81, pl. 32 (1985). *T. luciliae* Leeuwenberg, Agric. Univ. Wageningen Pap. 87. 5: 14 (1988), **syn. nov.** – Type: Peru, Ucayali, Aguaytia, *Woytkowski* 5345 (holotype MO; isotypes BR, C, F, P, S).

Anacampta pendula Mgf. in Brittonia 23: 438, fig. 2 (1971). *Bonafousia pendula* (Mgf.) Boiteau & Allorge, l.c. and in op. cit. 81, pl. 33, partly, excl. fruiting specimens. – Type: Brazil, Amazonas, Rio Urubu, between Cachoeira Iracema and Manaus-Itacoatiara Road, *Prance* et al. 5076 (holotype Z; isotypes K, NY, U, US).

Bonafousia prancei Allorge in op. cit. 342, fig. 2 and in op. cit. 79, pl. 31, partly, excl. fruiting specimens. *T. prancei* (Allorge) Leeuwenberg in op. cit. 22, **syn. nov.** – Type: Brazil, Rondônia, near Jaciparaná, *Prance* et al. 5171 (holotype NY; isotypes B, S, US, W, Z).

Shrub or small tree 0.40-3 m high. Branches brown, lenticellate; branchlets terete when fresh, not angular, sulcate when dried, puberulous or glabrous. *Leaves* shortly petiolate; petiole puberulous or glabrous, 2-20 mm long; (ocreae slightly widened into intrapetiolar stipules or not); blade papery when dried, elliptic, narrowly elliptic or narrowly ovate, variable in size, even in a single branchlet, 1.5-3.5 x as long as wide, 5-33 x 2-15 cm, acuminate at the apex, cuneate or rounded at the base, glabrous on both sides or sometimes puberulous at the base of the costa beneath, mostly pale and dull and often with scattered black dots beneath; with 10-21 pairs of rather straight secondary veins forming an angle of 60-85° with the costa; tertiary venation not conspicuous, reticulate. *Inflorescence* shortly pedunculate, 2 8 x 2 5 cm, 2 30 flowered, congested, often pendulous. Peduncle rather robust, puberulous or glabrous, 2-35 mm long; pedicels 0-5 mm long, puberulous or glabrous. Bracts sepal-like, about half as long as the sepals. *Flowers* fragrant, open during the day. *Sepals* red and white, pink or green (?), connate at the base for 0.1-0.5 mm, erect, ovate, broadly ovate or circular, 1-2.5 x as long as wide, 2-4 x 1.5-3 mm, rounded, puberulous or almost glabrous outside, ciliate or not, glabrous inside and with 4-8 colleters in 1 row in the middle at the base; colleters 0.3-1 x 0.1-0.3 mm. *Corolla* white, tube mostly pinkish, throat often (?) yellow, 10-30 mm long in the mature bud and forming a comparatively small broadly ovoid or subglobose head 0.1-0.25 of the bud length (2-4 x 2.5-3 mm) with a rounded or blunt apex, puberulous outside on the base of the lobes and often also on the apex of the tube for up to 2 mm or sometimes entirely glabrous, with a more or less interrupted belt of pubescence 2-8 mm wide inside, starting from 0-1 mm below the anther tails and ending 4 mm below to 2 mm above the apices of the anthers and mostly puberulous at the base of the lobes, otherwise glabrous; tube 3.3-10 x as long as the calyx, 1.3-2.6 x as long as the lobes, 9-28 mm long, narrow, almost cylindrical and 1.5-5 mm wide, slightly narrowed below and above the insertion of the stamens to 1.2-2.5 mm wide, not twisted; lobes obliquely elliptic or obovate, mostly more or less falcate, 0.4-0.76 x as long as the tube, 2-5 x as long as wide, 5-18 x 2-7 mm, rounded, auriculate at the left side of the base, undulate or not, upcurved, sometimes (?) spreading or slightly recurved. *Stamens* with apex 0.7-5 mm below mouth of corolla tube, 0.52-0.75 of the length of the corolla tube, (at 7-19 mm from the base), with anther tails 0.5-1 mm lower; anthers sessile, narrowly triangular, 3-5 x as long as wide, 3-5 x 1-1.2 mm, apex acuminate, sterile for 0.3-0.5 mm, sagittate at the base. *Pistil* glabrous, 6.5-20 mm long; ovary ovoid, 1.5-2.5 x 1.2-2.2 x 1.2-2 mm, of 2 separate carpels connate at the base by a disk-like ring 0.8-1 mm high; style most-

Fig. 69. *Tabernaemontana coriacea.* **1,** habit (x 2/3); **2,** leaf (x 2/3); **3,** flower (x 1); **4,** opened corolla (x 2); **5,** calyx with pistil (x 4); **6,** fruit (x 2/3). 1 from Sanaiotti 4 April 1989; 2 from Forero et al. P7144; 3-6 from Corner 88; 7 from Magnusson AC 090.387.

ly filiform, 4-14.5 mm long; pistil head 1-1.5 mm high, composed of a veil-like basal ring 0.2-0.3 x 0.8-1 mm, a 5-lobed stipitate depressed globe 0.2-0.4 x 0.6-0.8 mm and a cleft stigmoid apical part 0.2-0.3 x 0.1-0.2 mm. Ovules approximately 50-100 in each carpel. *Fruit* of 2 separate mericarps; mericarps green, violet or black, obliquely ellipsoid or pod-like, not dotted, 25-45 x 7-13 x 7-10 mm, mostly recurved, acuminate, with a lateral ridge at each side, approximately 10-15-seeded; wall thin; aril enveloping the seed. *Seed* narrowly oblong, 7-8 x 2.5-4 x 2-3 mm, with longitudinal grooves, with honeycomb-structure.

DISTRIBUTION: Northern Brazil and Peru.

ECOLOGY: Forest understorey on terra firme or on sandy islands in large rivers. Alt. low. Probably flowering and fruiting throughout the year.

Geographical selection of the approximately 100 specimens examined:

BRAZIL. Pará: Mun. Viseu, 6 km SSW of Canindé, *Balée* 1555 (WAG); Belém, *Martius* s.n. (M); S of Belém, Rio Acará, between Acará and Tome-acu, *Austin & Cavalcante* 4081 (MO); Portel, *Fróes* 32808 (MO); Pombal, *Black* 52-15485 (MO, Z); near Santarém, *Martius* Sept. 1819 (BP, L, M); ibid., *Spruce* 232 (AWH, BM, CGE, FI-W, G, GH, K, M, NY, P, TCD, W; phot. of G sheet GH, MO, NY, US; phot. of W sheet GH, MO, US), 675 (K, P); ibid., *Strudwick* et al. 3006 (F, GH, K, MO, WAG); ibid., Belterra, *Lobo* et al. 55 (MG, NY, WAG); Peninsula of Alter-do Chao, *Sanaiotti* 4 Apr. 1989 (WAG); ibid., *Magnusson* AC 090387 (WAG); Boa Vista, Tapajós R., *Monteiro da Costa* 134 (F, G, MO); Monte Alegre, *Lima* 53-1343 (K); Alto Tapajós, Missão Cururú, *Anderson* 10533 (F, K, NY, Z). Amazonas: Rio Aiary, Içana, *Froes* 21332 (US); Rio Urubu, above Cachoeira Iracema, *Prance* et al. 4970 (MO, NY, US); Rio Urubu, between Cachoeira Iracema and Manaus-Itacoatiara Road, *Prance* et al. 5076 (K, NY, U, US, Z, type of *Anacampta pendula*); Puraquequara, 40 km S of Manaus, *Prance* et al. 7199 (BR, K, NY, S, US, Z); near Manaus, *Ule* 5180 (F, G, HBG, K, L; phot. of G sheet GH, NY, US); ibid., *Coelho* INPA 4077 (INPA, MG, U); ibid., *Corner* 88 (MO); ibid., *Martius* 2761 (M); Mun. Manaus, road to Aleixo, *Krukoff* 7985 (A, BR, G, K, MO, NY, S, U); ibid., km 3, *Lasseign* P21188 (AAU, F, GH, NY, Z); near Rio Negro mouth, *Spruce* 1666 (AWH, BM, BP, BR, C, CGE, F, FI, G, GH, GOET, K, LD, MPU, NY, P, TCD, W, type of *T. submollis*); Rio Amazonas, between Coari and Rio Negro, *Poeppig* Mar. 1832 (F, W, type of *T. acutissima*); Mun. Borba, near Urucurituba, *Krukoff* 5968 (A, BM, BR, G, K, MO, NY, S, U, US); Nova Prainha, Rio Aripuanã, *Ramos* et al. INPA 62329 (INPA); Cauca Pirêra, *W. Rodrigues & Lima* 2286 (INPA); Rio Purus, S of Lago Prêto, *Prance* et al. 13682 (NY, U, US, Z); km 44 Humaitá-Porto Velho Road, *Lima* 485 (GH); Rio Curuquetê, near Cachoeira São Bento, *Prance* et al. 14329 (K, NY, U, Z); ibid., near Cachoeira San Antonio, *Prance* et al. 14225 (C, F, GH, K, NY, S, U, US, Z); Headwaters of Rio Curuquetê, near Igarapé João Bento, *Prance* et al. 14660 (C, F, G, K, NY, S, U, US, Z). Rondônia: Porto Velho Airfield, *Cordeiro* 582 (MG, Z); near Jaciparaná, *Prance* et al. 5171 (B, NY, S, US, W, Z, type of *Bonafousia prancei*); 15 km E of km 117 Porto Velho-Cuiabá Highway, near Santa Barbara, *Prance & Ramos* 7007 (K, MG, NY, S, U, US, Z); 20 km N of Ariquemes, *Forero & Wrigley* P7144 (INPA, K, M, NY, P, S, US, Z); Basin of Rio Madeira, 1 km N of Riberão, *Prance* et al. 6291 (F, NY, S, Z, paratype of *Anacampta pendula*); Bom Sossego Road, Yata, *Carreira* et al. 537 (INPA, LL, MG); 12 km NNE of Guajará-Mirim, *Prance* et al. 6662 (K, NY, US, Z, paratype of *Anacampta pendula*); Guajará-Mirim, *Sandeman* 2164 (K); 4 km from Vilhena, *Vieira* et al. 807 (NY, WAG). Mato Grosso: Rio Aripuaña, base of Salto dos Dardanelos, *Berg* et al. P18471 (K, NY, U, US, Z). Sin. loc., herb. *Link* s.n. (phot of lost B sheet G, GH, MO, NY, US, type); *Riedel* s.n. (G, P; phot. lost B sheet G, GH, MO, NY, US, type pf *T riedelii*).

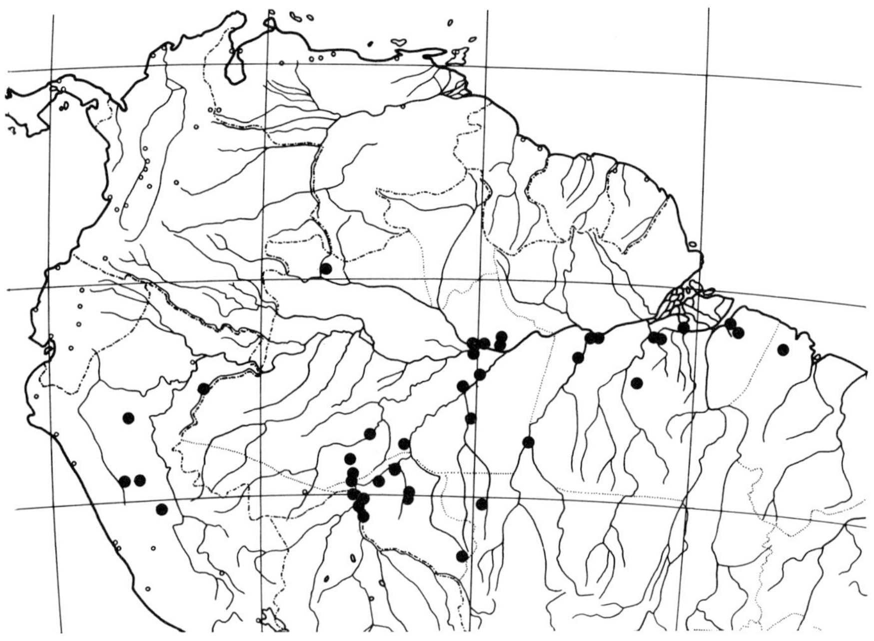

Map 35. *Tabernaemontana coriacea.*

PERU. San Martín: Santa Rosa de Davidcillo, km 72 Tarapoto-Yurimaguas Road, *Knapp* 7937 (WAG). Huanuco: road to Monzón, *Schunke* 1056 (ECON, F, NY, U, WAG); Río Javari, behind Angamo Garrison, *Lleras* et al. P17137 (K, MO, NY, S, US, Z). Ucayali: near Aguaytía, *Mathias & D.Taylor* 5089 (F, MO, US), 5124 (F, MO); ibid., *Woytkowski* 5345 (BR, C, F, MO, P, S, type of *T. brachyantha* Woods. non Stapf).

BOLIVIA. Pando: Río Madeira, opposite Abunã, *Prance* et al. 5699, partly, flowering specimens only (F, GH, K, NY, RB, S, US, Z, paratype of *Anacampta pendula*); S bank Río Abunã, 7-8 km above mouth, *Prance* et al. 6061 (F, K, NY, S, U, US, Z, paratype of *A. pendula and B. prancei*). Pasco: Puerto Bermudez, *Killip & A.C.Smith* 26489 (MO, NY, US).

Notes. Great confusion existed with *T. coriacea*, a species which is rather constant in almost all characters except for the length of the corolla tube. However, some dispute about its delimitation existed as can be seen in the list of synonyms.

The types of *T. acutissima, T. riedelii* and *T. submollis* have only slightly larger leaves than the type of *T. coriacea*. The inflorescences are often pendulous or recurved and therefore *Anacampta pendula* was reduced to a synonym of *T. prancei* by Leeuwenberg (1988). Markgraf (1971), Allorge (1985) and Leeuwenberg (1988) supposed that *T. prancei* (alias *Anacampta pendula*) has prickly fruits, but this opinion was based on three errors:

1. *Prance* et al. 5699, paratype of *Anacampta pedula*, had apparently been taken from two different plants, one with flowers and glabrous leaves with tertiary venation concolourous beneath and one with fruits with leaves sparsely pubescent beneath and tertiary venation darker than the rest of the blade beneath. The fruiting specimen turned out to belong to *T. vanheurckii*.

2. *Anderson* 10722 and *Prance* et al. 9008, both fruiting and with leaves sparsely pubescent and with darker venation beneath, had been identified as *Anacampta pendula* (= *Bonafousia pendula*) by both Markgraf and Allorge. These two specimens appear to belong to *T. linkii*.

3. A flower that could be identified as from *T. coriacea* and apparently belonging to another specimen, arrived in a pocket on the Z sheet of *Prance* et al. 9008, belonging to *T. linkii* as stated above.

Lleras et al. P17137, collected in Peru, perfectly fit in the above discussed species concept of *T. coriacea* and is the first specimens of this species collected outside Brazil (except for the following). The specimens Leeuwenberg (1988) named *T. luciliae* cannot be kept separate, as they are exactly like *T. coriacea* except for the length of the corolla tube which varies from 9 to 11 mm, while the variation of the length of the other specimens ranged from 11 to 28 mm.

The specimens with triangular branchlets identified as *T. coriacea* by Allorge belong to *T. angulata* according to the present author.

70. Tabernaemontana cumata Leeuwenberg, **sp. nov.** Fig. 70, p. 284; map 36; p. 288

Arbor parva. Folia petiolata anguste elliptica apice acuta vel obtusa margine undulata utroque latere glabra subtus opaca. Inflorescentia congesta pauciflora. Corolla alba extus glabra tubo brevi lobis obliquiter obovatis breviore. Stamina inclusa paulo supra medium corollae tubi inserta antheris sessilibus anguste triangularibus caudis rectis. Fructus muricatus spinis brevibus obtusis.

Type: Brazil, Amazonas, Manaus, Rio Tarumá-Mirim, *Ducke* 1290 (holotype UC; isotypes A, MO, NY, US).

Etymology: κυματιας, waving, as the leaves are strongly undulate, much more than in *T. undulata* which has much longer flowers and smooth fruits.

Small tree 5-13 m high. Trunk 10-35 cm in diameter. Branches dark brown, lenticellate; branchlets terete (?), deeply sulcate when dried, glabrous. *Leaves* petiolate; petiole glabrous, 3-15 mm long; (ocreae widened into intrapetiolar stipules); blade coriaceous when dried, narrowly elliptic, 2.3-4 x as long as wide, 10-27.5 x 2.5-11 cm, obtuse or acute at the apex, cuneate at the base, undulate, with a revolute margin, glabrous on both sides, shiny and with prominent reticulate venation above, dull, clearly dotted and with often obscure venation beneath, with 9-14 pairs of upcurved secondary veins forming an angle of 60-70° with the costa. *Inflorescence* shortly pedunculate, 2-3 x 2-3 cm, 5-8-flowered, congested. Peduncle robust, 2-12 mm long, glabrous; pedicels 3-5 mm long, glabrous. Bracts scale-like, numerous, about half as long as the sepals. *Flowers* open during the day. *Sepals* pale green (?), free, erect, or in dried flowers often spreading, ovate, 1.3-2 x as long as wide, 4-5 x 2.2-3 mm, rounded or obtuse, glabrous outside, ciliate, glabrous inside and with 2-4 colleters in 1-2 rows in the middle at the base; colleters variable in shape and size, 0.3-0.5 x 0.1-0.3 mm, lobed when broad. *Corolla* tube greenish-white, limb white, throat yellow, 10-11 mm long in the mature bud and forming a broadly ovoid or subglobose head 0.2-0.27 of the bud length (2-3 x 2-3 mm) with a blunt apex, glabrous outside, pubescent inside in a belt from 1.5 mm below to 0.5 mm above the insertion of the stamens; tube 2-2.5 x as long as the calyx, 0.8-0.9 x as long as the lobes, 10 mm long, almost cylindrical, 3 mm wide at the base and around the anthers, narrowed below and above the stamens to 1.5 mm wide, widened again at the mouth to 2 mm wide, not twisted; lobes obliquely obovate or nearly so, 1.1-1.2 x as long as the tube, 2.4-2.7 x as long as wide, 11-12 x 4-5 mm, rounded, spreading, recurved later. *Stamens* with apex 2 mm below mouth of corolla tube, inserted 0.55 of the length of the corolla tube, (at 5.5 mm from the base); anthers sessile, with tails 0.5 mm below insertion, narrowly

Fig, 70. *Tabernaemontana cumata.* **1,** habit (x 2/3); **2,** corolla (x 4); **3,** opened corolla (x 6); **4,** sepal inside (x 8); **5,** pistil (x 6); **6,** fruit (x 2/3). 1-5 from Mori et al. 21333; 6 from Prance et al. 22609.

triangular, 3 x 1 mm, apex acuminate, sterile for 0.3 mm, sagittate at the base. *Pistil* glabrous, 7 mm long, with apex about halfway along anthers; ovary ovoid, 2 x 1.5 x 1.5 mm; style rather thick, 3.7 mm long; pistil head 1.3 mm high, composed of a veil-like entire recurved ring 0.3 x 1.1 mm, a stipitate 5-lobed depressed globe 0.3 x 0.6 mm and a stigmoid apex 0.3 x 0.2 m. Ovules approximately 40 in each carpel. *Fruit* of 2 separate mericarps; mericarps green, obliquely ellipsoid, 20-30 x 10-17 x 10-15 mm, strongly recurved, obtuse at the apex, without ridges, densely covered with blunt excrescences; wall about 2 mm thick in dried fruits; aril enveloping the seed. *Seed* medium brown, obliquely ellipsoid, 8 x 4 x 3 mm, with longitudinal grooves, papillose.

DISTRIBUTION: Brazil (Mainly near Manaus).

ECOLOGY: Forest on terra firme. Alt. low. Flowering and fruiting collected in July and September.

Paratypes:

BRAZIL. Amazonas: Mun. Pres. Figueredo, near Vila Residêncial Atroarí, *Cid* et al. 8210 (WAG); Manaus, *Mello* INPA 4105 (MO); ibid., *Prance* et al. 22609 (AAU, F, NY, Z); Distr. Agropecuário, Res. 1501 (km 41), *Mori* et al. 21333 (NY, WAG); km 32 of Manaus-Itacoatiara Road, *W. Rodrigues* 1789 (INPA, MO); km 70 of same road, *Prance* et al 17251 (NY, Z); Manaus, AM-1, km 134, cidade Tavares, *W. Rodrigues* 7206 (ECON, INPA); Rio Cuieras, 2 km below mouth of Rio Brancinho, *Prance* et al. 17771 (GH, MO, NY, S, Z).

Notes. *Tabernaemontana cumata* has been confused with *T. muricata* as the fruits are very similar. Only completely developed flowers especially in herb. *Mori* et al. 21333 made a clear distinction possible. The two species differ as follows:

Corolla puberulous outside on the base of the lobes, tube 22-28 mm long; branchlets triangular in section; leaves entire .... **T. muricata**

Corolla glabrous outside, tube 10 mm long; branchlets deeply sulcate when dried, terete (?) when fresh; leaves undulate .... **T. cumata**

71. Tabernaemontana cuspidata Rusby, Descr. New Sp. S. Am. Pl. 83 (1920). – Type: Brazil, Mato Grosso, Madeira R. Falls, *Rusby* 2376 (holotype NY, phot. P).

Fig. 71, p. 287; map 36, p. 288

Homotypic synonym:

Anartia cuspidata (Rusby) Allorge in Mém. Mus. natn. Hist. Nat. IIB. 30: 62, pl. 25 (1985), partly excl. *Schunke* 1060 which belongs to *T. flavicans*.

Heterotypic synonyms:

Anacampta kuhlmannii Mgf. In Notizbl. Bot. Gart. Berlin 14: 162, 179 (1938). *Woytkowskia kuhlmannii* (Mgf.) Allorge in op. cit. 121, pl. 53, **syn. nov.** – Type: Brazil, Mato Grosso, Madeira-Mamoré, *Kuhlmann* 609 in RB21843 (holotype RB, not seen; phot. GH).

W. spermatochorda Woods. in Ann. Miss. Bot. Gard. 47: 74, fig. 1 (1960), **syn. nov.** – Type: Peru, Ucayali, Aguaytia, *Woytkowski* 5338 (holotype MO; isotypes F, P).

A. tenuicornuta Mgf. in Bradea 3 (11): 77 (1980). – Type: Brazil, Mato Grosso, Aripuana, near Humboldt Centre, road to Rio Juruema, *Manoel dos R. Cordeiro* 93 (holotype Z; phot. of holotype is pl. 53 in Allorge, op. cit.).

Shrub or small tree, 2-5 m high. Branches pale to dark brown, lenticellate; branchlets terete, glabrous. *Leaves* petiolate; petiole glabrous, 5-12 mm long; (ocreae widened into intrapetiolar stipules); blade papery when dried, elliptic or narrowly elliptic, 2.1-2.7 x as long as wide, 9-23 x 3.5-11 cm, apiculate or caudate at the apex, acumen 5-15 mm long and rounded, cuneate at the base, entire, glabrous on both sides, with scattered black dots beneath or not; with 9-16 pairs of upcurved secondary veins; tertiary venation reticulate. *Inflorescence* pedunculate, 5-8 x 3-5 cm, 2-12-flowered, lax. Peduncle robust, short, glabrous, 7-17 mm long; pedicels glabrous, 7-12 mm long. Bracts scale-like, about 1 mm long. *Flowers* open during the day (only one open seen, *Vieira* et al. 231). *Sepals* green, connate at the base for about 0.5 mm, erect, ovate, broadly ovate or subcircular, 1-1.6 x as long as wide, 5-6 x 3.5-5 mm, rounded, glabrous outside, not ciliate, glabrous inside and with 5-6 colleters in 1 row in the middle at the base; colleters 0.3-0.5 x 0.15-0.2 mm. *Corolla* yellow, 43 mm long in the mature bud and forming a subglobose head 0.18 of the bud length (8 x 8 mm) nearly as wide as the widest part of the tube, in immature buds mostly as wide as narrow part of the tube, glabrous outside, hirto-pubescent inside, with recurved hairs in a belt 1-2 mm wide immediately below the insertion of the stamens (also on the filament ridges among the tails); tube 8 x as long as the calyx, 6 x as long as the lobes, 40 mm long, almost cylindrical, widened above the base to 7-8 mm wide, gradually narrowed from the insertion of the stamens towards the throat to 5-6 mm wide, not or only slightly widened at the mouth to 6 mm wide, not twisted; lobes obliquely strap-shaped, 0.18 x as long as the tube, 2.3 x as long as wide, 7 x 3 mm, rounded, with an obtuse lobe about 2 x 2 mm at the base being often visible in the bud as a tooth (part covered in bud probably differently coloured), spreading. *Stamens* with apex 17 mm below mouth of corolla tube, inserted 0.42 of the length of the corolla tube, (at 17 mm from the base); anthers sessile, narrowly oblong, 6-6.5 x 1 mm, apex acuminate, sterile for 0.5-0.7 mm, sagittate at the base, with tails curved towards each other (at least in bud). *Pistil* glabrous, 20 mm long; ovary ovoid, 3 x 2 x 2 mm, with a disk-like ring 1 mm high; style filiform, 15.7 mm long; pistil head 1.2-1.3 mm high, composed of a basal ring 0.2-0.3 x 0.8-1.1 mm, a stipitate 5-lobed depressed globe 0.4 x 0.8-0.9 mm and a stigmoid apex 0.2-0.4 x 0.2-0.3 mm. Ovules approximately 30 in several rows in each carpel. *Fruit* of 2 separate mericarps; mericarps green, slenderly pod-like, 19-25 x 0.4-0.6 x 0.4-0.6 cm, acuminate at the apex, not ridged, smooth, torulose when dried, with 20-25 seeds in 1-2 rows; wall thin, about 0.5 mm thick in dried fruits; aril surrounding the seed. *Seed* dark brown, obliquely ellipsoid, 6-9 x 2-4 x 2-3 mm, with longitudinal grooves, dull, papillose; embryo 6.5 mm long or less; cotyledons ovate, 3.5 x 2 mm, obtuse at the apex, truncate at base; rootlet 3 x 0.7 mm.

DISTRIBUTION: Only known from a few collections from western South America.
ECOLOGY: Forest understorey on terra firme. Alt. 100-300 m. Probably flowering mainly October-November.

Specimens examined:
COLOMBIA. Amazonas: La Chorrera, *Sastre* 2247 (COL, G, P, US) & 3250 (COL, P).
BRAZIL. Amazonas: Rio Curuquetê, Providencia, *Prance* et al. 14637 (INPA, NY, U, Z). Rondônia: Vila Caneco-Mineraçao Jacundá, 106 km from Porto Velho, *Vieira* et al. 231 (GH, INPA, MO, NY, US, WAG); Rio Madeira Falls, *Rusby* 2376 (NY, phot. P, type); Pres. Marques (= Abuña), *Kuhlmann* 609 (RB, not seen; phot. GH, type of *Anacampta kuhlmannii*). Mato Grosso: Aripuana, near Humboldt Centre, Prance et al. 18251 (NY, Z, paratype of *A. tenuicordata*); Rio Aripuana, *Cordeiro* 93 (Z, type of *A. tenuicornuta*).
PERU. Ucayali: Aguaytia, *Woytkowski* 5338 (F, MO, P, type of *Woytkowskia spermatochorda*). Pasco: Oxapampa Prov., between Santa Rosa de Chives and crest of Cordillera San Matias, *D.N. Smith* et al. 2431 (WAG).

Fig. 71. *Tabernaemontana cuspidata,* **1,** habit (x 2/3); **2,** flower (x 1); **3,** opened corolla bud (x 4); **4,** calyx with pistil (x 4); **5,** fruit (x 2/3); **6,** seeds (x 2). 1-2 from Vieira et al. 231; 3-4 from Boom 4930; 5-6 from Cordeiro 93.

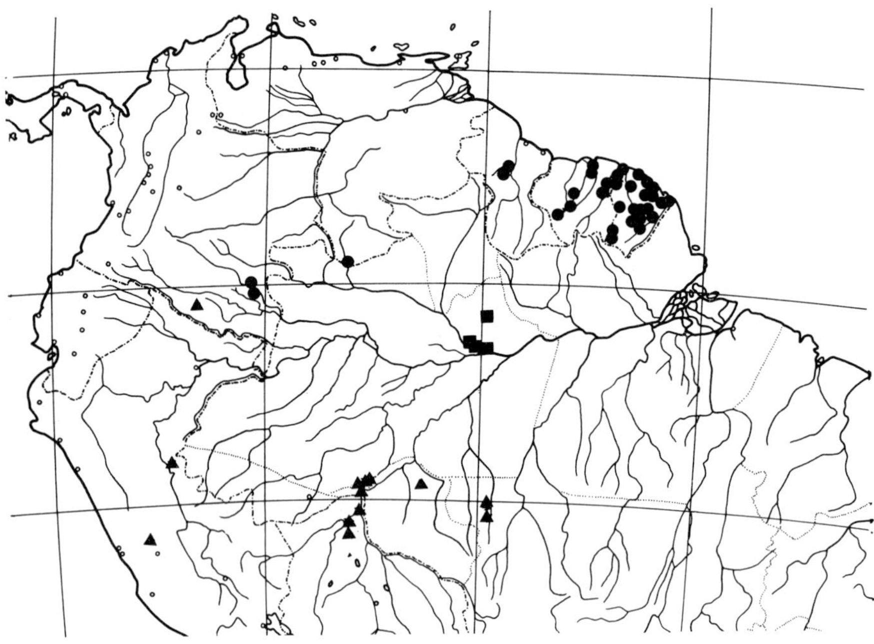

Map 36. ■. *Tabernaemontana cumata,* ▲. *T. cuspidata,* ●. *T. disticha.*

BOLIVIA. Beni: Cachuela Esperanza, Rio Beni, *G. Meyer* 354 (MO, U, Z); Riberalta, *Daly* et al. 6268 (NY, WAG); near Chácobo, *Boom* 4930 (NY, WAG), 5027 (NY, WAG).

Notes. All specimens examined are very similar in all characters and therefore it is very easy to combine them in a single species. The latter results in the reduction of *Woytkowskia* to a synonym of *Tabernaemontana* and the two species placed there by Allorge (1985) to synonyms of *Tabernaemontana cuspidata*. The immature flower bud is often indented at the apex, as is the case with *T. attenuata* which also has very short corolla lobes. Slender fruits 10 x as long as wide with seeds in 1-2 rows are known from *T. amygdalifolia* and the Asian *T. bufalina*. Although the fruits of *T. cuspidata* are much more slender, this remaining character is not considered sufficiently important to maintain *Woytkowskia* as a separate genus.

72. Tabernaemontana cymosa Jacq., Enum. Pl. Carib. 14 (1760); Select. Stirp. Amer. 39, t. 181, fig. 14 (1763), non Sessé & Mociño (1893). – Type: Icon. cit.

Fig. 72., p. 290; phot. 4, 5; map 37, p. 293

Homotypic synonyms:

Taberna cymosa (Jacq.) Miers, Apoc. S. Am. 62 (1878). *Peschiera cymosa* (Jacq.) Dugand in Caldasia 2: 299 (1943).

Heterotypic synonyms:

T. arcuata Ruiz & Pav., Fl. Per. 2: 22, t. 143 (1799), **syn. nov.** *Merizadenia arcuata* (Ruiz & Pav.) Miers, op. cit. 79. *P. arcuata* (Ruiz & Pav.) Mgf. in Notizbl. Bot. Gart. Berlin 14: 171 (1938). – Type: Peru, sin. loc., *Ruiz & Pavón* s. n. (holotype MA, not seen; isotypes BM, fruit only, G, veg.; phot. of MA sheets WAG; phot. of lost B sheet F, G, GH, MO, US).

T. psychotriifolia H.B.K., Nov. Gen. 3: 227 (1819). *P. psychotriifolia* (H.B.K.) Miers, op. cit. 42. – Type: Colombia, Bolívar, Magdalena R., Mompos, *Bonpland* 1516 (holotype P-BO; isotype P; phot. of holotype F, GH, US).
T. umbrosa H.B.K., op. cit. 226. *P. umbrosa* (H.B.K.) Miers, op. cit. 44. – Type: Venezuela, Sucre, near Bordones, *Bonpland* 1235 (holotype P-BO; phot. of holotype F, GH, US).
P. puberiflora Miers, op. cit. 43. – Type: Peru, near Tarapoto, *Spruce* 4245 (holotype BM; isotypes AWH, BR, CGE, E, G, GH, K, LE, P, TCD, W).
P. concinna Miers, op. cit. 44. *T .concinna* (Miers) Macbride in Publ. Field. Mus. Nat. Hist. Chicago, Bot. Ser. 13, 5: 402 (1959). – Type: Peru, San Martín, near Tarapoto, *Spruce* 4534 (holotype BM; isotypes AWH, BP, BR, C, CGE, E, G, GH, GOET, K, LD, LE, MPU, P, TCD, W; phot. of lost B sheet F, GH, MO, US).
T. buchtienii H. Winkler in Fedde, Repert. 7: 244 (1909). *P. buchtienii* (H. Winkler) Mgf. in Notizbl. Bot. Gart. Berlin 14: 171 (1938), **syn. nov.** – Type: Bolivia, La Paz, Charopampa, near Mapiri, *Buchtien* 1976 (holotype B†; lectotype US, designated here; phot, of lectotype WAG).
T. mapirensis Rusby in Bull. N.Y. Bot. Gard. 8: 115 (1912). – Type: Bolivia, La Paz, Mapiri, *R.S. Williams* 736 (holotype NY, not seen; isotypes BM, K; phot. of holotype P).

Shrub or tree, 2.50-25 m high. Trunk 2-40 cm in diameter; bark grey, rough, thick, fissured, lenticellate; wood yellowish. Branches pale or dark brown, lenticellate; branchlets probably terete when fresh, glabrous. *Leaves* of pair equal or unequal, (larger up to 3 x as long as other and more or less similarly shaped), petiolate; petiole 3-20 mm long; (ocreae slightly widened into intrapetiolar stipules); blade papery or membranaceous even when fresh, elliptic or narrowly elliptic, 2-3 x as long as wide, 5-25 x 2-12 cm, acuminate, bluntly apiculate or sometimes in some leaves obtuse at the apex, cuneate at the base or decurrent into the petiole, entire or undulate, glabrous and sometimes with scattered black dots on both sides, with 6-18 pairs of rather straight secondary veins forming an angle of 60-90° with the costa; tertiary venation reticulate. *Inflorescence* 4-7 x 4-12 cm, many-flowered, rather lax, corymbose. Peduncle 1-20 mm long, glabrous; pedicels glabrous, 5-10 mm long. Bracts numerous, slightly smaller than the sepals and almost similar, acute. *Flowers* fragrant, open during the day. *Sepals* green, subequal, connate at the base for 0-0.5 mm, recurved or not even when fresh (mostly recurved in specimens from Colombia and Venezuela, mostly not in those form from Peru and Bolivia), ovate or narrowly ovate, 1.25-2.3 x as long as wide, 2,5-4 x 1.5-2 mm, obtuse or acute, entire, glabrous outside, sometimes ciliate at lower half, glabrous inside and with 2-9 colleters in 1 row in the middle at the base; colleters 0.3-0.7 x 0.1-0.2 mm. *Corolla* yellow, orange or white and often with a yellow throat, or cream and turning orange-brown at anthesis, 14-16 mm long in the mature bud, skittle-shaped and with a large ovoid head wider than whole corolla tube 0.3-0.36 of the bud length (5 x 4-4.5 mm) with an obtuse apex, glabrous outside, pilose inside (partly in stripes) starting at the level of the insertion of the stamens to the apices of the anthers and ending at the level 0-0.4 of the length from the apices of the lobes; tube 2.3-5 x as long as the calyx, 0.6-1.3 x as long as the lobes, (8-)9-13 mm long, flask-shaped, mostly inflated above the base and 2-4.5 mm wide, narrowed above the stamens to 1.2-2 mm wide, not twisted; lobes obliquely oblong or dolabriform, 0.8-1.6 x as long as the tube, 1.5-3.2 x as long as wide, 7-16 x 4.5-8 mm, rounded, undulate, spreading. *Stamens* with apex 2-4.5 mm below mouth of corolla tube, inserted 0.19-0.33 of the length of the corolla tube (at 2-3 mm from the base); filament ridges thick; anthers sessile, narrowly triangular, 2.2-3 x as long as wide, 4-5 x 1.2 mm, apex acuminate, sterile for 0.3-0.5 mm, sagittate at the base,

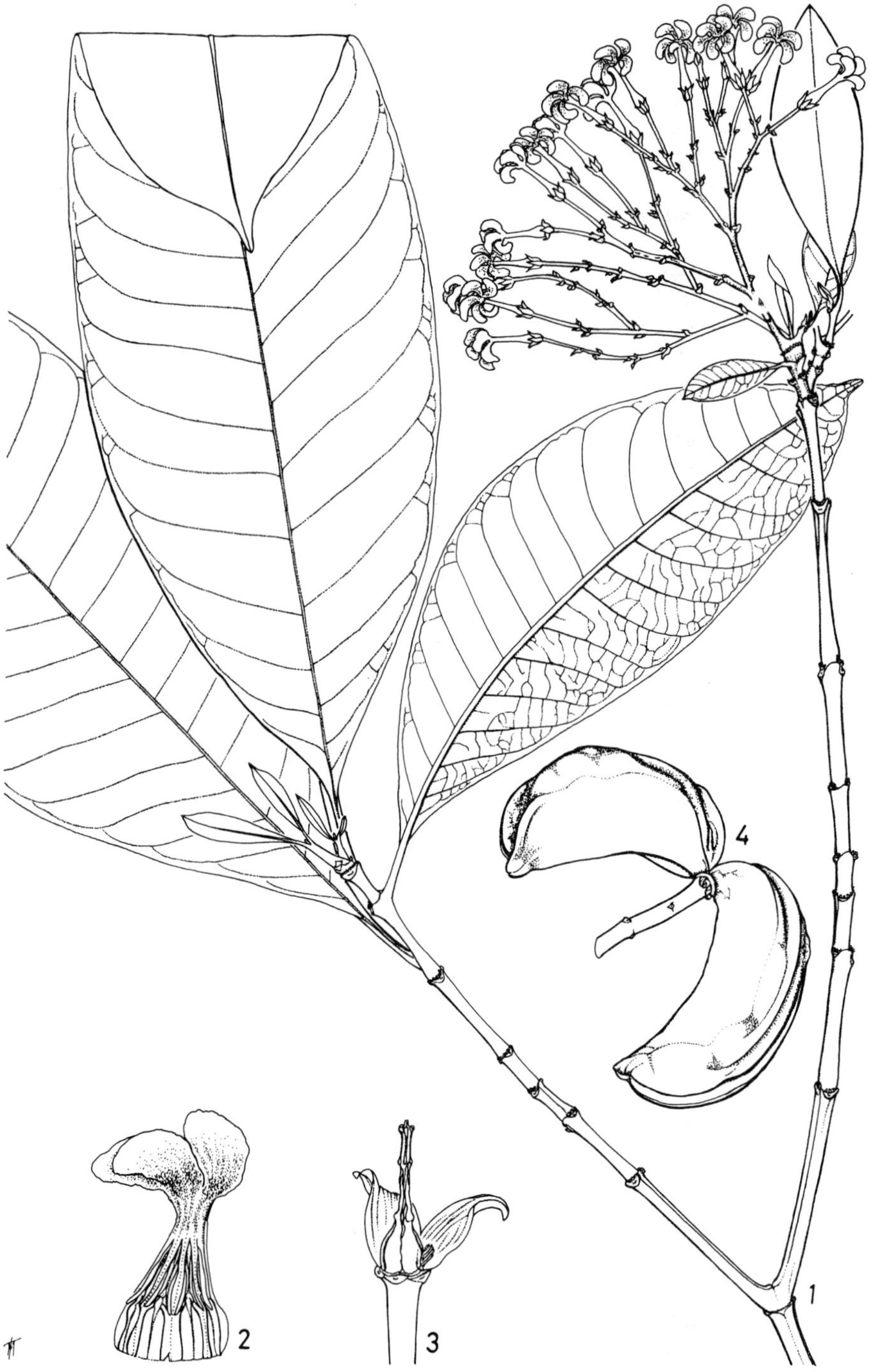

Fig. 72. *Tabernaemontana cymosa*. **1,** habit (x 2/3); **2,** opened corolla (x 2); **3,** calyx with pistil (x 4); **4,** fruit (x 2/3). 1-4 from Breteler 3434; 4 from Breteler 3904.

glabrous. *Pistil* glabrous, 4-5.2 mm long, with the apex almost halfway along the anthers; ovary ovoid or nearly so, 1.5-2.5 x 1-2 x 1-2 mm, gradually narrowed into the style, of two separate carpels; style very short and thick, 0.5-1.5 mm long; pistil head large, composed of an obscure undulate ring 0.3 x 0.6-1 mm, an almost cylindrical apically 5-lobed and laterally 5-ribbed central part 1-1.5 x 0.5-0.6 mm and a stigmoid apex, 0.2-0.3 x 0.2-0.3 mm. Ovules approximately 100-300 in each carpel. *Fruit* of 2 separate or occasionally partly united mericarps; mericarps brown or red, obliquely oblong, 30-90 x 15-50 x 15-40 mm, rounded at the apex, recurved, without or sometimes with faint ridges, slightly lumpy and slightly verrucose or tuberculate, approximately 40-50-seeded; wall 3-5 mm thick; aril orange or red. *Seed* dark brown or black, obliquely ellipsoid, 8-13 x 4-7 x 3.5-5 mm, with longitudinal grooves, papillose; embryo 6-8.5 mm long; cotyledons ovate, 1-1.25 x as long as wide, 2.5-4 x 2-4 mm, obtuse at the apex, cordate at the base; rootlet 3.5-5 x 0.7-0.8 mm.

DISTRIBUTION: Western South America.

ECOLOGY: Rather open or secondary forest. Alt. 0-400 m. Mainly flowering March-April in Colombia, Venezuela and Trinidad, and mainly September-October in Peru and Bolivia. No certainty about fruiting season.

Geographical selection of the approximately 160 specimens examined:

COLOMBIA. Atlántico: Piojó, *Dugand & Garcia Barriga* 2395 (US); Barranquilla, Bro. *Elias* 115 (MO, US); Manati, *Dugand* 586 (F). Bolívar: Torrecilla, near Turbaco, *Killip & A.C.Smith* 14653 (A, GH, US); near Bordones, *Bonpland* 1235 (P-BO; phot. F, GH, US, type of *T. umbrosa*); Las Piedras, *Cuadros* 1487 (COL, MO); Corozal, near Palmitos, *Romero* 9718 (AAU, F, MO, P); Mompos, left bank Río Magdalena, *Bonpland* 1516 (P, P-BO; phot. P-BO sheet F, GH, US, type of *T. psychotriifolia*); San Martin de Loba, *Curran* 83 (GH, S, US). Magdalena: Ciénaga, *de Romero* 163 (AAU); Tucurinca, *R. Romero* 1055 (MO, US). César: El Huerto, *G. Triana* 45 (COL); Poponte, *C. Allen* 943 (K). Guajira: Fonseca, *Haught* 4058 (F, MO, US).

VENEZUELA. Zulia: Headwaters of Río Guasare, *A. Gentry* 41142 (MO, U, WAG); near Perijá, *Tejera* 228 (US); between Agua Viva and Mene, *Oberwinkler* 14494 (M). Lara: between Sarare and El Altar, *Steyermark* et al. 109893 (Z). Portuguesa: 17 km NW of Tucupido, *Davidse* et al. 21428 (MO, WAG); Turén, *Pittier* 11740 (A, G, M, US); km 50-56 Guanare-Guanarito Road, *Liesner & González* 12708 (MO, WAG). Barinas: near Barinitas, *Breteler* 3160 (MO, NY, S, U, US, WAG, Z); Barrancas-Barinas Road, *Valverde* et al. 1193 (MO); between La Libertad and Agua Larga, *Bernardi* 1179 (K, MEL); Ticoporo F.R., *Breteler* 3434 (G, S, U, UPS, WAG), 3904 (U, US, WAG, Z); Caparo F.R., *Marcano-Berti* 2888 (MO). Carabobo: El Castaño, Las Trincheras-Puerto Cabello Road, *Pittier* 8812 (G, GH, US); Barbula Road, *Ll. Williams & Alston* 340 = *Ll. Williams* 10967 (BM, F, P); Valencia, *Pittier* 8386 (US). Guarico: between El Sombrero and Ortíz, *Pittier* 12383 (A, G, M, US). Anzoátegui: Río Querecual, SW of Bergantín, *Steyermark* 61482 (MO). Monagas: Caicara, *F.D. Smith* 210 (MO, US). Sucre: Peninsula of Paria, between Cangua and San Juan de las Galdonas, *Steyermark & Liesner* 120952 (MO); near Cristobal Colón (= Macuro), *Broadway* 326 (GH, US). Bolívar: between Moitacu and Río Aro, *Trujillo* 5817 (NY); Parque Caroní, San Félix-Puerto Ordaz, *Aristeguita* 5916 (VEN); km 17 Upata-San Félix Road, *Blanco* 239 (MO, NY); Cerro Bolívar, *Maguire & Wurdack* 35764 (US, VEN); between El Cristo and La Paragua, *A.L. Bernardi* 7366 (FI, G, K); La Prisión, *Ll. Williams* 11540 p.p. (A, F, MO, S, US; other part made here 11540a is *Stemmadenia grandiflora*), 11702 (F, K, US, VEN); Los Chacharos, Río Asa, *Steyermark* 86810 (MO); N of Río Hacha, 30 km of El Manteco, *Steyermark* 87044 (MO); between km 11 and 18.5 S of El Dorado, *Steyermark* 86603 (MO); NE of Canaima, *Agostini* 307 (US, VEN); Potrero Guaicaipuro, *Garófalo* et al. 439 (VEN);

La Paragua For.Res., Río Asa, *Blanco* 721 (US). Delta Amacuro: Sacupana, *Rusby & Squires* 85 (A, BM, E, F, G, GH, K, M, MO, UC, UPS, US, W, WU, Z); 5-14 km ESE of Los Castillos de Guyana, *Davidse & González* 16374 (MO, NY).

TRINIDAD. Maracas Bay, *Richards & Simmonds* Herb.Trin. 15433 (K, U); Mt St Benedict, *Adames & Baker* 313 (MO); near Tabaquite, *Britton* et al. 2599 (GH); Pt Radix, *Ramcharan* 381 (NY).

GUYANA. Rupununi Northern Savanna, Marakanata, *Goodland* 979 (US).

BRAZIL. Roraima: Serra Tepequem, *Prance* et al. 4596 (F, GH, K, S, U, US, Z); Ilha de Maracá, *Ratter* et al. 5784 (K, U); Bôa Vista, Caracarai, *Fróes* 23216 (MO); Mun. São Paulo de Olivença, near Palmares, *Krukoff* 8359 (A, BM, BR, G, K, MO, P, S, U). Acre: Cruzeiro do Sul, Maas et al. P13166 (INPA, MG, NY, U, US, Z); ibid., N bank of Rio Jurua, *Prance* et al. 2916 (F, K, NY, S, US, Z); near mouth of Rio Macauhan, tributary of Rio Yaco, *Krukoff* 5784 (A, BM, G, K, LE, M, S, U, UC, US); Mun. Rio Branco, Area do Parque Zoobotânico da UFAC, *Cid* et al. 10108A (WAG); Seringal São Francisco, *Ule* 9702 (G, K, L); Mun. Brasiléia, km 31 of road to Rio Branco, *Cid* et al. 10231 (WAG).

ECUADOR. Napo; Km 65 Hollin-Loveto, *Palacios* 2191 (WAG); Pastaza: 50 km SSE of Curaray, *Espinoza & Coba* 401 (US).

PERU. Loreto: Maynas Prov., Indiana, Explorama Res., *Vásquez & Jaramillo* 12873 (WAG); Lower Huallaga R., *Ll. Williams* 4273 (F). Ucayali: Pucallpa, *Sandeman* 3313 (K); Distr. Campo Verde, Purmas, km 11 road to Nueva Requena, *Arana* 25 (K). Huánuco: Distr. Honoria, San Miguel de Semuya, *Suárez* 26 (K); near Tingo Maria, Perez 28 Sept. 1966 (K); Distr. Rupa-Rupa, Cerro Aeropuerto, *Vásquez* 42 (F, P, US, WIS). San Martín: near Shapaja, *Belshaw* 3177 (BH, GH, K, LL, MO, UC, US); Bellavista, Río Huallaga Valley, *R. Ferreyra* 4848 (MO); Juanjui, *Klug* 3854 (BM, F, GH, K, MO, S, US, WIS); Distr. Tocache Nuevo, E of Puerto Pizana, *Schunke* 6513 (F, GH, MO, Z); Fundo San Rafael, *Schunke* 4421 (F, G, GH, K, MO, P, US, WIS); between Majo and Saposoa, *Woytkowski* 5495 (BR, F, MO, P, S, US); near Tarapoto, *Spruce* 4245 (AWH, BM, BR, CGE, E, G, GH, K, LE, P, TCD, W, type of *Peschiera puberiflora*) & 4534 (AWH, BM, BP, BR, C, CGE, E, G, GH, GOET, K, LD, LE, MPU, P, TCD, W; phot. of lost B sheet F, GH, MO, US, type of *P.concinna*). Huánuco: Prov. Puerto Inca, Dantas, *Flores & Tello* 1749 (MO), 2321 (MO). Pasco: Oxapampa, near Paujil, *Forster* et al. 8947 (MO, P). Cuzco: Río Madre de Dios, 1 hour from Puerto Maldonado, *Nunez & Monice* 5376 (MO). Madre de Dios: Manu National Park, Cocha Cashu Station, *Forster* 5552 (U); Tambopata Prov., Las Piedras, Albergue Cuzco Amazónica, *Timaná & Astete* 1693 (WAG); Río Tambopata, *A. Gentry* et al. 45688 (MO, WAG); 39 km SW of Puerto Maldonado, *S.F. Smith & Shuhler* 389 (F, K, P, US, WAG). Andes, sin. loc., *Ruiz & Pavón* s.n. (BM, fruit only, G veg., MA, not seen; phot. of lost B sheet F, G, GH, MO, US; phot. of MA sheets WAG, type of *T. arcuata*).

BOLIVIA. Pando: Prov. Nicolás Suárez, near Porvenir, *Casas & Susanna* 8368 (MO); near Campoana, *Casas & Susanna* 8300 (MO); Río Madre de Dios, Nueva Etea, Nee 31764 (WAG). Beni: Cachuela-Esperanza Road, Río Beni, *G. Meyer* 351 (MO, U, Z); ca 18 km S of San Joaquín, *Balick* et al. 1438 (WAG); Prov. Ballivian, Espíritu Santo, *Beck* 2550 (WAG); Rurrenabaque, *Rusby* 816 (GH, K, US); Río Beni, above confluence with Río Quiquibey, 3.5 hours upstream from Rurrenabaque, *Daly* et al. 6597 (WAG); San Borja, *Tercoros & Chore* 365 (MO); Estancia Porvenir, 50 km E of Río Maniqui (San Borja), *Solomon* 14707 (MO, WAG); between Trinidad and Misiones de Guarayos, *Werdermann* 2586 (MO, S). La Paz: Tumupasa, *R.S. Williams* 416 (BM, K, US); Mapiri, *Buchtien* 1425 (US) & 1976 (US, phot. WAG, type of *T. buchtienii*); ibid., *R.S. Williams* 736 (BM, K, P, type of *T. mapirensis*). Cochabamba: Prov. Carrasco, Proyecto Valle del Sacta, *D.N. Smith* et al. 13714 (G). Santa Cruz:

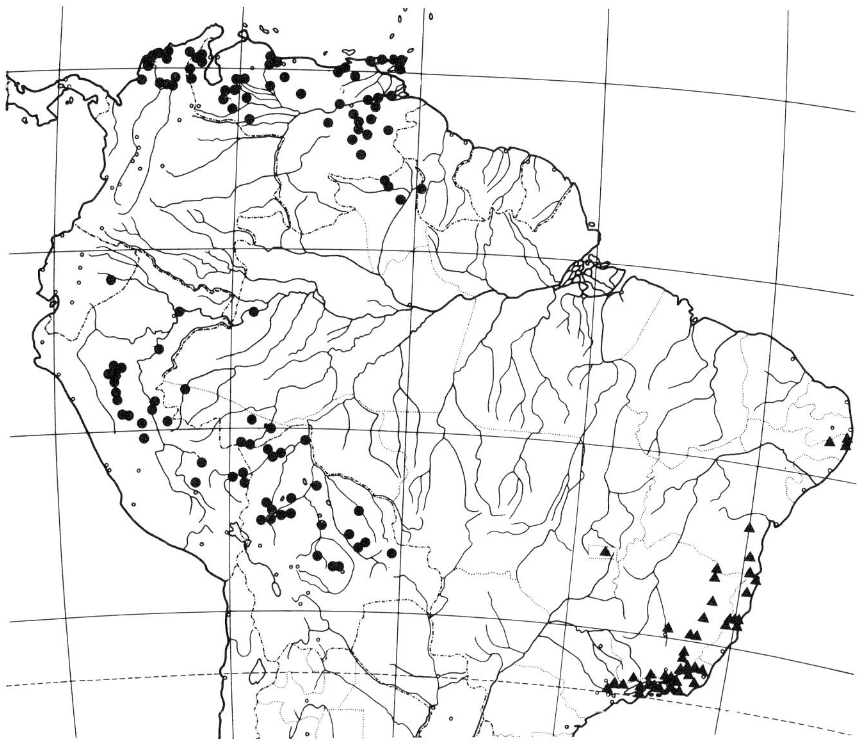

Map 37. ●. *Tabernaemontana cymosa,* ▲.*T. laeta.*

Prov. Guarayos, 2 km NE of Perseverancia, Nee 38738 (WAG); Acención de Guarayos, *Beck* 12245 (HBG); Concepción, *Krapovickas* & *Schinini* 31951 (Z); Lagua Brava, 10 km NW of San Ramón, *Bettella* 26 (WAG); between San Rafael and Velasco, *E. Schmidt* 100 (HBG, M); Prov. Sara, Bosque de Buena Vista, *Steinbach* 6505 (A, BM, G, K, S); Prov. Ichilo, ca 30 km SE of Buena Vista, *Solomon* 14212 (MO, WAG).

Notes. The sepals are mostly recurved in the specimens collected in Colombia and Venezuela and mostly erect in the specimens collected elsewhere. As no other character support the distinction of *T. arcuata* and *T. buchtienii* from *T. cymosa*, they are not kept up here.

73. Tabernaemontana disticha A. DC., Prod. 8: 362 (1844); Leeuwenberg in Journ. Ethnopharmacology 10: 10 (1984), partly), excl. syn. *T. oblongifolia*. – Type: French Guiana, sin. loc., *Martin* 49 (holotype G-DC, phot. F, MO, NY, US; isotype FI-W).
Fig. 73, p. 295; phot. 6; map 36, p. 288

Homotypic synonyms:

Taberna disticha (A. DC.) Miers, Apoc. S. Am. 64 (1878). *Anacampta disticha* (A. DC.) Mgf. in Notizbl. Bot. Gart. Berlin 14: 162 (1938). *Bonafousia disticha* (A. DC.) Boiteau & Allorge in Bull. Soc. Bot. France 130, Lettres 4-5: 340 (1983); Mém. Mus. natn. Hist. Nat. IIB, 30: 106, pl. 46 (1985).

Shrub or small tree 0.60-5 m high. Trunk 1-4 cm in diameter; bark dark brown, smooth or shallowly fissured. Branches pale grey-brown, lenticellate; branchlets terete, glabrous. *Leaves* petiolate; petiole glabrous, 2-7 mm long; (ocreae slightly widened into intrapetiolar stipules); blade coriaceous when fresh, subcoriaceous and often more or less ochraceous when dried, elliptic or narrowly elliptic, 2.1-5 x as long as wide, 5-23 x 1.5-9 cm, acuminate at the apex, cuneate at the base and often decurrent into the petiole, entire, glabrous on both sides, with scattered black dots beneath; with 5-15 pairs of upcurved secondary veins forming an angle of 70-85° with the costa; tertiary venation inconspicuous. *Inflorescence* shortly pedunculate, 3-5 x 3-5 cm, 3-10-flowered, congested. Peduncle robust, glabrous, 2-12 mm long; pedicels 3-7 mm long, glabrous. Bracts sepal-like, about 0.3 x as long as the sepals. *Flowers* fragrant, erect, open during the day. *Sepals* pale green or greenish, connate at the base for 0.3-0.5 mm, erect, broadly ovate or almost circular, 1-1.6 x as long as wide, 2-4 x 1.5-3 mm, rounded, glabrous outside, mostly ciliolate, glabrous inside and with 3-5 colleters in 1 row in the middle at or shortly above the base; colleters 0.3-0.7 x 0.1-0.2 mm. *Corolla* tube white or greenish with pink or violet base, or tube pink with violet or purple base, limb white and throat orange or yellow or entire limb orange or yellow, basal coloration usually (?) in short lines, 14-30 mm long in the mature bud and forming a comparatively small very broadly ovoid head 0.1-0.14 of the bud length (2-3 x 3-4 mm) with a blunt apex, puberulous outside on the part of the lobes not covered in bud and on the distal part of the tube, pubescent inside, usually in stripes, on the filament ridges from 0-1 mm below the anther tails to the insertion of the stamens or up to 1 mm above apices of anthers and puberulous at the base of the lobes; tube 7-10 x as long as the calyx, 1.4-2.3 x as long as the lobes, 15-30 mm long, almost cylindrical, 2-5 mm wide above the base, below and above the anthers slightly narrowed or not to 1.5-3 mm wide, mostly somewhat widened around them to 2-4 mm wide, also slightly widened at the throat to 2-5 mm wide, not twisted; lobes obliquely and narrowly obovate, laterally curved, 0.5-0.7 x as long as the tube, 2.2-3.8 x as long as wide, 8-15 x 3-6 mm, rounded, undulate, upcurved. *Stamens* with apex 3.5-8.5 mm below mouth of corolla tube, inserted 0.4-0.7 of the length of the corolla tube, (at 7-16.5 mm from the base); anthers sessile, with tails 1 mm below insertion, narrowly triangular or narrowly oblong, 4.5-6 x as long as wide, 4.5-6 x 0.8-1 mm, apex acuminate, sterile for 0.3-0.7 mm, sagittate at the base, glabrous. *Pistil* glabrous, 9-18 mm long; ovary ovoid, 2.5-3 x 1.2-2 x 1-2 mm, carpels united by a basal thickening, ring-shaped and 1-1.5 mm high; style filiform, 5.5-13.6 mm long, persistent; pistil head 1.2-1.4 mm high, composed of an entire basal ring 0.2 x 0.8-1.1 mm, a stipitate 5-lobed depressed globe 0.2-0.3 x 0.6-0.8 mm and a stigmoid apical part 0.2-0.3 x 0.2 mm. Ovules approximately 50-100 in each carpel. *Fruit* of 2 separate mericarps; mericarps green (immature ?), obliquely ellipsoid, not dotted, 25-45 x 15-16 x 15 mm, mostly slightly recurved, acuminate at the apex, with a lateral ridge at each side, smooth, approximately 20-seeded; wall thin; aril probably only at the hilar side. *Seed* medium brown, obliquely ellipsoid, about 6 x 3 x 2 mm, with longitudinal grooves, papillose.

DISTRIBUTION: Venezuela, Colombia, the three Guianas and northern Brazil.
ECOLOGY: Forest understorey. Alt. 0-700 m. Flowering mainly February-March and August-September. A fruiting season could not be deduced from collections extant.

Geographical selection of the approximately 125 specimens examined:
COLOMBIA. Vaupés: Río Apaporis, Soratama, *Schultes & Cabrera* 19598 (MO, US) & 19618 (COL, GH, MO, U, US); Jinogojé, Río Piraparaná mouth, *Schultes & Cabrera* 19792 (MO, US).

Fig. 73. *Tabernaemontana disticha*. **1,** habit (x 2/3); **2,** flower (x 2); **3,** opened corolla x 2); **4,** pistil (x 6); **5,** sepals inside (x 8); **6,** fruit (x 2/3). 1 from Prévost 216; 2-5 from Leeuwenberg 11615; 6 from de Granville 5231.

VENEZUELA. Amazonas: Neblina Base Camp, Río Mawarinuma, *Davidse & J.S.Miller* 26939 (MO); 1.5 km S of Cerro de la Neblina, *Liesner & Funk* 16436 (MO).

GUYANA. Tumatumari, *Gleason* 106 (NY); Potaro-Siparuni Region, Eagle Mt, *McDowell & Gopaul* 3468 (US, WAG).

SURINAME. km 121 of Railway, *Florschütz* 1816 (NY, U); Saramacca R., near Camp 2, *Maguire* 24139 (A, K, MO, NY, U); Emmaketen, *Daniels & Jonker* 1175 (U, UC, US); Nassau Mts, *Lanjouw & Lindeman* 2362 (K, NY, U); Lely Mts, *Lindeman* et al. 240/75 (K, MO, U, US, WAG); Manlobi Mt, *Wessels Boer* 209 (U); Wilhelmina Gebergte, Zuid R., *Irwin* et al. 57648 (NY); ibid., SE of Julianatop, Near headwater of West R., *Irwin* et al. 55089 (F, K, NY, U, US); Tafelberg, *Maguire* 24540 (A, K, MO, NY, U, US).

FRENCH GUIANA. Paul-Isnard Region, *Cremers* et al. 11279 (P, U); ibid., Mt Lucifer, *de Granville* 5231 (CAY, P); Mana R., Sabbat Fall, *F. Hallé* 533 (NY, P); Trinité Mts, *de Granville* 6078 (B, BR, CAY, P, U, US, WAG); Piste St Elie, 16 km SW of Sinnamary, *Prévost* 216 (CAY, P, WAG); km 13 of same road, *Leeuwenberg* 11652 (CAY, P, WAG); Mt Trésor, 22 km from Roura Ferry, *Sastre* et al. 8000 (CAY, U, US); Nouragues Station, Approuague R. basin, *de Granville* et al.11056 (P, U, US); 10 km S of Cayenne, *Skog* et al. 5653 (CAY, MO, U, US, WAG); Montsinéry For., *Jacquemin* 2585 (CAY, P); Cayenne, *Martin* 49 (FI-W, G-DC, type); ibid., *von Rohr* 108 (BM, C, MEL); Mt Mahury, E of Cayenne, *Leeuwenberg* 11615 (MO, P, USF, WAG); Ile de Cayenne, Cabassou, *Sastre* 6414 (CAY, MO, P); Rorata Lake, *Prévost* 166 (CAY, P); Mt des Trois Pitons, *Cremers* 6986 (CAY, P, WAG); Camopi R., Mt Alikéné, *Sastre* 285 (CAY, NY, P); Mts Bakra, *de Granville* 4130 (CAY, P, WAG); Sommet Tabulaire, ca 50 km SE of Saül, *Cremers* 6438 (CAY, P, WAG); Emerillons Mts, Cremers 6687 (CAY, P, WAG); Petite Ouaqui R., *de Granville* 1881 (CAY, MO, NY, P, U); Mt Galbao, *Skog* et al. 7369 (CAY, P, U, US, WAG); Mt Bellevue de l'Inini, *de Granville* et al. 7697 (BM, BR, CAY, G, K, U); Upper Litani R., Antekumpata, *Moretti* 567 (CAY, P, WAG); Emerillons Road, km 15, Degrad Claude, *de Granville* 2263 (CAY, P).

BRAZIL. Pará: Tumuc Humac, 1 km W of Mitaraka Nord, *de Granville* 1452 (CAY).

74. Tabernaemontana flavicans Willd. ex Roem. & Schult., Syst. Veg. 4: 797 (1819); Leeuwenberg in Agric. Univ. Wageningen Pap. 87, 5: 10 (1988). – Type: Brazil, sin. loc. *Hoffmannsegg* s.n. (holotype B-W 5195; isotype FI-W).

Fig. 84, p. 298; map 38, p. 301

Homotypic synonym:

Anartia flavicans (Willd. ex Roem. & Schult.) Miers, Apoc. S. Am. 82 (1878); Allorge, Mém. Mus. natn. Hist. Nat. IIB. 30: 66, pl. 27 (1985).

Heterotypic synonyms:

T. oblongifolia A. DC., Prod. 8: 368 (1844), partly, as for lectotype; Leeuwenberg in Journ. Ethnopharmacology 10: 12 (1984). *Bonafousia oblongifolia* (A. DC.) Miers, op. cit. 50. *A. oblongifolia* (A. DC.) Mgf. in Notizb. Bot. Gard. Berlin 14: 165 (1938). – Type: Brazil, Bahia, Ilhéus, *Blanchet* 2358 (lectotype G-DC; isolectotypes BM, BP, BR, FI-W, G, K, LD, NY, P, W; phot. of G-DC sheet MO, US).

T. olivacea Muell. Arg. in Martius, Fl. Bras. 6, 1: 75 (1860). *B. olivacea* (Muell. Arg.) Miers, op. cit. 52. *A. olivacea* (Muell. Arg.) Mgf., l.c. – Type: Venezuela, Amazonas, near San Carlos, Río Negro, *Spruce* 3114 (lectotype W, designated by

Leeuwenberg (1988); isolectotypes AWH, BM, BP, BR, C, FI, G, GH, GOET, K, LD, NY, P, TCD).

B. latifolia Miers, op. cit. 50. – Type: Brazil, Pará, near Santarém, *Spruce* 236 (holotype BM; isotypes CGE, K, M, TCD, W).

Taberna disparifolia Miers, op. cit. 63. – Type: Peru, San Martín, near Tarapoto, *Spruce* 4611 (holotype BM; isotypes AWH, BP, BR, C, E, F, FI, G, GH, GOET, K, LD, LE, MPU, NY, P, W).

A. glabrata Miers, op. cit. 81. – Type: Brazil, Rio de Janeiro, Barra do Facão, *Martius* s.n. (holotype BM; isotype K).

Shrub or small tree, 0.30-12 m high. Trunk up to at least 20 cm in diameter; bark longitudinally fissured, rough, lenticellate. Branches grey, or pale or dark brown, with fissured bark, with mostly paler lenticels; branchlets terete, often sulcate and angular when dried, glabrous. *Leaves* of a pair equal or sometimes unequal (larger up to 2 x as long as other and comparatively not narrower), petiolate; petiole 2-13 mm long; (ocreae not widened into intrapetiolar stipules); blade subcoriaceous or papery when dried, elliptic, narrowly elliptic, or sometimes ovate or narrowly ovate, 1.8-4.3 x as long as wide, 2.8-17 x 0.6-6.5 cm, acuminate or caudate at the apex, cuneate at the base or decurrent into the petiole, entire, sometimes undulate, glabrous and sometimes with scattered black dots on both sides, with 4-12 pairs of secondary veins; tertiary venation inconspicuous. *Inflorescence* shortly pedunculate, 3-5(-9) x 2-4 cm, few-flowered, rather lax. Peduncle glabrous, 5-10 mm long; pedicels 3-10 mm long, glabrous. Bracts sepal-like but smaller, persistent. *Flowers* sweet-scented, open during the day. *Sepals* pale green or sometimes (?) white, connate at the base for 0.1-0.5 mm, erect, ovate, 1-2 x as long as wide, 2-5 x 1.5-4 mm, rounded or obtuse, entire, glabrous outside, not ciliate, glabrous inside and with 4-8 colleters in a single row in the middle at the base; colleters 0.4-1 x 0.1-0.2 mm. *Corolla* white with a yellow throat, creamy or pale yellow, rather thin, 25-45 mm long in the mature bud and forming an ovoid head about 0.2-0.4 of the bud length (6-14 x 4-9 mm) with an acute or obtuse apex, glabrous outside, pubescent or pilose-pubescent inside from 10-18 mm above the base to the insertion of the stamens; tube 5-12 x as long as the calyx, 0.76-2.7 x as long as the lobes, 20-35 mm long, almost cylindrical, 2-5 mm wide at the base, above it slightly widened to 3-6 mm wide, about halfway again narrowed to 1.5-3 mm wide, widened around the anthers to 4-6 mm wide, slightly narrowed at the throat to 3-5 mm wide, not twisted; lobes obliquely oblong or obovate, 0.4-1.3 x as long as the tube, 1.6-3.4 x as long as wide, 16-38 x 6-22 mm, rounded, obscurely auriculate at the left side of the base, not undulate, recurved. *Stamens* with apex 3-5 mm below mouth of corolla tube, inserted 0.6-0.8 of the length of the corolla tube (at 13-25 mm from the base); anthers sessile, narrowly triangular, 2.7-4 x as long as wide, 5-6 x 1.5-2 mm, apex acuminate and sterile for 0.3-0.8 mm, sagittate at the base; glabrous. *Pistil* glabrous, 13-27 mm long, with the apex halfway along anthers, with an abscission layer at the apex of the ovary and therefore style and pistil head shed with the corolla; ovary ovoid, 3-4 x 1.5-2.5 x 1.5-2 mm, acuminate, of 2 separate carpels, with an obscure disk-like base; style filiform, 9-22 mm long; pistil head composed of an undulate basal ring, 0.2 x 1.2-1.3 mm, an almost obovoid central part 0.7-1 x 0.6-0.8 mm, with 5 distal ellipsoid lobes about 0.4-0.5 x 0.2-0.3 mm, and a stigmoid apex 0.2-0.3 x 0.1 mm. Ovules approximately 40-50 in each carpel. *Fruit* of 2 separate mericarps; mericarps orange, darker inside, obliquely ellipsoid, 25-45 x 15-30 x 15-20 mm, apiculate, slightly recurved at least at the apex, with faint ridges, 3-ca.20-seeded; wall rather thin, ca. 1-2 mm thick in dried fruits; aril orange, enveloping the seed. *Seed* dark brown, obliquely ellipsoid, 7-9 x 3.5-5 x 3.5-4 mm, with longitudinal grooves, dull, papillose; embryo 5-5.5 mm long; cotyledons elliptic, 1.5-1.8

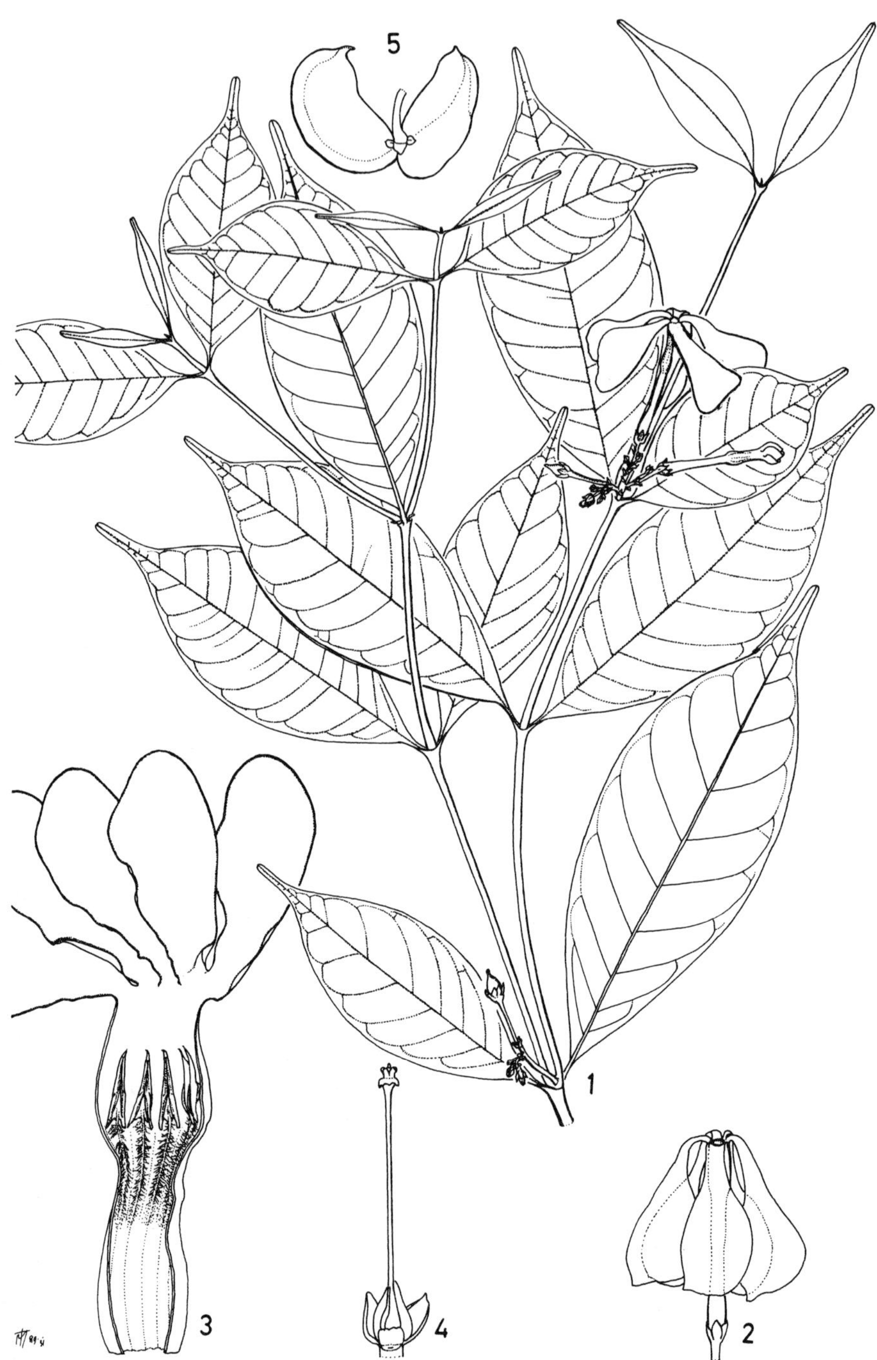

Fig. 74. *Tabernaemontana flavicans*. **1,** habit (x 2/3); **2,** flower (x 1); **3,** opened corolla (x 2); **4,** calyx with pistil (x 2); **5,** fruit (x 2/3). 1-2 from Plowman et al. 9870; 3-4 from Austin et al. 7113; 5 from Cid et al. 6724.

x as long as wide, 2.2 x 1.2-1.5 mm, rounded at the apex and at the base; rootlet 2.8-3.5 x 0.5-0.7 mm.

DISTRIBUTION: Venezuela, Peru, Brazil.

ECOLOGY: Humid or dry forests, mainly in understorey; or low scrub. Alt. 300-950 m. Flowering in Venezuela mainly in May, in Brazil mainly from September to December and in Peru mainly from July to September. A fruiting season could not be deduced from the extant collections.

Geographical selection of the approximately 210 specimens examined:

COLOMBIA. Cauca: Gorgona Island, "St. George" Exp. 566 (K). Amazonas: La Chorrera, Sastre 3038 (COL, P). Cundinamarca: Dindal, Mun. Caparrapi, *Garcia Barriga* 7641 (COL).

VENEZUELA. Bolívar: Río Paraguaza, *Steyermark* 131699 (MO). Amazonas: Dept. Atures, ca 5 km NE of Galipero, *O.Huber* 5242 (MO, Z); 8 km W of Ytuaje, *Holst & Liesner* 3156 (MO, WAG); 35 km S of Puerto Ayacucho, *Steyermark & O.Huber* 113861 (K, VEN); 22 km S of Puerto Ayacucho, *Davidse & O.Huber* 15170 (MO, US, Z); Isla Ratón, *Morillo* et al. 7354 (VEN); Upper Caño Picure, *O. Huber* 4036 (VEN); Caño Cupaven, *Wurdack & Adderley* 42803 (NY, US); W of Cerro Cucurital, *Guanchez* 1298 (MO); Tamatama, *Ll. Williams* 15140 (F, G, MO, US, VEN); Near San Carlos, Río Negro, *Spruce* 3114 (AWH, BM, BP, BR, C, FI, G, GH, GOET, K, LD, NY, P, TCD, W, lectotype of *T. olivacea*).

BRAZIL. Mun. São Gabriel de Cachoeira, Rio Xié, *Farney* et al. 1778 (WAG); Rio Cauaburí, near Pico da Neblina, *N.T. Silva & Brazão* 60616 (MO); Carauari, *A.S.L. Silva* et al. 504 (Z); near Ega (= Tefé), *Poeppig* 2744 (G, GOET, MO, NY, P, W; phot. of lost B sheet GH, MO, NY, US, paratype of *T. olivacea*); Mun. Humaitá, near Livramento, *Krukoff* 6980 (A, K, MO, NY, S, U); Manaus, *Ducke* RB 21616 (RB); Mun. Pres. Figueredo, Atroarí, *Cid* et al. 6724 (K, WAG); 30 km S of Borba, near Rio Maparí, *Zarucchi* et al. 2855 (WAG); Rio Urubú, São Francisco, *Fróes* 25500 (MO); Maués, *N.T. da Silva* 4504 (NY, WAG). Pará: Rio Trombetas, *Ducke* RB 21795 (RB); ibid., Porteiro, *Campbell* et al. P22490 (NY, U, US, Z); Tapajós National Park, *M.G. Silva & Rosário* 3913 (MG, NY, WAG); Serra do Cachimbo, Mun. Itaituba, *M.N. Silva* 359 (WAG); Rio Cupari, *Black* 47-2115 (MO); near Santarém, *Spruce* 236 (BM, CGE, K, M, TCD, W, type of *Bonafousia latifolia*); ibid. Pindobal-Porto Novo Road, *Maciel & Cordeiro* 220 (U); Obidos, Spruce 378 (K, P); Alter do Chão-Pindobal Road, near Pindobal, *Maciel & Cordeiro* 282 (NY, U); Taperinha, *Ginzberger* 332 (WU); Santarém, road to Alter do Chão, *Vilhena* et al. 216 (NY, WAG); Prainha, *Trail* 515 (K); Mun. Almeirim, Mt Dourado, *M.J. Pires & N.T. Silva* 1218 (WAG), 1464 (WAG); Mt Dourado-Munguba Road, *N.T. Silva* 1505 (MO, NY, US, Z); Gurupá, *Killip & A.C. Smith* 30583 (MO, NY, US); Rio Tocantins, *Ducke* RB 21781 (RB); 12 km N of Tucuruí, *Plowman* et al. 9870 (MG, NY, WAG); Igarapé Cametaú, opposite Itupiranga, *Daly* et al. 1564 (MO, US, WAG); Tomé-Açu, Acará, *M.G. Silva & Pinheiro* 5120 (NY); Belém, *Pereira* 3308 (INPA, MG, RB, Z); ibid., *Burchell* 9636 (BR, GH, K, P); ibid., *Martius* 2581 (BP, L, M, W); km 95 Belém-Brasília Road, *Kuhlmann & Jimbo* 309 (K, MO, U, US); Ilha do Mosqueiro, near Belém, *Killip & A.C. Smith* 30438 (MO, NY, US); Cajutuba, *Monteiro da Costa* 266 (MO); Curuçá, Abade, *M.G. Silva* 4057 (NY, Z); near Paragominas, *Prance & N.T.* Silva 58705 (K, NY, U, US, Z); 17 km S of Ligaçao do Pará, *Plowman* et al. 9407 (F, GH, MG, MO, NY, US, WAG). Amapa: Macapá, *Nonato* 245 (MG); E of Porto Grande, *Austin* et al. 7113 (MG, MO, NY, US, WAG); Ferreira Gomes, *Fróes & Black* 27666 (MO, NY, Z); Coastal Region, near km 48, *J.M. Pires & Cavalcante* 52059 (K, NY, S, US, Z). Maranhão: Maracaçumé R. region, Boa Esperança, *Fróes* 1852 (A, G, K, MO, NY, P); 3 km NW of Lago do Junco, *Daly* et al. 134 (GH, MO, US, WAG); sin. loc. *Don*

27 (BR, CGE). Pernambuco: Amaraji, *Paiva* 1494 (US); Rio Formosa, *Falcão* et al. 1203 (RB). Alagoas: São Miguel dos Campos, *Paiva* 3325 (MO). Bahia: Jacobina, *Blanchet* 95 (BM, G, P); Lamarão do Passé, *Noblick & Lemos* 3388 (WAG); Salvador, *Martius* s.n. (M); km 7 Serra Grande-Itacaré Road, *de Carvalho* et al. 3588 (WAG); Itabuna, *Fróes* 12692/57 (A, NY); Ilhéus, *Blanchet* 2358 (BM, BP, BR, FI-W, G, G-DC, K, LD, NY, P, W; phot. of G-DC sheet MO, US, type of *T.oblongifolia*); Bela Vista, Rio Una basin, *Fróes* 12693/58 (A, NY); Santa Luzia-Canavieras Road, *Pinheiro* 1689 (P, Z); 14 km W of Porto Seguro, *Euponino* 389 (P); near Est. Ecol. of Pau-Brasil, *S. Mori* et al. 10814 (K, NY, P); Mt Pascoal, *Belém & Pinheiro* 2704 (NY, Z); km 10 Cumuruxatiba-Prado Road, *L.A. Mattos Silva* et al. 2127 (WAG); sin. loc., *Blanchet* 1471 (G), 2314 (FI-W). Espirito Santo: Mun. Linhares, 2 km S of Alegre, *Hatschbach & Guimãraes* 46951 (C, US); Ponta do Tubarão, *Pabst* 9058 (Z). Minas Gerais: Est. Biol. of Caratinga, *Andrade & Malopes* 528 (P). Rio de Janeiro: Mun. Marica, Area proteçao Ambiental, *Araujo & M.C.A.Pereira* 7731 (GUA); Mun. Macae, Cabuinas, Araujo & Maciel 7082 (GUA); Petropolis, *Goés & Costantino* 997 (RB); near Rio de Janeiro, *Glaziou* 14068 (BR, G, K, P); ibid., *Kuhlmann* 97 (G, GH, MO, NY); Guanabara, *Sucre* 1661 (HBG); Barra do Facão, *Martius* s.n. (BM, K, type of *Anartia glabrata*). Mato Grosso: Mun. Santa Terezinha, Fazenda Primavera, *J.B. Fereira dos Santos* INPA 117.751 (INPA); Serra Cobrinha, *W. Thomas* et al. 4411 (WAG); Mun. São Felix do Araguaia, Fazenda Jamaica, *Cid* et al. 6479 (WAG); km 257 Xavantina-Cachimbo Road, *Philcox* 3129 (K, NY, P, U); Corrego do Porco, Rio Suiazinho, *Harley* et al. 10830 (K, NY, P, U); Xavantina, road to São Felix, *Fonseca* 358 (NY, Z); Mun. Colider, Serra do Cachimbo, *N.T. Silva* et al. 16 (WAG); Mun. Sinop, road to Fazenda Atlântica, *W. Thomas* et al. 4033 (F, K, US, WAG); Chapada dos Guimarães, *Oliveira* 71 (HBG, MU, UC, Z); Burity, NE of Cuiabá, *Collenette* 175 (K, NY); km 59.5 of Rio Teles Pires-Porto dos Gaúchos Road, *W. Thomas* et al. 3989 (K, MO, US, WAG); Santa Cruz (= Barra do Bugres), *S. Moore* 289 (BM, NY, WU); Aripuaña, *Berg* et al. P19939 (AAU, F, NY, Z); ibid., near Centro Científico, *Andrade* 3295 (US). Rondônia: Guaporé, *Vieira* et al. 975 (NY, US, WAG), 1037 (GH, MO, NY, US, WAG). Sin. loc., *Hoffmannsegg* s.n. (B-W 5195, FI-W, type); *Riedel* s.n. (CGE, G, LE, P).

PERU. Loreto: Yanayacu, *Vásquez & Jaramillo* 11169 (WAG); Prov. Requena, Ucayali R., *L. Bernardi* 16225 (G, WAG); Arboretum Jenaro Herrera, *Spichiger & Encarnación* 1084 (G, MO); Lagunas, Prov. Alto Amazonas, *McDaniel & Rimachi* 16430 (MO); Shucushuyacu, *Vásquez & Jaramillo* 2427 (MO, WAG); Yurimaguas, *Killip & A.C. Smith* 29053 (MO, NY, P, US); Balsapuerto, *Klug* 2856 (A, BM, G, GH, K, MO, NY, S, US). Ucayali: near Aguaytia, *Croat* 20978 (F, MO, Z); Granja del Sr. Barrera, Prov. Coronel Portillo, *Schunke* 5408 (GH, K, NY, U, US). San Martín: Pongo de Cainarachi, *Klug* 2709 (A, BM, G, GH, K, MO, NY, S, US); km 17.5 Tarapoto-Yurimaguas Road, *Knapp* 8285 (F, MO, WAG); near Tarapoto, *Spruce* 4611 (AWH, BM, BP, BR, C, E, F, FI, G, GH, GOET, K, LD, LE, NY, P, W, type of *Taberna disparifolia*); E of Puerto Pizana, *Schunke* 6519 (GH, MO, Z); SE of airport of Tocache, *Schunke* 3865 (F, G, GH, K, MO, NY, US); Quebrada de Ishichimi, *Schunke* 10009 (GH, MO, U, USF, WAG); km 2 Caserío de Nuevo Progreso-Río Uchiza, *Schunke* 3226 (G, GH, K, MO, NY, P, US, WIS). Huánuco: Tingo María, *Asplund* 13182 (S); ibid., *Woytkowski* 5328 (BR, MO, P, S); Maquizapa, *Schunke* 1060 (ECON, F, NY, U, WAG); Monzón, *Woytkowski* 1523 (MO, P). Madre de Dios: Tambopata Res., 30 km S of Puerto Maldonado, *Young & Stratton* 102 (MO, WAG); near confluence of Río Tambopata and Río La Torre, *S.F. Smith & Shuhler* 183 (US, WAG); km 41 Puerto Maldonado-Quince Mill Road, *A. Gentry* et al. 19766 (F, MO).

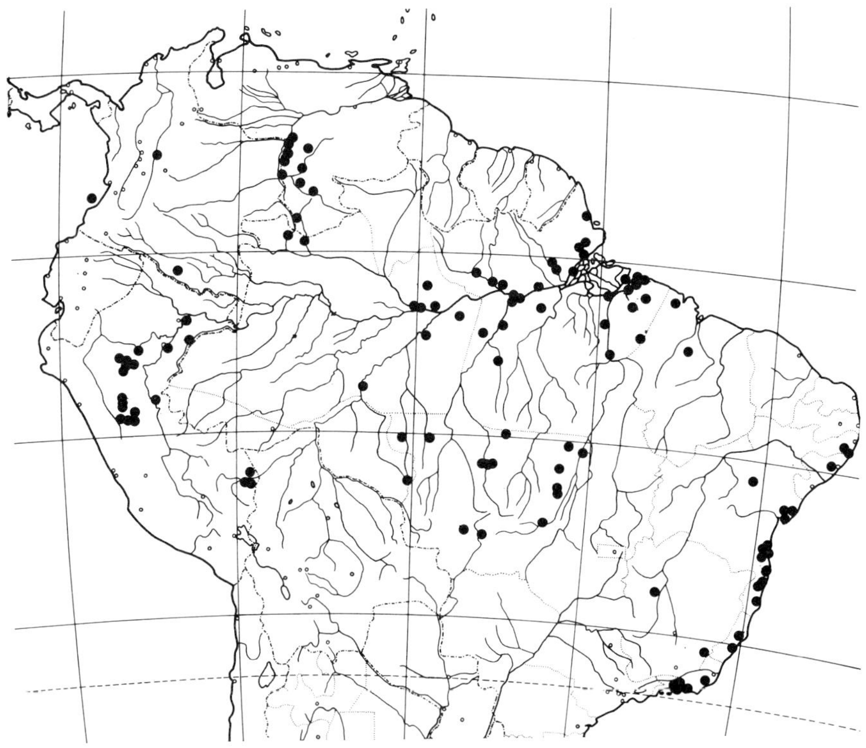

Map 38. *Tabernaemontana flavicans.*

CULT. Brazil, Pará, Belém, Botanical Garden of Mus. Goeldi, *Markgraf* 3804 (MO, RB).

75. Tabernaemontana heterophylla Vahl, Eclog. amer. 2: 22 (1798). – Type: French Guiana, Cayenne, *J. von Rohr* s.n. (holotype C-VAHL; isotypes BM, C, MEL, P).
Fig. 75, p. 303; photo. 7, 8, 9; map 39, p. 308

Homotypic synonyms:

Peschiera heterophylla (Vahl) Miers, Apoc. S. Am. 38 (1878). *Stenosolen heterophyllus* (Vahl) Mgf. in Pulle, Fl. Surinam 4, 1: 455 (1937).

Heterotypic synonyms:

P. tenuiflora Poepp., Nov. Gen. et Sp. 3: 70, t. 280 (1845). *T. tenuiflora* (Poepp.) Muell. Arg. in Martius, Fl. Bras. 6, 1: 76 (1860), non Miq. (1862), **syn. nov.** – Type: Brazil, Amazonas, Rio Aragona, between Coary and Rio Negro, *Poeppig* Mar. 1831 (holotype W; isotype F).

P. diversifolia Miq., Stirp. Surinam. Sel. 164 (1851). – Type: Suriname, sin. loc., *Hostmann* 9 (lectotype U, designated here; isolectotypes BM, CGE, G, K, M, NY, S, W; phot. of M sheet F, GH, MO, NY, US).

T. stenoloba Muell. Arg. in Linnaea 30: 407 (1860), **syn. nov.** *P. stenoloba* (Muell. Arg.) Miers, op. cit. 38. *S. stenolobus* (Muell. Arg.) Mgf. in Notizbl. Bot. Gart. Berlin 14: 177 (1938); Allorge in Mém. Mus. natn. Hist. Nat. IIB, 30: 163, pl. 73 (1985). – Type: Peru, near Cuchero, *Poeppig* anno 1829 (lectotype W, designated by Allorge in 1985).

P. cuspidata Miers, op. cit. 37. – Type: Colombia, Santander del Sur, Barranca Bermeja, *Weir* 76 (holotype BM; isotype K).

T. unguiculata Rusby in Mem. N.Y. Bot. Gard. 7: 324 (1927), partly, as for type. – Type: Bolivia, La Paz, Huachi (= San Miguel de Huachi), *O.E. White* 461 (holotype NY; isotypes GH, US; phot. of US sheet WAG).

S. eggersii f. *glabra* Mgf., op. cit. 177, 183, **syn. nov.** – Type: Ecuador, Manabi, Hacienda El Recreo, *Eggers* 15110 (holotype B†; lectotype M, designated here; isotypes BM, C, F, GH, K, L, LD, MA, MO, NY, P, S, US; phot. of US sheet WAG).

S. eggersii f. *pubescens* Mgf., l.c, **syn. nov.** – Type: Ecuador, Guayas, Bulao, *Eggers* 14326 (holotype B†; lectotype M, designated here; isotypes A, L, US, WU; phot. of US sheet WAG).

S. holothuria Mgf. in Ann. Miss. Bot. Gard. 61: 899 (1974). *T. holothuria* (Mgf.) Leeuwenberg in Agric. Univ. Wageningen Pap. 87, 5: 12 (1988), **syn. nov.** – Type: Panama, Darién, Tumaganti, *Duke* 14149 (holotype Z; isotypes MO, NY).

S. grandifolius Mgf. in Acta Bot. Venezuel. 10: 250 (1975), **syn. nov.** – Type: Venezuela, Zulia, Sierra de Perijá, between la Misión de los Angeles de Tukuku and Pishikakao, *Steyermark* et al. 105789 (holotype Z).

P. laevifructa Allorge in Bull. Soc. Bot. Fr. 130, Lettres 4-5: 344, fig. 6 (1983); in Mém. Mus. natn. Hist. Nat. IIB, 30: 155, pl. 69 (1985), partly, as for type only, **syn. nov.** – Type: Venezuela, Aragua, Rancho Grande, headwaters of Río Limón, *Pittier* 12124 (holotype US; isotype G). (The paratypes belong to *T. amygdalifolia*).

Shrub or small tree, 0.50-7 m high. Trunk 1-20 cm in diameter; bark grey-brown, smooth; wood white. Branches pale brown, lenticellate; branchlets terete when fresh, glabrous, puberulous, or sometimes pubescent. *Leaves* of a pair equal or unequal (smaller down to 0.2 x as long as the other and comparatively wider), shortly petiolate or sessile; petiole glabrous, 1-12 mm long, (ocreae slightly or not widened into intrapetiolar stipules); blade herbaceous when fresh, membranaceous or thinly papery when dried, elliptic, narrowly elliptic or, especially in smaller leaves of a pair ovate, 1.5-4.5 x as long as wide, 2-25 x 1-11 cm, acuminate or caudate at the apex, cuneate, rounded or in smaller leaves of a pair often cordate and at least in some leaves of a pair uequal-sided at the base, entire or undulate, sometimes bullate, with scattered black dots and glabrous on both sides or pubescent, especially beneath, with 5-15 pairs of rather straight or upcurved secondary veins forming an angle of 50-70° with the costa; tertiary venation more or less reticulate. *Inflorescence* shortly pedunculate, 2-7 x 2-7 cm, 1-10-flowered, lax. Peduncle slender, glabrous or pubescent, 1-40 mm long; axis covered with bracts and pedicel scars at anthesis; pedicels glabrous or pubescent, 2-10 mm long. Bracts numerous, more or less sepal-like and 0.3-1 x as long as them. *Flowers* fragrant, open during the day. *Sepals* pale green, subequal, connate at the base for about 0.3 mm, erect or recurved at the apex even when fresh, ovate to strap-shaped 1.5-7 x as long as wide, 1.5-7 x 0.8-1.3 mm, acute or obtuse, entire, glabrous or pubescent outside, occassionally ciliate, glabrous inside and with 1-9 colleters in 1-2 rows in the middle at the base; colleters 0.2-0.7 x 0.1-0.2 mm. *Corolla* white, creamy or sometimes pale yellow, often with a greenish tube, 13-26 mm long in the mature bud and forming a comparatively small ovoid or narrowly ovoid head 0.27-0.35 of the bud length (3-8 x 1.5-2 mm) with an acute or less often obtuse apex, glabrous or pubescent outside, pilose inside in a belt starting 1 mm below the insertion of the stamens to 1 mm below apices of anthers and ending at the mouth or the base of the lobes; tube slender, 2.5-9 x as long as the calyx when sepals straightened out, 1-2.7 x as long as the lobes, 10-21 mm long, almost cylindrical, widest around the stamens and 1.5-3 mm wide, narrowed above to 0.7-2 mm wide, twisted 0-1 turn

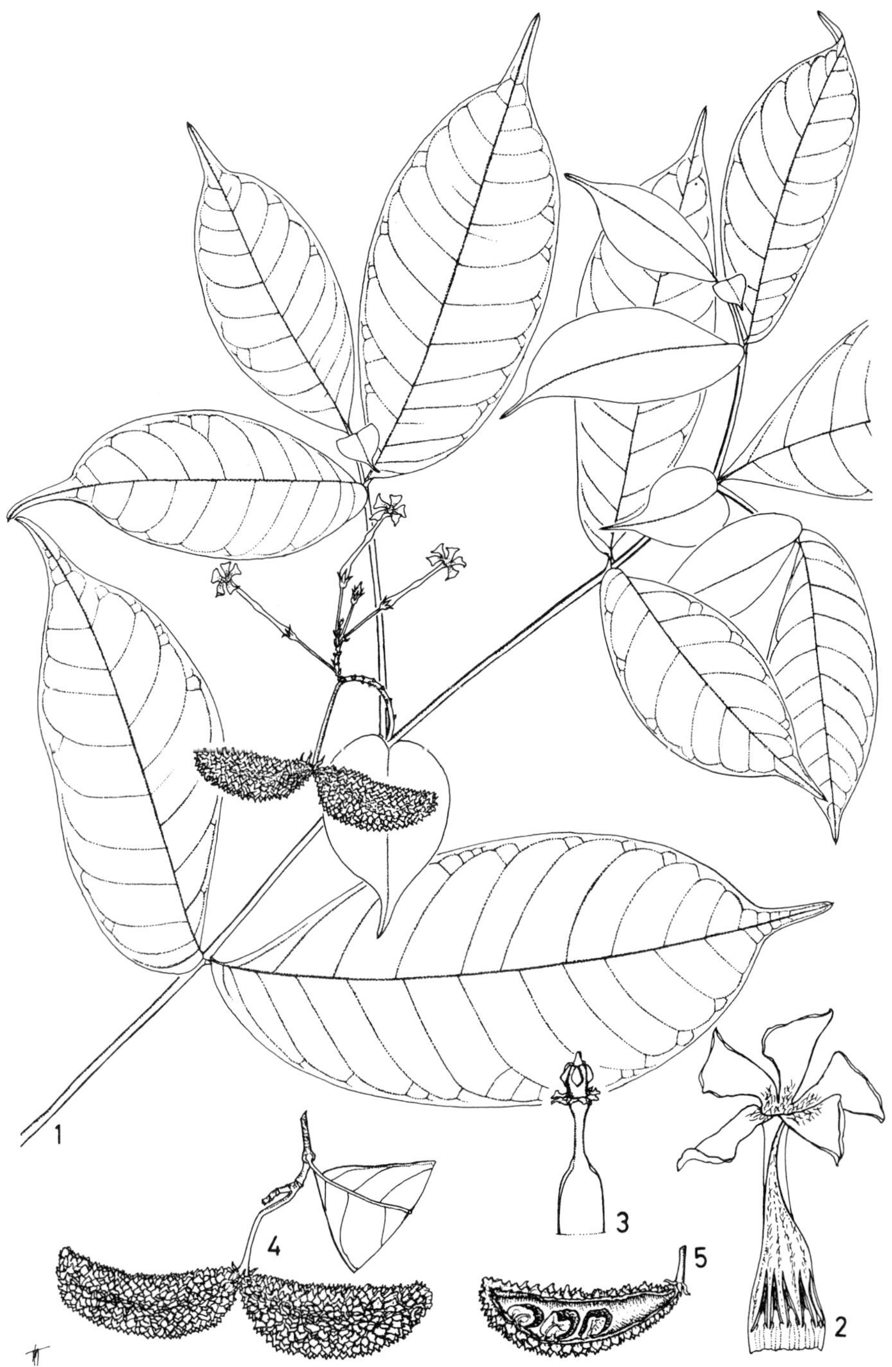

Fig. 75. *Tabernaemontana heterophylla.* **1,** habit (x 2/3); **2,** opened corolla (x 2); **3,** pistil (x 8); **4,** fruit (x 2/3); **5,** fruit valve with seeds partly covered by aril (x 2/3). 1-5 from Leeuwenberg 11611.

mostly just below the mouth; lobes dolabriform or narrowly elliptic and falcate, 0.37-1 x as long as the tube, 1.8-5 x as long as wide, 5-11 x 1.5-5 mm, acute or acuminate, spreading. *Stamens* deeply included, inserted 0.1-0.25 of the length of the corolla tube (at 1.5-3 mm from the base); anthers sessile, narrowly triangular, 3-5 x as long as wide, 3.5-6 x 1-1.2 mm, apex acuminate, sterile for 0.3-1 mm, sagittate at the base, glabrous. *Pistil* glabrous, 3-7 mm long, with apex about halfway along anthers; ovary ovoid 1.3-2 x 0.8-1.3 x 0.8-12 mm, acute or obtuse, of 2 basally connate carpels, with or without an obscure disk-like thickening; style short, 0.6-3.5 mm long; pistil head composed of a 10-lobed ring 0.1-0.3 x 0.6-1 mm, a stipitate apically 5-lobed depressed globe 0.2-0.6 x 0.5-0.7 mm and a stigmoid apex, 0.2-0.3 x 0.1-0.2 mm. Ovules approximately 30-40 in each carpel. *Fruit* of 2 separate mericarps; mericarps orange, yellow or red, obliquely ellipsoid, fusiform or pod-like, obtuse to acuminate, 25-60 x 10-20 x 10-20 mm, straight or recurved, without ridges, with blunt soft prickles all over, sometimes only on abaxial side or occasionally entirely smooth, 3-20-seeded; wall 2-3 mm thick; aril orange, eveloping hilar side of the seed only. *Seed* dark brown or black, obliquely ellipsoid, 7-10 x 4-5 x 2.5-4 mm, with longitudinal grooves, papillose; embryo 5.8-6.5 mm long; cotyledons ovate, 1.2 x as long as wide, 2.5-3.3 x 2-2.8 mm, rounded at the apex, cordate or rounded at the base, rootlet 2.8-3.5 x 0.5-0.7 mm.

DISTRIBUTION: From Panama to Peru and northern Brazil.

ECOLOGY: Forest understorey or scrub, not on river banks. Alt. 0-1650 m. Flowering and fruiting throughout the year.

Geographical selection of the approximately 450 specimens examined:

COSTA RICA. Alajuela: Río Naranjo, *Tonduz* 7645 (BR).

PANAMA. Panamá: Mts above Torti Arriba, *Folsom* et al. 6570 (MO, USF); Past Río Ipetí, along Panamerican Highway, about 4-5 km E, *Folsom & Collins* 1678 (MO); 8 km E of Ipetí, *Hamilton & D'Arcy* 1397 (MO). San Blas: Ailigandi, *Hammel & D'Arcy* 5016 (MO, USF). Darién: Río Chucunaque, *Duke* 15534 (MO); Tumaganti, *Duke* 14149 (MO, NY, Z, type of *Stenosolen holuthuria*); Río Pirre, *Duke & Bristan* 8266 (MO, paratype of *S. holothuria*); ca 18 km E of Pocuro, *Hammel* et al. 16470 (WAG); N of Punta Guayabo, *Knapp & Mallet* 3152 (MO, WAG); Serrania de Pirre, *McPherson* 12249 (WAG).

COLOMBIA. Choco: near Sautata, *Duke* 15398 (MO); Mun. Ríosucio, Los Catios National Park, *Forero* et al. 1644 (COL, NY); W of Mutatá, *A. Gentry & H. Leon* 20233 (COL, MO, Z); Bahía Solano, *Killip & H. Garcia* 33561 (A, BM, COL, MO, US). Antioquia: Punta de Piedra, *Haught* 4826 (COL, MO, US); Villa Arteaga, *Cuatrecasas & Willard* 26174 (COL, MO, US); 12 km N of Mutatá, *Callejas* et al. 5778 (WAG); Caucheras Path, *Bernal & Galeano* 524 (COL); Mun. Cáceres, Hacienda Catatumbo, *Fonnegra & Roldán* 2361 (WAG); 4 km NW of Medellín, *Callejas* et al. 3607 (WAG); near Planta Providencia, *Shepherd* 702 (COL, MO, WIS); Independencia-Santa Rita Road, *Romero-Castañeda* 1579 (COL); San Julian, *J. Triana* 3396. 10 (COL); Mun. San Luís, Cañon del Río Claro, *Cogollo* 1935 (MO); Río Claro, road to Cairo, *Rentería* et al. 2758 (MO); El Refugio Res., Río Claro, 30 km W of Puerto Triunfo, *Stein & McDade* 3339 (MO); Río Nus, *Lehmann* 7911 (K); Mun. Puerto Berrío, Alicante Path, *Callejas* et al. 9283 (WAG); El Río-Segovia, *Rentería* et al. 2327 (COL); Casabe, *Rentería* et al. 1939 (COL, MO). Carabobo: Mun. Tierralta, Río Esmeralda, *Bernal* et al. 1183 (COL); Monte Líbano-San Pedro Road, *Romero-Castañeda* 1773 (COL, MO). Bolívar: Ayapel, *Romero-Castañeda* 1679 (COL). Norte de Santander: Reyes, Río Sardinata, *García-Barriga* 18247 (US); Río Cubugón, *Cuatrecasas* 13280 (MO, US). Santander del Sur: near Barranca Bermeja, *Haught* 2026 (COL, MO, NY, US); ibid., *Weir* 76 (BM, K, type of

Peschiera cuspidata); 20 km from Montoya, Sitio Capote, *Camacho Durán* 429 (WAG); near Puerto Berrio, *Haught* 1663 (MO, US); Virolín, *Díaz* 1500 (COL). Boayaca: NW of Bogota, *Lawrance* 306 (A). Risaralda: Mun. Santa Rosa, El Crucero, *Idrobo* 9962 (U). Meta: Mun. Mesetas, Gaviotes-Río Leiva, *Callejas & Marulanda* 5967 (MO, WAG); near Villavicencio, *Alston* 7667 (BM, NY, P, S); Río Ocoa, *Dugand* 3114 (COL, MO, US); Sierra de la Macarena, *Philipson* et al. 1954 (BM, COL, MO, S, US) & 2240 (BM, COL, MO, US). Vaupés: San José del Guaviare, *Cuatrecasas* 7387 (COL); Río Inírida, *Fernández* 2255 (COL, US); Mitú, *Zarucchi* 1203 (COL, GH, WAG); Cauca: Belalcázar, *Lehmann* Aug. 1891 (K). Valle: Río Dígua, Quebrada de San Juan, *Cuatrecasas* 23831 (MO, S, U, US). Amazonas: Río Caquetá, La Pedrera, *Schultes & Cabrera* 17690 (GH, MO, US); Amacayacu National Park, *Rudas* et al. 3051 (MO); Puerto Nariño, *Plowman* 3218 (GH, K, S, US). Nariño: Tumaco, *Romero-Castañeda* 5163 (COL); Mun. Espriella, Quebrada de Sabaletas, *H. León* et al. 1483 (U). Putumayo: San Antonio del Río Guamues, *Plowman* 2105 (GH, K, S, US); Umbría, *Klug* 1767 (A, NY, S, US).

VENEZUELA. Zulia: ca 11 km S of Casigua, *Bunting* et al. 7253 (P); Ariguaisa, *Liesner & González* 13064 (MO, WAG); ca 13 km NE of the intersection of the Maracaibo-La Fria Highway and Río Aricuaisá, *Davidse* et al. 18321 (MO, WAG); S of Río Catatumbo, *Liesner & González* 13202 (MO, WAG); near Ensenada Congo, near Maracaibo Lake, *de Bruijn* 1388 (WAG); Sierra de Perijá, between Misión de los Angeles and Pishikakao, *Steyermark* et al. 105789 (Z, type of *Stenosolen grandifolius*); Río Las Piñas, *Zambrano & Gutierrez* 1573 (MO). Mérida: between Caño Zancudo and La Azulita, *Breteler* 4571 (MO, NY, S, U, US, WAG, Z). Trujillo: La Ceiba, *Reed* 919 (MO, NY, US). Tachira: La Pocina, *Steyermark* et al. 120292 (MO); ca 6 km SW of Coloncito, *Steyermark* et al. 120504 (MO); ca 35 km SSE of San Cristóbal, *Liesner & González* 10869 (MO, WAG); 10 km S of El Piñal, *Davidse & González* 21639 (MO). Apure: San Camilo F.R., *Steyermark* et al. 101454 (M, Z). Barinas: Santa Barbara de Barinas, *Valverde & Peña* 1140 (MO); Mun. La Pedraza, Río Camatuche, *Stergios & Taphorn* 8733 (MO); Highway 5, between Río Acequia and junction with road to Ciudad Bolivia, *Steyermark & Rabe* 96497 (NY, US); Ticoporo F.R., *Breteler* 3532 (NY, U, WAG); Distr. Pedraza, above El Agarrobo, *van der Werff & Ortíz* 5830 (WAG). Aragua: Pittier National Park, *Weitzmann & Boom* 88 (GH, K, MO, US, WAG); Rancho Grande, headwaters of Río Limón, *Pittier* 14124 (G, US, type of *Peschiera laevifructa*). Yaracuy: NE of Nirgua, W of Valencia, *Croat* 54656 (MO, WAG). D.F.: between Colonia Tovar-Junquito Road and Hacienda El Limón, *Steyermark & Nevling* 95902 (GH, M, WAG); near Colonia Tovar, *Fendler* 2570 (GH); Cerro Naiguatá, *Steyermark* 92142 (K, MO, NY, US). Miranda: Los Guayabitos, *Aristeguieta* 2954 (NY, US). Anzoategui: Puerto La Cruz, *Pittier* 8073 (GH, US). Amazonas: San Pedro de Orinoco, *Morillo* 7411 (VEN); Dept. Atures, between Paso el Diablo and Caño de Culebra, *Steyermark* 122341 (MO, WAG); Siquita, *Morillo* 7478 (VEN); Patacame, *Morillo* 7531 (VEN); Dept. Atabapo, SE of Santa Bárbara del Orinoco, *Guanchez & Mercado* 1881 (MO); Yureba Fall, *Delascio & Guánchez* 10686 (MO); Santa María de los Guaicas, *Guanchez* 470 (MO, Z); Dept. Río Negro, Río Siapa, *Liesner & Delascio* 21933 (MO); between Ocamo and Mavaca, *Aristeguieta & Lizot* 7389 (MO, NY, US); Caño Amarillo, *A.L. Bernardi* 6379 (G); Tamatama, *Ll. Williams* 15159 (A, US); between La Esmeralda and Ugueto, *Croizat* 841 (NY, VEN). Bolívar: Mun. Raul Leoní, 85 km SSE of Entre Ríos, *Aymard & Fernández* 7286 (MO); 5 km E of Las Chicharras, *Steyermark* 89327 (K, NY, U, US). Bolívar-Delta Amacuro border: Río Grande o Toro, *Breteler* 3875 (WAG). Delta Amacuro: Dept. A.Diaz, 60 km NE of El Palmar, *Aymard* 5413 (NY); E of El Palmar, *A. Gentry & Berry* 14934 (AAU, MO, NY); Río Cuyubini, *Steyermark* 87631 (K, NY, U, US).

GUYANA. NW Distr., Waini R., *de la Cruz* 3594 (F, GH, MO, NY, US); Mabaruma Compound, *Archer* 2239 (K, US); Wanama R., *de la Cruz* 3890 (F, GH, MO, NY); NW of Baramita, *McDowell* et al. 4311 (US, WAG); Pomeroon Distr., Moruka R., *de la Cruz* 950 (NY); Supernaam Creek, Essequibo R., *S.A. Harris* TP 469 (K); Cuyuni-Mazaruni Region, Aurora, *Gillespie & Tiwari* 2121 (P); Rupununi Distr, Kuyuwini R., *Jansen-Jacobs* et al. 2916 (U, WAG); Kamuni Creek, Groete Creek, Essequibo R., *Maguire & Fanshawe* 22904 (A, NY); Penal Settlement, *Hitchcock* 17138 (GH, K, NY, US); near Bartica, *Sandwith* 274 (K, NY, U); below the Kaieteur, Potaro R., *Jenman* 927 (K); Corantyne R., above Cow Falls, *McDowell & Gopaul* 2407 (US, WAG); Rupununi area, Surama, *Acevedo* et al. 3320 (US, WAG); Kanuku Mts, *A.C. Smith* 3424 (A, B, G, K, MO, NY, P, S, U, US, W); ibid., *Jansen-Jacobs* et al. 293 (CAY, K, MO, U, US, WAG, WIS); Essequibo R., *Robert Schomburgk* I, 3 (BM, CGE, E, FI-W, G, K, L, NY, P, SING, TCD, U, UPS, US, W); Corantyne R., For. Dept. 6485 (K, MICH, NY, S).

SURINAME. Corantyne R., near Apoera, *Heyde* 423 (U, US); Matapi, *Florschütz* 2264 (NY, U); Mamaboen Creek, *Stahel* 4628 (L, U); near Tibiti, *Lanjouw & Lindeman* 1574 (K, NY, U); near Paramaribo, *Hostmann & Kappler* 1632 (FI-W, G, MO, NY, P, S, U, W, paratype of *Peschiera diversifolia*); Charlesburg, *Maguire & Stahel* 22745 (K, MO, NY, U, US); SW of Paramaribo, *Kramer & Hekking* 2079 (C, U); near Republiek, *Mennega* 324 (U); Berg en Dal, *Wullschlaegel* 317 (BR, NY, U, W); Slootwijk, *Soeprato* 9H (U); Jodensavanne, *Schulz* LBB 8462 (U); Akintosoela, *Voordouw* 21 (WAG); Bakhuis Mts, *Florschütz & Maas* 2671 (NY, U); Wilhelmina Gebergte, 4 km S of Julianatop, *Irwin* et al. 55300 (K, MO, NY, U, US); ibid., Zuid R., *Irwin* et al. 55967 (NY); Saramacca R., near Jakob Kondre, *Pulle* 132 (L, U); Upper Suriname R., near Goddo, Exp. to Wilhelmina Gebergte 2 (U, Z); Sara Creek, Brokopondo Distr., *van Donselaar* 1100 (U); Lely Mts, *Lindeman* et al. 498/76 (U); Tapanahoni R., *Versteeg* 579 (U); La Rencontré, *Focke* 788 (U, paratype of *P. diversifolia*), 1215 (U); sin. loc., *Hostmann* 9 (BM, CGE, G, K, M, NY, S, U, W; phot. of M sheet F, GH, MO, NY, US, lectotype of *P. diversifolia*).

FRENCH GUIANA. St. Jacques Creek, *Wachenheim* 39 (P); Godebert, *Wachenheim* 129 (BM, K, NY, P, US); Acarouany, *Sagot* 3 (P), 390 (BM, BR, G, K, MPU, NY, P, S, W), 390bis (BM, S); S of Sinnamary, km 16 Piste St.Elie, *Leeuwenberg* 14048 (WAG); Montagnes Françaises, *Sastre* et al. 8102 (CAY, US); Paul Isnard Region, Décou Décou Mts, *Feuillet* 397 (CAY); Mt Lucifer, *de Granville* 5279 (CAY, P); Cayenne, *Martin* s.n. (G-DC, K; phot. of G-DC sheet MO, NY, US, cited as *Peschiera echinata* by A. De Candolle); ibid., *J. von Rohr* s.n. (BM, C, C-VAHL, MEL, P, type); Remire, *Leeuwenberg* 11641 (CAY, MO, P, USF, WAG); Mt Mahury, *Leeuwenberg* 11611 (CAY, P, WAG); Mt Atachi Bacca, Inini R. Region, *de Granville* et al. 10674 (P, U, US); Grand Inini R., Saï Creek, *Oldeman* 3569 (CAY, K, P, WAG); Approuague R., *de Granville* 8265 (CAY, P); Camopi R., Ouassaye Fall, *Oldeman* 2601 (CAY, NY, P).

BRAZIL. Roraima: near Auaris, *Prance* et al. 9734 (COL, GH, NY, S, U, US, Z); Serra dos Surucucus, *Prance* et al. 10027 (GH, K, NY, S, U, US, Z). Amazonas: Rio Negro, Japurá, *Fróes* 22324 (NY, U); Mun. São Paulo de Olivença, Rio Solimoes basin, *Krukoff* 7848 (K, MO, NY); Bôca do Acre, *Prance* et al. 2346 (NY); near São Paulo de Olivença, *Spruce* 2101 (BM, GH) & 2106 (AWH, BM, BR, E, G, GH, K, NY, P, TCD, W); ibid., *Vogel* 344/1964 (MJG); near Camanaus, *Prance* et al. 15928 (K, MO, NY, S, U, US, Z); Karapaná, *Schultes & López* 9195 (GH, K, US); Rio Tucano, *Ewel* 103 (MY); Serra do Curicuari, *Nascimento* 758 (WAG); Rio Cauaburí, Rio Maturacá, *N.T. Silva & Brazão* 60666 (MO); Mun. Novo Japurá, Rio Acanauê, *Cid & Lima* 3541 (WAG); Lago do Mapari, *Amaral* et al. 417 (MO, WAG); Matambá, *Gourd* XX.1115R (K); Rio Purus, opposite Lábrea, *Prance* et al. 13470 (K, NY, S, U,

US, Z); Ega (= Tefé), *Poeppig* Dec. 1831 (W; phot. GH); Rio Castanho, *Cardona* 1381 (MO, US); Pico Rondon, *Pipoly* et al. 6872 (WAG); Mun. Humaitá, near Tres Casas, *Krukoff* 6417 (A, BM, F, G, K, MO, NY, S, U, US); Rio Aragona, between Coary and Rio Negro, *Poeppig* Mar. 1831 (F, W, type of *Peschiera tenuiflora*); Mun. Novo Aripuanã, near Gaúchos, *Cid* 6034 (K, WAG); Manaus, Igarapé do Binda, *Coêlho* INPA 6304 (MO); Mun. Borba, near Urucurituba, *Krukoff* 5949 (A, BM, G, K, MO, NY, S, U, US); Rio Canumã, *S.R. Hill* et al. 12854 (K, WAG); Rio Urubú, Sucurijú, *Fróes* 25427 (MO); Serpa (= Itacoatiari), *Trail* 513 (K); near Maués, *Campbell* et al. P22089 (NY, U, US, Z), 22129 (K, MO, NY, Z); ibid., *Zarucchi* et al. 3134 (WAG); Rio Jamundá (= Nhamundá), *Black & Ledoux* 50-10859 (MO). Acre: Rio Macauhan mouth, *Krukoff* 5348 (G, GH, K, NY, S, U, US); Porangaba, *Maas* et al. P12975 (NY, Z). Rondônia: Rio Jatuarana source, *Krukoff* 1643 (A, BM, G, K, MO, NY, P, S, U, UC, US, WIS); near São Lorenço, Rio Madeira basin, *Prance* et al. 8891 (GH, NY, P, S, U, US, Z); Rio Madeira, 2 km above Matumparaná, *Prance* et al. 5672 (NY, Z); W of Ariquemes, *Frame* et al. 207B (F, GH, K, USF); Rio dos Pacaás Novos, *Anderson* 12185 (WAG); Mineraçao Taboca, *Vieira* et al. 367 (MO); São Manoel-Amazonas, *Hoehne* 5126 (US). Pará: Mun. Oriximiná, Rio Paru do Oeste, *Cid* et al. 2369 (GH, NY, US, WAG); Rio Cachorro, *Cid* et al. 8025 (K, WAG); Rio Trombetas, Tapagiha, *Ducke* RB 21782 (RB); Oriximiná-Óbidos Road, near Cuminá-Mirim, *Cid* et al. 2528 (NY, US, WAG); Beira do Rio Cupari, *Black* 48-2202 (MO); Serra dos Carajás, *Sperling* et al. 6393 (WAG).

ECUADOR. Esmeraldas: San Lorenzo, *Játiva & Epling* 708 (NY); Río Cayapa, Zapallo Grande, *Kvist & Asanza* 40887 (QCA); ibid., *Barford* et al. 48123 (AAU, WAG). Imbabura: Santiago Providence, *Lowell* 505 (NY). Los Rios: Hacienda Clementina, *Harling* 448 (S); near Canton Vinces, *Mexia* 6626 (BM, MO, UC, US); Río Palenque Field Station, halfway between Quevedo and Santo Domingo, *A. Gentry* 9876 (GB, MO, S, Z). Manabi: Hacienda El Recreo, *Eggers* 15110 (BM, C, F, GH, K, L, LD, M, MA, MO, NY, P, S, US; phot. of US sheet WAG, type of *Stenosolen eggersii* f. *glabra*). Cotopaxi: Río Guapara, 20 km NW of El Corazón, *Sparre* 17339 (S). Pastaza: Lorocachi, *Jaramillo* et al. 31383 (AAU, WAG). Guayas: near Guayaquil, *Ruiz & Pavón* s.n. (G); Cerro Azul, W of Guayaquil, *Asplund* 19014 (NY, S); Bulao, S of Guayaquil, *Eggers* 14326 (A, L, M, US, WU; phot. of US sheet WAG, type of *S. eggersii* f. *pubescens*). El Oro: Zaracay-Las Piedras Road, *Harling* et al. 15608 (GB, USF).

PERU. Loreto: near Yurimaguas, *Ll. Williams* 3920 (F); ibid., *Poeppig* D.2079 (US, W; phot. of W sheet F, GH, paratype of *T. stenoloba*). Prov. Maynas, near Paco Caño, *Diaz & Jaramillo* 1566 (MO); Gamitanacocha, Río Mazán, *Schunke* 379 (A, MO, NA, NY, UC, US); NW of Santa Maria de Nanay, *Schunke* 2462 (COL, G, NY, US); 21 km S of Iquitos, *Pipoly* et al. 12123 (WAG); Rio Napo, Bella Vista, Croat 20632 (MO); Momoncillo, *Revilla* 302 (AAU, MO, NY); near Mazán, *McDaniel* 15212 (F, MO); Quebrada Tahuayo, above Tamishiyaco, *Croat* 19699 (AAU, K, MO, NY, UC, US, USF, WIS, Z); near Sucusarí, *Pipoly* et al. 13303 (WAG); near Pebas, *Plowman* et al. 6529 (F, GH, MO). San Martín: E of Puerto Pizana, *Schunke* 6479 (F, GH, Z); Río Sión, *Schunke* 3479 (F, NY, US); km 17.5 Tarapoto-Yurimaguas Road, *Knapp* et al. 7900 (MO). Madre de Dios: Manu National Park, Cocha Cashu Station, *Forster* et al. 7099 (F); Atalaya, *Forster & Wachter* 7250 (MO); Río Tambopata, *Evrard* 9629 (BR, WAG); Tambopata Res., *Barbour* 5712 (MO, USF). Not localized: near Cuchero, *Poeppig* anno 1829 (W, lectotype of *T. stenoloba*).

BOLIVIA. Pando: W bank of Río Madeira, opposite Abunã, *Prance* et al. 8643 (NY). La Paz: Huachi (= San Miguel de Huachi), *O.E. White* 461 (GH, NY, US, phot. of US sheet WAG, type of *T. unguiculata*).

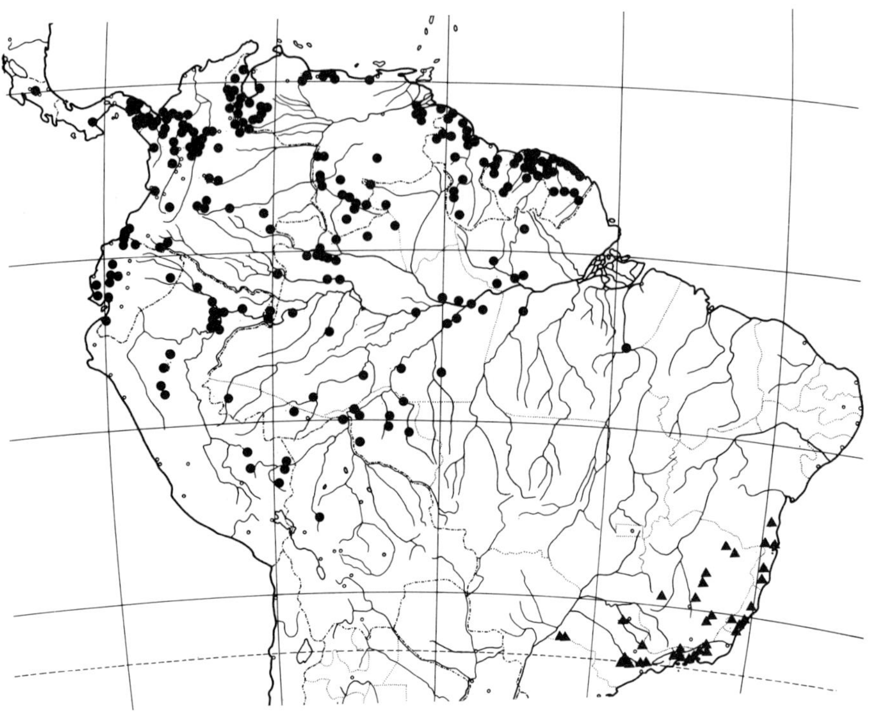

Map 39. ●. *Tabernaemontana heterophylla*, ▲. *T. hystrix.*

CULT. Netherlands, Wageningen Botanical Garden, seedling of herb. *Breteler* 3875 from Venezuela, *Bos* 1546 (WAG) and *Leeuwenberg* 11041 (WAG).

Notes. In *T. heterophylla* characters such as the size and number of the prickles on the fruit (they are even absent in a few specimens collected in Venezuela), the length of the sepals, and the size and the shape of the leaves vary independently. Several names have had to be reduced to synonymy.

76. Tabernaemontana hystrix Steud, Nom. ed.2, 2: 658 (1840). – Type: Vellozo, Fl. Flum. 3: t. 17 (1829). Fig. 76, p. 310; map 39, p. 308

Basionym and homotypic synonym:

T. echinata Vell., op. cit. 105 and 3: t.17, non Aubl.(1775). *Peschiera hystrix* (Steud.) A. DC., Prod. 8: 360 (1844).

Heterotypic synonyms:

T. collina Gardn. in Hooker, London Journ. Bot. 1: 178 (1842). – Type: Brazil, Rio de Janeiro, Serra dos Orgãos, *Gardner* 74 (holotype BM; isotypes CGE, FI, FI-W, G, G-DC, K, NY, P, S, TCD, US, W; phot. of G-DC sheet GH, MO, US; phot. of NY sheet P).

T. fuchsiifolia A. DC., op. cit. 365. *P. fuchsiifolia* (A. DC.) Miers, Apoc. S. Am. 34 (1878) (both as *fuchsiaefolia*). – Type: Brazil, Rio de Janeiro, *Lhotzky* anno 1832 (holotype G-DC; phot. P; phot. of lost B sheet GH, MO, US).

T. gaudichaudii A. DC., l.c., **syn. nov.** *P. gaudichaudii* (A. DC.) Miers, op. cit. 40. – Type: Brazil, Rio de Janeiro, *Gaudichaud* 540 (lectotype G-DC, designated here; isolectotypes FI-W, G, P; phot. of lost B sheet GH, MO, US).
T. lundii A. DC., l.c. *Peschiera lundii (*A. DC.) Miers, op. cit. 36. – Type: Brazil, Rio de Janeiro, *Lund* 883 (lectotype G-DC, designated here; phot. MO, US; isolectotype C).
T. gracilis Muell. Arg. in Martius, Fl. Bras. 6, 1: 82 (1860), non Benth. (1841). *P. gracillima* Miers, op. cit. 41. *T. gracillima* (Miers) Pichon in Not. Syst. ed. Humbert 13: 237 (1948). – Type: Brazil, sin. loc., *Riedel* s.n. (holotype B†; lectotype G, designated here; isotypes GH, P, U; phot. of lost B sheet GH, US).
T. bracteolaris Muell. Arg. in op. cit. 83. – Type: Brazil, Minas Gerais, sin. loc., *Saint-Hilaire* B 19 (lectotype P, designated here).
P. granulosa Miers, op. cit. 37. – Type: Brazil, Rio de Janeiro, Serra dos Orgãos, Vargem Bridge, *Miers* 4028 (lectotype BM, designated here).
P. solandri Miers, op. cit. 46. – Type: Brazil, Rio de Janeiro, *Solander* s.n. (holotype not seen but cited from BM by Miers, 1878).

Shrub or small tree 1-7 m high. Branches pale or dark brown, lenticellate; branchlets terete, glabrous or sometimes pubescent. *Leaves* of a pair equal or unequal (larger up to 2 x as long as other and not comparatively narrower), petiolate; petiole glabrous or sometimes pubescent, 2-10 mm long; (ocreae not widened into intrapetiolar stipules); blade thinly papery when dried, elliptic or narrowly elliptic, 2.5-5 x as long as wide, 2-12.5 x 1-4 cm, acuminate at the apex, cuneate at the base or decurrent into the petiole, glabrous on both sides or sometimes, especially on the veins, pubescent beneath, often with scattered black dots beneath, with 7-15 pairs of upcurved secondary veins, forming an angle of 60-80° with the costa; tertiary venation reticulate, conspicuous or not. *Inflorescence* shortly pedunculate, 2-5 x 2-5 cm, 3-15-flowered, rather lax. Peduncle glabrous or sometimes pubescent or puberulous, 2-10 mm long; pedicels glabrous or sometimes pubescent or puberulous, 4-12 mm long. Bracts sepal-like and 0.3-1 x as long as them. *Flowers* fragrant, open during the day, often many at a time, sometimes on bare trees. *Sepals* pale green (?), connate at the extreme base, ovate, oblong, or narrowly so, 1.5-4 x as long as wide, 3-5 x 1.2-2 mm, acute or subacute, recurved for about half of their length or only at the apex, glabrous or minutely puberulous on both sides, not or only obscurely ciliate, with (0-)3-11 colleters inside in 1 row in the middle at the base; colleters 0.3-0.5 x 0.1-0.2 mm. *Corolla* white or yellow, 11-16 mm long in the mature bud, skittle-shaped and forming a comparatively small ovoid or more often narrowly ovoid head 0.27-0.43 of the bud length (3-6 x 1-3 mm) with an obtuse apex, glabrous outside or sometimes puberulous on the tube, pilose-pubescent inside with indumentum more dense and usually with longer hairs towards the apex, from 1.5-3.2 mm above the base (= 0-2 mm below the insertion of the stamens to the base of the lobes or even up to halfway them; tube 2.1-4 x as long as the calyx (when sepals straightened out, if not (3-)4-8 x); 1.5-2.4 x as long as the lobes, 8-15 mm long, narrowly flask-shaped, 2.5-3.5 mm wide around the anther bases (= above the base), narrowed above to 1.5-2 mm wide, not twisted; lobes dolabriform, 0.4-0.7 x as long as the tube, 1-2 x as long, as wide, 4-8 x 3-6 mm, neither auriculate, nor undulate, spreading. *Stamens* with apex 2-7 mm below mouth of corolla tube, inserted 0.2-0.37 of the length of the corolla tube (at 2.5-4 mm from the base); anthers sessile, narrowly triangular, 4-5 x as long as wide, 4-5.5 x 1-1.5 mm, apex acuminate, sterile for 0.5-0.9 mm, sagittate at the base (tails straight), glabrous. *Pistil* glabrous, 4-5.8 mm long, with apex nearly halfway along anthers; ovary ovoid, 1.5-2.5 x 1-1.5 x 1-1.5 mm, of 2 separate carpels, rather gradually narrowed into the style; style short, 0.5-2.5 mm long; pistil head

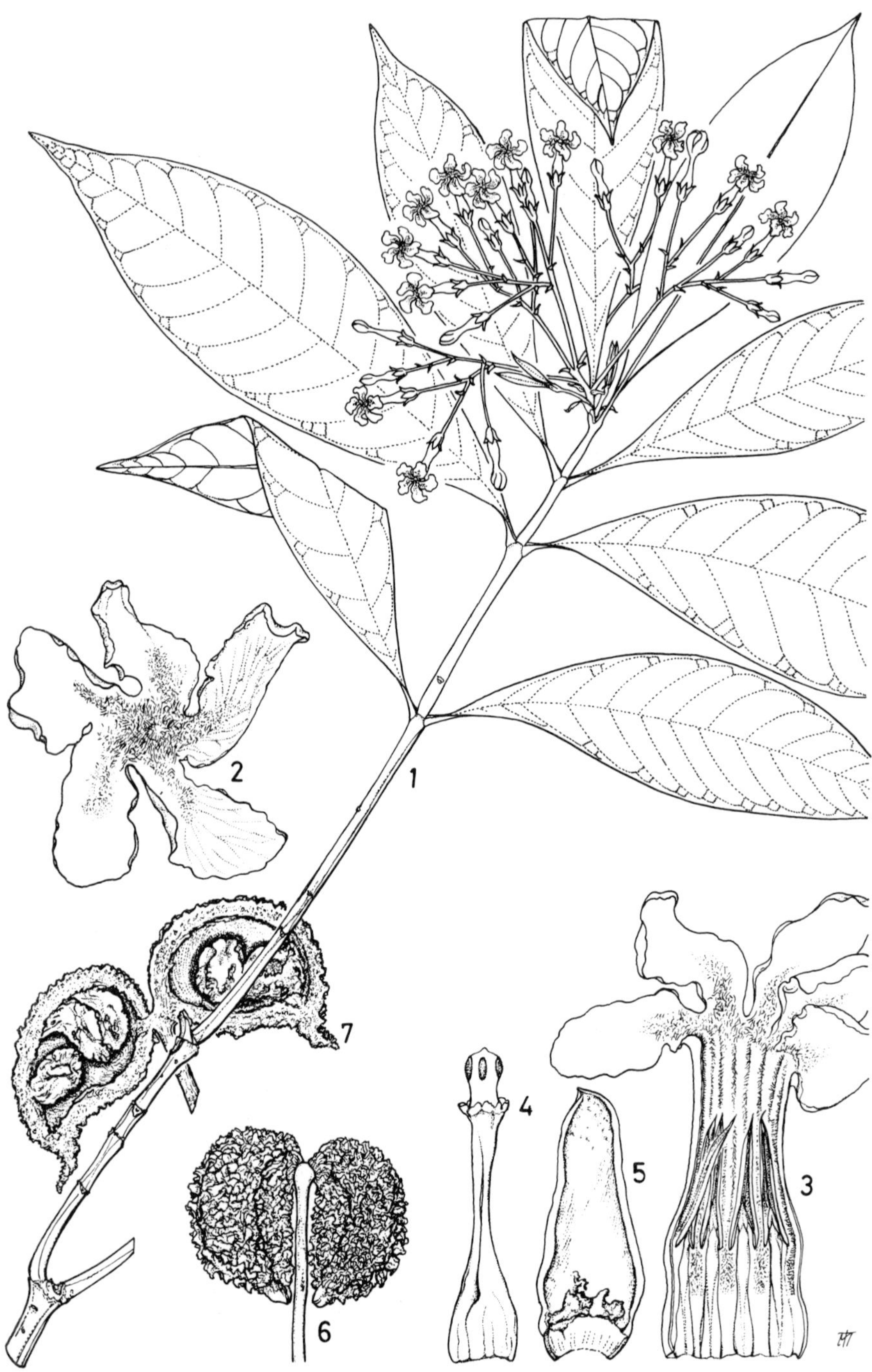

Fig. 76. *Tabernaemontana hystrix.* **1,** habit (x 2/3); **2,** flower above (x 4); **3,** opened corolla (x 4); **4,** pistil (x 8); **5,** sepal inside (x 12); **6,** fruit (x 2/3); **7,** section of fruit (x 2/3). 1 from Pereira 7184; 2-5 from Mexia 5185; 6 from Dorsett et al. 361; 7 from Wilkes US 65742.

1.3-1.7 mm high, composed of a basal ring of 5-10 acute lobes, a cylinder 1.1-1.5 x 0.5-0.8 mm, constricted in the middle, with 5 longitudinal grooves and 5 rounded apical lobes rendering it there 0.5-1.2 mm wide, and a bilobed stigmoid apex 0.2 x 0.2 mm. Ovules approximately 70-100 in each carpel. *Fruit* of 2 separate mericarps; mericarps yellow, with acute more or less wart-like short prickles, obliquely ellipsoid or reniform, 20-45 x 10-25 x 10-15 mm, rounded or acute at the apex, mostly recurved, not ridged, approximately 10-20-seeded; wall 2-4 mm thick in dried fruits; aril red. *Seed* dark, brown, obliquely ellipsoid, 8-10 x 4-5 x 3-3.5 mm, with longitudinal grooves, papillose.

DISTRIBUTION: Southern Brazil.

ECOLOGY: Open places, not in forest. Alt. 0-700 m. Flowering from September to January with a peak in October. Fruiting known only from February and April.

Geographical selection of the approximately 90 specimens examined:

BRAZIL. Bahia: Banco Central-Itajuípe Road, *Pinheiro* 1307 (Z); Mun. Belmonte, Gregório Bondar, *T.S. dos Santos* 4328 (WAG); Itapebí, *Belém & Pinheiro* 2967 (WAG); km 31 Prado-Itamaraju Road, *L.A. Mattos Silva & Brito* 695 (U, WAG); Nova Viçosa, *Baumgratz* et al. 160 (RB); Rio Gongogi, *Curran* 235 (GH, US); sin. loc., *Blanchet* 6006 (G). Minas Gerais: Caatinga, *Magalhães* 15244 (MO, Z); Itaobim, *Magalhães* 15671 (MO, Z); ibid., *Mautone* et al. 1399 (RB); Arraial das Merces, *Gardner* 5010 (BM, CGE, FI-W, K, W); near San Antonio de Itambé, Anderson et al. 35709 (NY); Serra d'Estrella, *Weddell* 876 (G, P, paratype of *T. affinis*); Lagoa Santa, *Dorsett* et al. 361b (US); road to Barroso, *Mexia* 5185 (A, BH, BM, G, GB, GH, K, LL, MO, S, U, UC, US, WIS, Z); Viçosa, *Mexia* 5140 (A, BH, BM, F, G, GB, GH, K, LL, MO, S, U, UC, US, WIS, Z) & 5140a (MO, UC); ibid., *Ramalho* 1602 (RB); sin. loc., *Clausen* 87 (AWH, P) & 325 (P, paratype of *T. affinis*); *Saint-Hilaire* 32 (P, paratype of *T. affinis*) & B 19 (P, lectotype of *T. bracteolaris*). Espírito Santo: Preguiças, *Mattos Filho* RB 90035 (RB); Mun. Jaguaré, Agua Limpa, *Hatschbach & Guimarães* 46972 (C, F, G, MU, US, Z); Linhares F.R., *I.A. Silva* 259 (MO); Br 5, near Vitória, *Trinta* 1058 = *Fromm* 2134 (Z); Mun. Conceição do Castelo, *Hatschbach* 48641 (C, G, HBG, MU, S, US, WAG, Z). Rio de Janeiro: Nova Friburgo, *Clausen* 47 (G, P); Serra dos Orgãos, *Gardner* 74 (BM, CGE, FI, FI-W, G, G-DC, K, NY, P, S, TCD, US, W; phot. of G-DC sheet GH, MO, US; phot. of NY sheet P, type of *T. collina*); ibid., Vargem Bridge, *Miers* 2065 (BM, paratype of *Peschiera granulosa*), 4028 (BM, lectotype of *P. granulosa*); Mt Corcovado, *Miers* 4654 (BM, paratype of *P. granulosa*); Petropolis, *Glaziou* 3052 (BR, C, K, P); km 13 road to Teresópolis, Baixada, *Pereira* 7184 (K, Z); Morro Flamengo, *Miers* 3013 (BM, K, TCD); Jacarepaguá, *Carauta* 438 (GUA, WAG); Serra do Coelho, *Glaziou* 8797 (C, K, P); near Margaritiba, *Pohl* 5107 (BM, M, W, paratype of *T. affinis*); Rio Funil, *Kuhlmann* 4046 (K, US, Z); sin. loc., *Gaudichaud* 97 (P, paratype of *T. affinis*) & 539 p.p. (G-DC) & 540 (FI-W, G, G-DC, P; phot. of lost B sheet GH, MO, US, lectotype of *T. gaudichaudii*); *Lhotzky* anno 1832 (G-DC; phot. P; phot. of lost B sheet GH, MO, US, type of *T. fuchsiifolia*); *Lund* 882 (C), 883 (C, G-DC; phot. of G-DC sheet MO, US, lectotype of *T. lundii*); *Martius* 103 (BR, P) & 172 (M) & s.n. (BP, L, W, named *T. fuchsiaefolia* by Muell.Arg.) & s.n. (M, paratype of *T. bracteolaris*); *Pohl* 1840 (BP, FI, K, W). São Paulo: Parque do Estado, *Handro* SP 37593 (SP); Tijuco, *Nonnenprediger* 67 (MEL); Mun. Votorantin, Serra de São Francisco, *V.F. Ferreira* 3204 (GUA); Sarapui, *Yano* 20 (SP); Itatiaya, Rio Taquaral, Est. Biol. 1364 (Z); Sandovalina, *Dambrós* 263 (RB); Rio Cuyabá, *Riedel* 1191 (LE); sin. loc., *Lund* 884 (G-DC; phot. US, paratype of *T. lundii*). Sin. loc., *Riedel* s.n. (G, GH, P, U,; phot. of lost B sheet GH, US, type of *T. gracilis* Muell.Arg. non Benth.) & s.n. (G, GH, M,

P, S, U, UPS, W, distributed as *T. fuchsiaefolia*).
CULT. Brazil, Santa Catarina, Horto Florestal I.N.P., *Gevieski* 15 (M, S, UC, Z).

Notes. *Tabernaemontana hystrix* is closely allied to *T. catharinensis*, a species with a similar ecology. They resemble each other by the leaves, the recurved sepals and the fruits. They may be distinguished from each other by the characters given in couplet 16 of key 1.

The specimens examined, including the types of the name and the synonyms, are remarkably similar in almost all characters. The smooth fruit mentioned by De Candolle for *T. gaudichaudii* has not been preserved. It may have belonged to a completely different species.

77. Tabernaemontana laeta Mart. in Flora 20, Beibl. 2: 98 (1837). – Type: Brazil, sin. loc., *Martius* 104 (holotype M; isotypes BR, FI-W, G-DC, K, L, W; phot. of G-DC sheet MO, US). Fig. 77, p. 314; map 37, p. 293

Homotypic synonym:

Peschiera laeta (Mart.) Miers, Apoc. S. Am. 35 (1878).

Heterotypic synonyms:

T. laeta var. *pubiflora* Muell. Arg. in Martius, Fl. Bras. 6, 1: 79 (1860). *P. florida* Miers, op. cit. 41. – Type: Brazil, Minas Gerais, Paraibuna, *Gardner* 5011 (holotype W; isotypes BM, CGE, F, FI, FI-W, G, GH, K, P, US). The BM sheet is the holotype of *P. florida*.

T. laeta var. *minor* Muell. Arg. in op. cit. 80. – Type: Brazil, Minas Gerais, sin. loc., *Saint-Hilaire* B1 515 (holotype P).

T. laeta var. *densa* Muell. Arg., l. c. – Type: Brazil, Rio de Janeiro, near Japuahiva (= Japuíba), near Angra dos Reis, *Pohl* 1847 (lectotype W, designated here; isolectotypes BR, CGE, FI, K).

T. spixiana Mart. ex Muell. Arg. in op. cit. 78, partly, as for lectotype. *P. spixiana* (Mart. ex Muell. Arg.) Miers, op. cit. 36. – Type: Brazil, Rio de Janeiro, sin. loc., *Martius* 359 (lectotype M, designated here; phot. GH, MO, US).

T. breviflora Muell. Arg. in op. cit. 79. *P. breviflora* (Muell. Arg.) Miers, op. cit. 45. – Type: Brazil, Rio de Janeiro(?), between Campos and Vitória, *Sellow* 461 (holotype B†; phot. GH, MO, US; the K sheet *Sellow* 416 may be an isotype).

Shrub or small tree, 3-10 m high. Trunk up to at least 10 cm in diameter; bark pale brown. Branches pale or dark brown, lenticellate; branchlets terete, glabrous. *Leaves* petiolate; petiole glabrous, 5-17 mm long; (ocreae not widened into intrapetiolar stipules); blade papery when dried, elliptic or narrowly elliptic, (2-)2.5-3.3(-4) x as long as wide, 3-16 x 1-6.5 cm, acuminate at the apex, cuneate at the base, entire, glabrous on both sides, without black dots, with 7-16 pairs of upcurved secondary veins forming an angle of 50-90° with the costa; tertiary venation reticulate beneath. *Inflorescence* pedunculate, 3-9 x 3-10 cm, many-flowered, rather dense. Peduncle 1-25 mm long, glabrous; pedicels 3-12 mm long, glabrous. Bracts sepal-like, 0.5-1 x as long as them. *Flowers* fragrant, open during the day. *Sepals* pale green(?), connate at the base for 0-0.5 mm, erect or slightly spreading in dried flowers, ovate or broadly ovate, 1-1.6 x as long as wide, 1.5-3 x 1.5-2 mm, obtuse or acute, glabrous outside, not ciliate, glabrous inside and with 1 row of 2-9 colleters in the middle at the base; colleters variable in shape and size, 0.2-0.6 x 0.1-0.4 mm, lobed when broad. *Corolla* white, with or without (?) pale yellow throat, 11-17 mm long in the mature bud and forming a comparatively large ovoid head 0.3-0.47 of the bud length (3.5-7 x 2-5 mm), 2-3 x as wide as narrow part of tube with a blunt apex, glabrous outside, pubescent or pilose-pubescent grading to pilose upwards inside in stripes alternating with the filament ridges from 0-1.5 mm below the insertion of the stamens and from there in a belt to the base of the lobes (0.1-0.35 of their length); tube 3.2-7.7 x as long as the calyx, 0.7-1.2 x as long as the lobes, 8-12 mm long, flask-shaped, 2.5-3.5 mm wide above the base, narrowed around the anthers and even more above to 1.2-2 mm wide, widened again at the mouth to 2-3 mm wide, not twisted; lobes obliquely elliptic, mostly more or less falcate, 0.84-1.4 x as long as the tube, 1.3-2.8 as long as wide, 8-15 x 4-8 mm, rounded, neither auriculate, nor undulate, spreading. *Stamens* with apex 1.5-3.2 mm below mouth of corolla tube, inserted 0.3-0.45 of the length of the corolla tube, (at 2.5-4.5 mm from the base), anthers sessile with tails 0.5 mm below insertion; narrowly triangular, 3-5 x as long as wide, 3.5-5 x 1-1.5 mm, apex acuminate, sterile for 0.4-1 mm, sagittate at the base. *Pistil* glabrous, 4.5-6.5 mm long, with apex about halfway along anthers; ovary ovoid, 1.5-2.5 x 1-1.5 x 1-1.5 mm; style short, 1-2.5 mm long; pistil head 1.1-1.5 mm high, composed of a basal ring of 10 acute spreading or upcurved lobes, 0.1-0.3 x 0.8-1 mm, a stipitate 5-lobed depressed globe 0.3-0.5 x 0.6-0.8 mm and a cleft stigmoid apical part 0.2-0.3 x 0.2 mm. Ovules approximately 50-60 in each carpel. *Fruit* of 2 separate mericarps; mericarps green, orange-yellow inside, obliquely ellipsoid, 35-50 x 25-30 x 25-30 mm, recurved, rounded or less often apiculate at the apex, with a faint ridge at each side, densely covered with blunt excrescences mostly wider than long, approximately 15-40-seeded; wall 3 mm thick in dried fruits; aril enveloping the seed(?). *Seed* black, obliquely ellipsoid, 11-12 x 6-7 x 4 mm, with longitudinal grooves, dull, papillose; embryo 8.5-9 mm long; cotyledons ovate, 1.6-2 x as long as wide, 4.5-5 x 2.5 mm, rounded at the apex, cordate at the base; rootlet 4.5 x 0.7-0.8 mm.

DISTRIBUTION: Southeastern Brazil.

ECOLOGY: Secondary forest. Alt. low. Flowering mainly from September to November. Fruiting probably a few months later.

Geographical selection of the approximately 230 specimens examined:

BRAZIL. Pernambuco: Vicência, *Teixeira* 2557 (US); Tapera, *Pickel* 1816 (WIS); Res. Ecol. Tapacura, *Chioppeta* 5807 (WAG). Paraiba: Rio Paraiba, *Prince Maxim. Vidensis* Oct. 1815 (BR). Bahia: Between Feira de Santana and Vitória da Conquista, *R.S. Santos* HB 27986 (Z); Muritiba, *Blanchet* s.n. (G); Itapebí, *Belém & Pinheiro* 2867 (F, Z); Porto Seguro F.R., *V. de Souza* 262 (WAG); ca 5 km W of Itamaraju, *Mori* et al. 10739 (K, NY, P); Mun. Prado, Sul Baiana, *Mori* et al. 10653 (P, WAG). Minas Gerais: 8 km W of Pedra Azul, *Harley* et al. 25169 (K, WAG); Mun. Medina, between Itaobim and Pedra da Onça, *Magelhães* 15646 (MO, Z); Mun. Teofilo Otoni, *Magelhães* 15725 (MO); Mun. Governador Valadares, Rio Doce, *Magelhães* 4432 (MO); Lago Santa: *Warming* Nov. 1885 (C); Mun. Marliéria, Parque Estudial do Rio Doce, *Heringer* 18613 (SP, US); Est. Biol. Caratinga, *Andrade & Malopes* 624 (P); Viçosa, *Mexia* 4200 (A, BH, BM, G, GB, GH, K, MO, NY, S, U, UC, US, WIS, Z); Fazenda Bom Jardim, Cambusca, Mexia 5250 (BH, BM, G, GB, GH, K, LL, MO, S, U, UC, US, WIS, Z); Miraí-Muriaé Road, *Rubens* 162 (RB); Paraibuna, *Gardner* 5011 (BM, CGE, F, FI, FI-W, G, GH, K, P, US, W, type of *T. laeta* var. *pubiflora* and *Peschiera florida*); Mt Itatiaia, *Wettstein & Schiffner* Sept. 1901 (WU); sin. loc., *Saint-Hilaire* B1 515 (P, type of *T. laeta* var. *minor*), B 523 (P). D.F.: Brasilia, *Heringer* et al. 5710 (K). Espirito Santo: 6-8 km from Itaúnas, *Martinelli & Soderstrom* 9708 (US); Mun. Nova Venecia, Serra de Lima, *Duarte* 8992 (RB); São Mateus, *Bondar* 19 Oct. 1946 (GH, MO, US); NW of Colatina, *Shepherd* et al. 5865 (F); Linhares, *Hatschbach & Guimarães* 46922 (BR, F, G, HBG, MU, US, Z). Rio de Janeiro: between Campos and Morro do Côco, *Trinta* 1027 = *Fromm* 2103 (M, Z); near Penedo, Sítio Palmital, *Lanna* 1236 (WAG); Mun. Macae, Fazenda Jurubatiba,

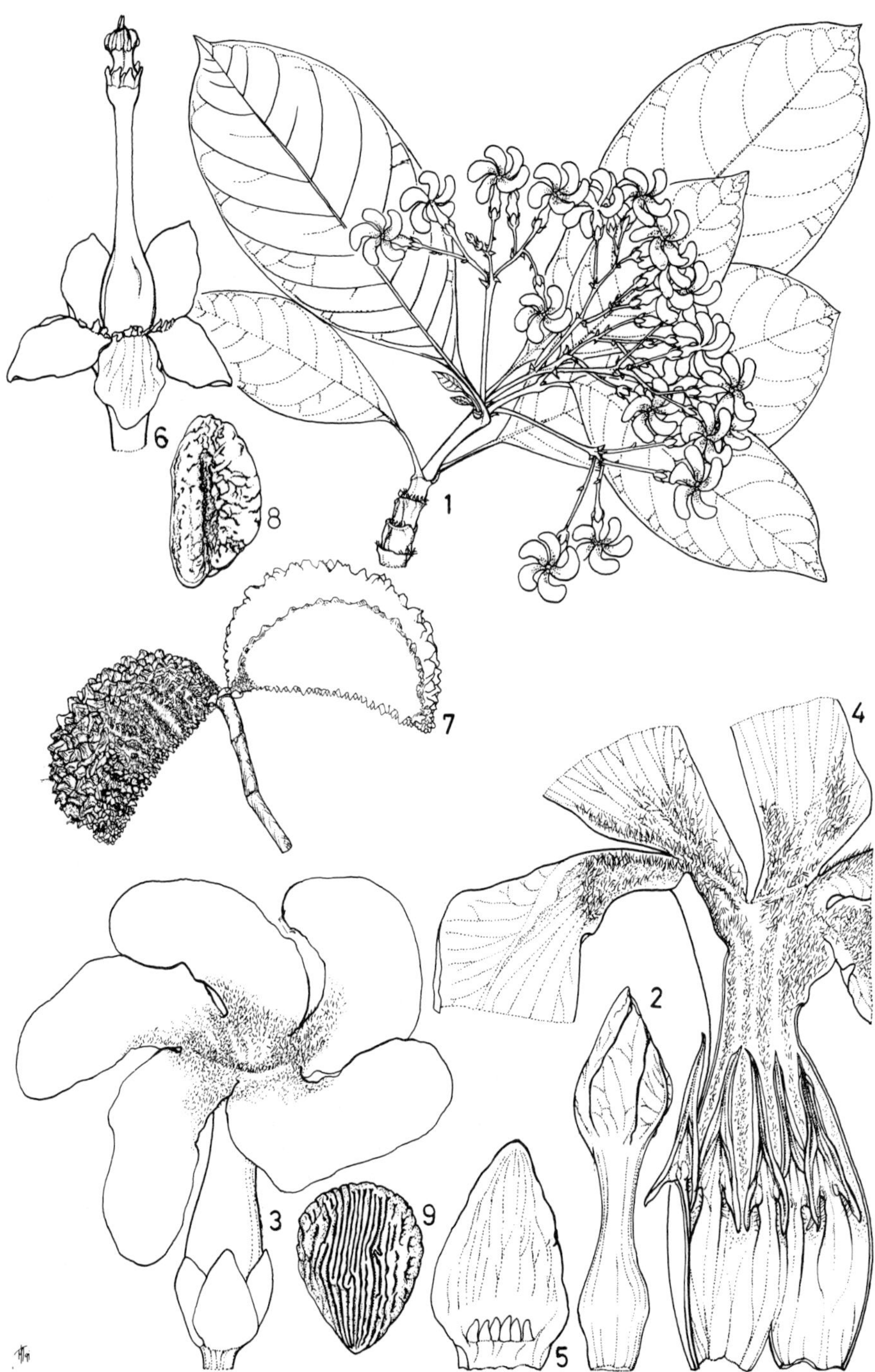

Fig. 77. *Tabernaemontana laeta.* **1,** habit (x 2/3); **2,** corolla bud (x 4); **3,** flower (x 2); **4,** opened corolla (x 6); **5,** sepal inside (x 8); **6,** calyx with pistil (x 8); **7,** fruit (x 2/3); **8-9,** seeds (x 2). 1-2 from Hatschbach et al. 46922; 3 from Mori et al. 10653; 4-6 from Pabst 4417; 6-9 from Araujo et al. 7510.

Araujo 6526 (GUA); Larangueiras, *Glaziou* 1610 (BR, C, K, P); Mun. Miracema, road to Palma, *Pirani* et al. 256 (SP); Corcovado, *Gardner* 5546 (BM); ibid., *Vauthier* 76 (G-DC, GH, L, MPU, P, W); border of Mun. Niteróbi and Mun. Maricá, Pico Alto Moirão, *Farney* 368 (K); Santa Teresa, *Dusén* 5117 (BM, G, S, US); ibid., *Glaziou* 4078 (C, K, P, US); Sebastião de Lacerda, *Iglesias* RB 50507 (MO, RB, WAG); Guanabara, Ilha do Governador, *Pabst* 4417 (M, Z); near Rio de Janeiro, *Gardner* 75 (BM, FI-W, GH, K, P, TCD, US, W); Sebastionopolis, *Martius* 201 (M); Paracambi, *Carauta* et al. 5987 (GUA); Mangaratiba, *Carauta* 5396 (WAG); near Japuhiva (= Japuíba), near Angra dos Reis, *Pohl* 1847 (BR, CGE, FI, K, W, lectotype of *T. laeta* var. *densa*); sin. loc., *Glaziou* 15213 (BM, BR, C, K, MA, MO, P, US); *Martius* 359 (M; phot. GH, MO, US, lectotype of *T. spixiana*); *Miers* 2830 (BM) & 2971 (BM) & 3013 (TCD) & 3014 (BM, K); *Saint-Hilaire* A 506 (P) & A1 569 (P) & C2 200 (P). São Paulo: Quelez, *Bamps* 5017 (BR, NY); km 180 São Paulo-Rio de Janeiro Road, *Bamps* 16 Dec. 1975 (BR); Mun. Ubatuba, Ubatuba-Taubaté Road, *Gibbs* et al. 4602 (F); Cachoeira Paulista, *Yano* 1187 (SP); Campinas, *Movaes* 14 (IBSC). Sin. loc., *Martius* 104 (BR, FI-W, G-DC, K, L, M, W; phot. of G-DC sheet MO, US); *Riedel* s.n. (G, GH, P, S, U, UPS, W); *Sellow* 416 (K) & 461 (phot. of lost B sheet GH, MO, US) & 677 (BM) & 771 (BM, K).

CULT. D.F., Brasilia, Est. Biol. UNB, *Heringer* 18424 (SP, US); Rio de Janeiro, Horto Florestal, *Antenor* 96 (GH, RB, WAG); ibid., *Tatto* et al. 722 (RB, S).

Notes. Some specimens of *T. solanifolia* have leaves with blade decurrent into the very short petiole and have at times been confused with *T. laeta*. The former, however, are pubescent, while *T. laeta* is always glabrous.

T. laeta is easily confused with *T. linkii* (q.v.).

78. Tabernaemontana lagenaria Leeuwenberg, **sp. nov.**

Arbor ramis lenticellatis. Folia petiolata lamina tenui elliptica vel anguste elliptica apice acuminata basi cuneata utroque latere glabra. Inflorescentia multiflora. Flores odorantes. Sepala viridia basi connata recurvata ovata vel oblonga obtusa utroque latere puberula. Corolla alba vel tubo viridi et lobis albis extus glabra intus pro parte pubescens tubo lagena simili. Stamina inclusa antheris sessilibus anguste triangularibus. Fructus mericarpis duobis oblique ellipsoideis plus minusve muricatis.

Typus: French Guiana, Upper Camopi R., Mt Belvédère, *de Granville* 7000 (holotypus P; isotypi CAY, G, MG, MO, US, WAG). Fig. 78, p. 317; map 40, p. 318

Tree 10-30 m high. Trunk 15-50 cm in diameter. Branches pale or dark brown, lenticellate; branchlets terete, glabrous. *Leaves* petiolate; petiole glabrous, 3-12 mm long; (ocreae widened into intrapetiolar stipules); blade membranaceous or thinly papery when dried, elliptic or narrowly elliptic, 2.3-2.8 x as long as wide, 4-15 x 1.5-6 cm, acuminate at the apex, cuneate at the base, glabrous on both sides, dotted beneath or not, with 7-12 pairs of upcurved secondary veins forming an angle of 50-60° with the costa; tertiary venation reticulate, rather conspicuous. *Inflorescence* pedunculate, 3-6 x 3-6 cm, 15-ca 50-flowered, rather lax. Peduncle 6-27 mm long, glabrous; pedicels slender, 4-7 mm long, glabrous. Bracts sepal-like, 0.5-3 x as long as them, deciduous or not. *Flowers* fragrant, open during the day. *Sepals* green, connate at the base for 0-0.5 mm, recurved, ovate or oblong, 1.6-3.4 x as long as wide, 2.5-4 x 1-2 mm, (part not recurved 0.5-2 mm long, obtuse, puberulous on both sides, ciliate, with 1 row of 2-7 colleters inside in the middle at the base; colleters 0.2-0.5 x 0.1-0.3 mm, lobed when broad. *Corolla* white or tube greenish and limb white, 8-10 mm long in the mature bud and forming a comparatively large broadly ovoid head 0.37-0.4 of the bud length (3-4 x 2-3 mm) with a blunt apex, glabrous outside, pubescent inside from 0-1 mm above the insertion of the stamens to 1-4 mm from the base of the lobes; tube 3-10 x as long as the calyx, when sepals recurved (as in fresh flowers), 2-4.7 x as long as, when sepals straightened out, 0.54-0.8 x as long as the lobes, 5-7 mm long, flask-shaped, 2-2.5 mm wide above the base, narrowed above the anthers to 1-1.5 mm wide, again widened at the mouth to 1.5-2 mm wide, not twisted; lobes obliquely oblong and more or less falcate, 1.3-1.8 x as long as the tube, 2.2-2.5 x as long as wide, 5-12 x 2-5.5 mm, slightly auriculate at the left side of the base, undulate or not, spreading. *Stamens* with apex 1-1.5 mm below mouth of corolla tube, inserted 0.4 of the length of the corolla tube, (at 2.2-3 mm from the base); anthers sessile, with tails 0.5 mm below insertion, narrowly triangular, 2.5-2.7 x as long as wide, 3-3.3 x 1.1-1.2 mm, apex acuminate, sterile for 0.3-0.5 mm, sagittate at the base. *Pistil* glabrous, 3-4.5 mm long; ovary ovoid, 1.5-2 x 0.8-1.2 x 0.8-1.2 mm; style short, 0.5-1.5 mm long; pistil head composed of a basal ring 0.1-0.3 x 0.7-1 mm consisting of 10 suberect blunt lobes, a stipitate 5-lobed depressed globe 0.3-0.4 x 0.5-0.6 mm and a stigmoid apical part 0.1-0.2 x 0.2 mm. Ovules approximately 40 in each carpel. *Fruit* of 2 separate mericarps; mericarps green (?), obliquely ellipsoid, 25-45 x 15-25 x 15-25 mm, recurved, rounded or apiculate at the apex, not ridged, covered with broad blunt excrescences, approximately 10-20-seeded; wall 2 mm thick in dried fruits; aril enveloping the seed. *Seed* brown, 11-13 x 4-5 x 3-4 mm, with longitudinal grooves, papillose.

DISTRIBUTION: French Guiana, Brazil and Peru.

ECOLOGY: Rain forest, not on river banks. Alt. low. Flowering mainly in October. A fruiting season could not be deduced from the collections studied.

Paratypes:

FRENCH GUIANA. Tampoc R., Awali Fall, *Moretti* 651 (CAY, WAG); 11 km from Mt Belvedère, *Sabatier & Prévost* 1674 (P); Trois Sauts, *Grenand* 978 (CAY, P); ibid., *Jacquemin* 1895 (BR, CAY, P, U, WAG) & 2350 (BR, CAY, P).

BRAZIL. Amazonas: Upper Rio Jurua, Mun. Eirunepe, Parana do Ouro Preto, *Fróes* 21706 (LL, MO, US, WIS); Lago Tefé, *Plowman* et al. 12606 (F, GH, K, MO, NY, US, WAG). Acre: km 40 Rio Branco-Santa Rosa Road, *Lowrie* et al. 437 (GH, MO, NY, US, WAG, WIS); near Sena Madureira, *Prance* et al. 7555 (F, GH, K, P, S, US, Z); Rio Iaco, 3 km above confluence with Rio Purus, *Prance* et al. 7868 (K, M, S, US, Z); Mun. Brasiléia, Serringal Poromgaba, *Daly* 7007 (WAG) & 7008 (WAG); Mun. Cruzeiro do Sul, INCRA F.R., *Cid* et al. 5131 (NY, WAG).

PERU. Ucayali: km 4 Pucallpa-Huánuco Road, *Magín* 82 (F, GB, K, UPS, US, WAG, WIS) & 113 (NY, US, WIS). Madre de Dios: Prov. Tambopata, km 12.5 Maldonado-Quicemil Road, *Acosta* 17 (F, K, MO).

Notes. *Tabernaemontana lagenaria* is closely allied to *T. catharinensis* and *T. cymosa* in almost all characters. The two first species are distinguished as follows:

Corolla tube 5-7 mm long; mature bud with a broadly ovoid head 3-4 x 2-3 mm with a blunt apex; sepals recurved for 0.5-0.8 of their length. French Guiana, northern Brazil and Peru .... **T. lagenaria**

Corolla tube 6-14 mm long; mature bud with a narrowly ovoid head 4-9 x 2-4 mm, with an acute or obtuse apex; sepals recurved at the apex only. Eastern and southern Brazil, Bolivia, Paraguay and northern Argentina .... **T. catharinensis**

T. cymosa has corolla lobes pubescent all over or nearly so, while the other species have flowers with pubescence only at the base of the corolla lobes.

Fig. 78. *Tabernaemontana lagenaria.* **1,** habit (x 2/3); **2,** flower bud (x 6); **3,** flower (x 4); **4,** opened corolla (x 8); **5,** pistil (x 18); **6,** sepal inside (x 14); **7,** fruit (x 2/3). 1 and 3 from de Granville 7000; 2 and 4-6 from Cid et al. 5131; 7 from Moretti 651.

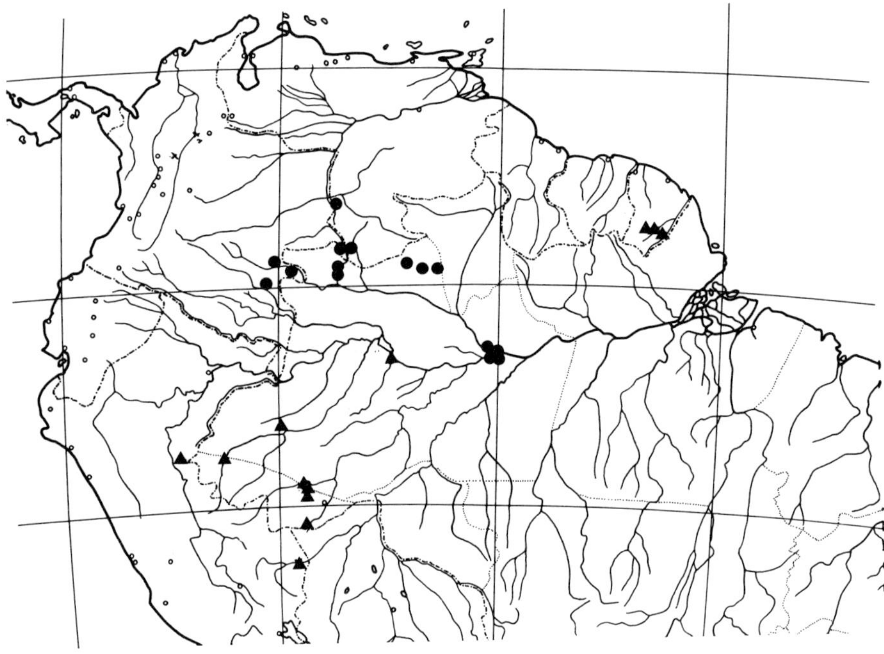

Map 40. ▲. *Tabernaemontana lagenaria*, ●. *T. maxima*.

Aublet (1775) described and illustrated material of *Tabernaemontana* under the name *T. echinata*. His description was based on leaves of *T. undulata* and fruits of *T. heterophylla*, while his plate is of a flowering branch, with erroneous ramification into 3 instead of 2 branchlets, of *T. siphilitica* and of a separate fruit of *T. heterophylla*. The epithet echinata refers to the prickly fruits from the *T. heterophylla* element. The fruits of the 2 other species are smooth. Two specimens collected by Aublet are preserved in the Rousseau herbarium Vol. 6: 224 at Paris, a sterile branch of *T. undulata* (under the name "*T. latifolia*", first sheet) and a branch with flower buds of *T. siphilitica* (under the name "*T. laurifolia*", second sheet).

Although Mme Allorge (1985) correctly observed Aublet's confusion, she increased it by using Aublet's plate as lectotype of the name *T. echinata* and by naming specimens belonging to *T. lagenaria* (*Grenand 978*, *Jacquemin* 1895 and *Moretti* 651) and *T. linkii* (*de Granville* 4488 and *Oldenburger* 406) as such.

As the name *T. echinata* has priority over all other names involved, as it had hardly and always wrongly been used since its publication and as it has been a basis of a great confusion, the present author proposes to consider it as a nomen rejiciendum. Moreover, all three species seen by Aublet are well known under the present names.

79. Tabernaemontana laurifolia L., Sp. Pl. 210 (1753), non Ker-Gawl (1823), nor Blanco (1837); Stearn in Journ. Arn. Arb. 52: 617 (1971). – Type: Sloane, Voy. Jam. 2: t. 186. f.2 (1714), designated by Stearn. Typotype: Jamaica, St. Catherine, Cobre R., Santiago de la Vega, Herb. *Sloane* 6: 58 (BM).

Fig. 79, p. 320; map 41, p. 321; map 44, p. 336

Heterotypic synonym:

Taberna laurina Miers, Apoc. S.Am. 63 (1878). – Type: Jamaica, sin. loc., *Shakespear* s.n. (holotype BM; isotypes G, K).

Shrub or small tree 2-9 m high. Trunk about 5-20 cm in diameter; bark light brown. Branches pale brown, lenticellate; branchlets terete, glabrous. *Leaves* long-petiolate; petiole glabrous, 10-40 mm long; (ocreae widened into intrapetiolar stipules); blade papery or coriaceous when dried, elliptic or narrowly elliptic, 1.2-3 x as long as wide, 4-21 x 2-10 cm, obtuse or shortly and bluntly acuminate at the apex, cuneate or rounded at the base, entire or sinuate and/or undulate, with a revolute margin, glabrous on both sides, without black dots, with 6-13 pairs of mostly upcurved secondary veins forming an angle of 60-90° with the costa; tertiary venation reticulate. *Inflorescence* many-flowered, pedunculate, once branched into 2 racemes, when young 2 x 2 cm, when completely developed up to 7 x 5 cm. Peduncle 5-25 mm long, robust, glabrous; raceme axes densely covered with bract and pedicel scars; pedicels 5-17 mm long, glabrous. Bracts sepal-like and 0.3-0.5 x as long. *Flowers* fragrant, open during the day. *Sepals* pale green (?), connate at the base for about 0.5 mm, erect, ovate, broadly ovate or subcircular, 1-1.3 x as long as wide, 2-3.5 x 1.5-3 mm, rounded, glabrous outside, not ciliate, glabrous inside and with 5-7 colleters in 1-2 rows in the middle at the base; colleters 0.3-0.8 x 0.1-0.2 mm. *Corolla* yellow, greenish-yellow or less often cream, 14-22 mm long in the mature bud and forming a comparatively large subglobose head 0.23-0.3 of the bud length (4.5-5 x 4.5-5 mm) with a rounded apex, glabrous outside, pilose inside for 1 mm on filament ridges among anther tails; tube 5-8.5 x as long as the calyx, 1.4-1.9 x as long as the lobes, 13-19 mm long, almost cylindrical, 1.5-3.5 mm wide above the base, narrowed below the anthers and often also above to 1-2.2 mm wide, around them 2-3 mm wide, not twisted; lobes obliquely oblong, 0.54-0.7 x as long as the tube, 2.9-3.5 x as long as wide, 7-13 x 2-4.5 mm, rounded, auriculate at the left side of the base, not undulate, spreading. *Stamens* with apex from 1 mm below to 0.5 mm above mouth of corolla tube, inserted 0.7-0.8 of the length of the corolla tube, (at 9.5-15.5 mm from the base); anthers sessile, with tails 0.5 mm below insertion, 2.5-3.2 x as long as wide, 3-3.5 x 1-1.5 mm, apex acuminate, sterile for 0.3-0.5 mm, sagittate at the base with straight tails, glabrous. *Pistil* glabrous, 10.5-16.5 mm long, with apex almost halfway along anthers; ovary ovoid, 2-3 x 1.5 x 1.5 mm, with a disk-like thickening 0.5-1 mm high; style filiform, 6.5-12 mm long; pistil head composed of a basal ring 0.3-0.5 x 1-1.2 mm, comprising 10 suberect rounded lobes, a stipitate 5-lobed depressed globe 0.4-0.5 x 0.8-0.9 mm and a stigmoid apex 0.2 x 0.2 mm. Ovules approximately 80 in each carpel. *Fruit* of 2 separate mericarps; mericarps green (?), obliquely ellipsoid to pod-like, 30-50 x 12-15 x 12-15 mm, acuminate at the apex, with lateral ridges, approximately 20-40-seeded; wall about 1 mm thick in dried fruits; aril surrounding the seed. *Seed* dark brown, obliquely ellipsoid, 5-6 x 2.5-3.5 x 2-3 mm, with longitudinal grooves, dull, papillose, embryo 4.2-5 mm long; cotyledons broadly ovate, 1.7-2 x 1.5 mm, rounded at the apex, subcordate at the base; rootlet 2.5-3.2 x 0.6-0.7 mm.

DISTRIBUTION: Jamaica and Grand Cayman Island.
ECOLOGY: Coastal scrub on limestone. Alt. 0-30 m. Flowering and fruiting throughout the year.

Geographical selection of the approximately 30 specimens examined:
GRAND CAYMAN. near Grape Tree Point, *Proctor* 11990 (BM, GH, IJ); NW of East End Village, *Correll* 51006 (NY); Georgetown, *N. Chevalier* 268 (F, USF, WIS); sin. loc., *Howard & Wagenknecht* 15031 (A).
JAMAICA. St. James: between Montego Bay and Round Hill Bluff, *Britton* 2450 (NY). Hanover: Booby Cay, *Proctor* 33036 (BM, IJ); near Lances Point, *Proctor*

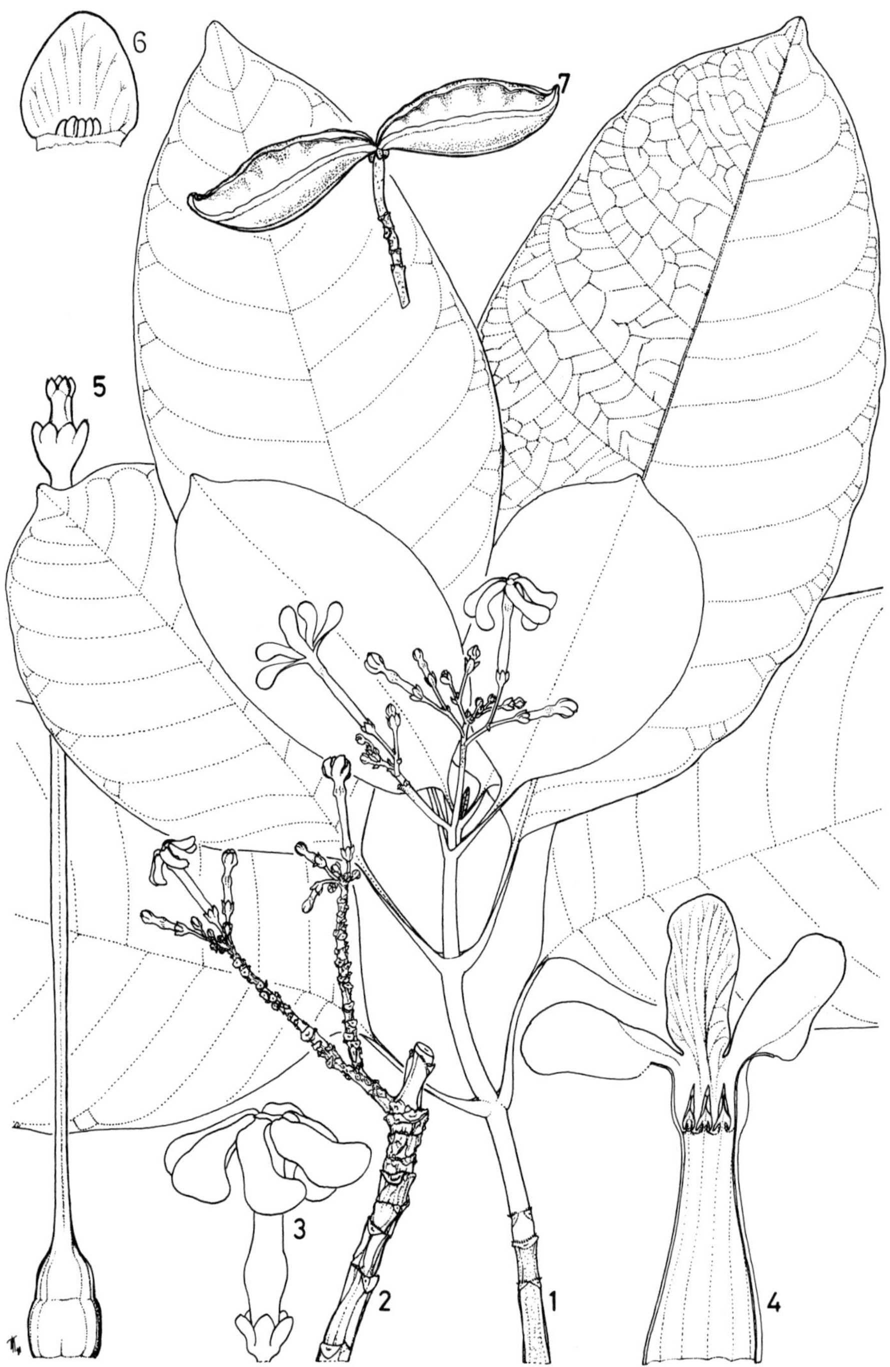

Fig. 79. *Tabernaemontana laurifolia.* **1,** habit (x 2/3); **2,** branch with old inflorescence (x 2/3); **3,** flower (x 2); **4,** opened corolla (x 2.2); **5,** pistil (x 8); **6,** sepal inside (x 8); **7,** fruit (x 2/3). 1 from Webster et al. 8506; 2-3 from Correll 51006; 4-6 from Howard & Wagenknecht 15031; 7 from Proctor 8411.

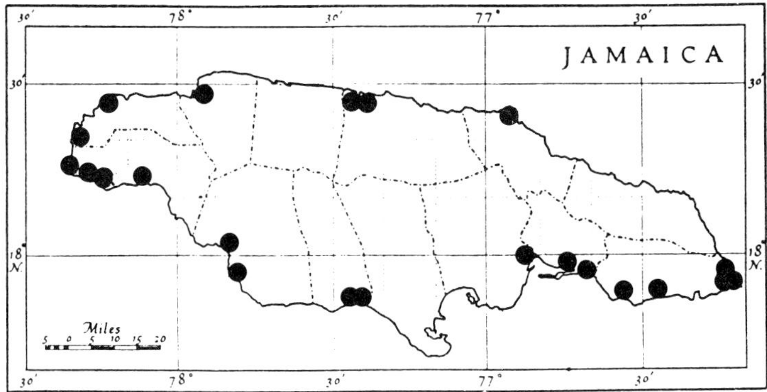

Map 41. *Tabernaemontana laurifolia.*

23971 (BM, LL, NY, U). Westmoreland: Negril, *Adams* 10975 (BM, M); Brighton, *Webster* et al. 8506 (BM, G, S); Little Bay, *Stearn* 393 (A, BH, BM); 2.5 km NW of Savanna-La-Mar, *Proctor* 11108 (BM, IJ). St. Elizabeth: near Black River, *Harris* 9933 (BM, F, K, NY, US); Parottee Distr., *Stearn* 887 (A, BM). Manchester: near Gut R. mouth, *Stearn* 645 (BM) = *Proctor* 11949 (IJ); Two Rivers, *Adams* 11151 (M). St. Ann: Fort Point, Discovery Bay, *Stearn* 728 (BM); Queen's Highway, *Stearn* 316 (A, BH, BM). St. Catherine: Cobre R., Santiago de la Vega, *Herb. Sloane* vol. 6: 58 (BM, typotype). St. Mary: Galina Point, N of Port Maria, *Stearn* 779 (A, BM). St. Andrew: Kingston, *McNab*(?) Oct. 1838 (E); behind the Quarry, near the Ferry, *Harris* 10388 (BM, F, K, NY, US). St. Thomas: Yallahs, *Barkley* & *Gentles* 3 (IJ); Roselle, *Proctor* 15567 (BM, IJ); near Grant's Penn, *Stearn* 692 (A, BH, BM); between Holland Bay and Point Morant, *Hespenheide* 1372 (GH, LL, MO, UC, US); Morant Point, *Proctor* 8411 (BM, UC, US); ibid., *Webster* & *K.A. Wilson* 5234 (A, BM, G, IJ, S); near Holland Bay, *Purdie* July 1843 (K). Sin. loc., *Aiton* s.n. (G-DC); LINN 304.3 (LINN; phot. F); *Shakespear* s.n. (BM, G, K, type of *Taberna laurina*).

80. Tabernaemontana linkii A. DC., Prod. 8: 364 (1844); Leeuwenberg in Agric. Univ. Wageningen Pap. 87, 5: 13 (1988). – Type: Brazil, sin. loc., herb. *Link* s.n. (holotype B†); Brazil, Pará, *Siber* in *Hoffmannsegg* s.n. (neotype B-W 5190, designated by Leeuwenberg in 1988; phot. GH, MO, US; isoneotypes BR, FI-W).

Fig. 80, p. 323; map 42, p. 325

Basionym and homotypic synonym:

T. multiflora Link ex Roem. & Schult., Syst. 4: 431 (1819), non Sm. (1817). *Peschiera linkii* (A. DC.) Miers, Apoc. S. Am. 47 (1878).

Heterotypic synonyms:

T. muricata Willd. ex Roem. & Schult., op. cit. 797, non Link ex Roem. & Schult., op. cit. 431. – Type: Brazil, sin. loc. *Siber* in *Hoffmannsegg* s.n., see above.

T. benthamiana Muell. Arg. in Martius, Fl. Bras. 6, 1: 80 (1860), not p. 70, **syn. nov.** *Peschiera multiflora* Spruce ex Miers, op. cit. 45. *P. benthamiana* (Muell. Arg.) Mgf. in Notizbl. Bot. Gard. Berlin 14: 171 (1938). – Type: Brazil, Pará, near Obidos, *Spruce* 235 (holotype B†; lectotype G; isotypes BM, FI-W, K, M, TCD, W; phot. of lost B sheet GH, MO, US; phot. of G sheet GH, NY, US). The BM sheet is the holotype of *P. multiflora.*

P. ochracea Miers, op. cit. 42. – Type: Brazil, Pará, near Santarém, *Spruce* 234 (holotype BM; isotypes CGE, K, M, TCD, W).
T. myriantha Britton ex Rusby, Descr. 300 New Sp. S. Am. Pl. 84 (1920), **syn. nov.** *P. myriantha* (Britton ex Rusby) Mgf. l.c. *P. benthamiana* var. *myriantha* (Britton ex Rusby) Allorge in Mém. Mus. natn. Hist. Nat. IIB, 30: 154 (1985). – Type: Bolivia, Pando, Junction of Rivers Beni and Madre de Dios, *Rusby* 2377 (holotype NY; isotypes BM, E, F, G, GH, K, MO, P, US, W, WIS; phot. of US sheet WAG).
T. stenantha Mgf. in Notizbl. Bot. Gart. 10: 1037 (1930), **syn. nov.** *P. stenantha* (Mgf.) Mgf. in op. cit. 14: 171 (1938). *P. benthamiana* var. *stenantha* (Mgf.) Allorge, l.c. – Type: Brazil, Pará, Serra de Parintins, *Ducke* RB 21802 (holotype B†; phot. G, GH, MO, US).

Shrub or small tree, 2-15 m high. Trunk 10-40 cm in diameter or less. Branches pale or dark brown, lenticellate; branchlets terete, glabrous. *Leaves* petiolate; petiole glabrous or sparsely pubescent, 5-20 mm long; (ocreae not widened into intrapetiolar stipules); blade subcoriaceous when dried, elliptic or narrowly elliptic, 2-3.5 x as long as wide, 5-17 x 2-7 cm, acuminate at the apex, cuneate at the base, entire, glabrous on both sides or more or less pubescent beneath, with or without scattered black dots, with 6-17 pairs or upcurved secondary veins forming an angle of 60-80° with the costa; tertiary venation conspicuous and reticulate beneath or hardly visible. *Inflorescence* shortly pedunculate, 3-8 x 3-10 cm, 10-50-flowered, rather lax. Peduncle glabrous, 3-20 mm, long; pedicels 3-10 mm long, glabrous or pubescent. Bracts sepal-like and similarly sized. *Flowers* fragrant, open during the day. *Sepals* pale green(?), equal or unequal and larger up to 1.5 x as long as smaller, connate at the base for 0-0.5 mm, erect, ovate or broadly ovate, 0.8-2(-2.7) x as long as wide, 1.5-3(-5.5) x 1.2-2 mm, smaller from 1 x 1.2 mm, obtuse or rounded, glabrous on both sides, mostly not ciliate, with 1 row of 1-8 colleters inside in the middle at the base; colleters 0.3-0.5 x 0.1-0.4 mm, lobed when broad. *Corolla* pale yellow or yellowish-orange and turning darker at anthesis, rarely white, 10-21 mm long in the mature bud and forming a comparatively rather narrow ovoid or narrowly ovoid head 0.33-0.5 of the bud length (2-8 x 1-4 mm) with a blunt apex, glabrous outside, not ciliate at lobes, pilose inside in stripes alternating with filament ridges 0.5-1.5 mm long below the insertion of the stamens and from there in a belt to the base of the lobes (up to 0.4 of their length); tube (3-)4.4-11 x as long as the calyx, 0.8-2.9 x aslong as the lobes, 9-17 mm long, almost cylindrical, narrow, 1.8-3 mm wide above the base, narrowed around the anthers and even more above to 1-2 mm wide, widened again at the mouth to 2-3 mm wide, not or less often twisted 0.1-0.5 turn from above the anthers or only at the throat; lobes obliquely elliptic to dolabriform, mostly more or less falcate, 0.35-1.2 x as long as the tube, 1.2-2 x as long as wide, 4-14 x 2-9 mm, rounded to acute, neither auriculate nor undulate, spreading. *Stamens* with apex (2.8-)3.7-11.5 mm below mouth of corolla tube, inserted 0.13-0.4 of the length of the corolla tube, (at 1.7-4.3 mm from the base); anthers sessile, with tails 0.5 mm below insertion, narrowly triangular, 3-5 x as long as wide, 3.5-4-5 x 0.8-1.2 mm, apex acuminate, sterile for 0.5-1 mm, sagittate at the base. *Pistil* glabrous, 2.5-5.5 mm long; ovary ovoid or subglobose, 1-2.5 x 1-1.5 mm x 0.8-1.5 mm; style short or very short, 0.1-1.4 mm long; pistil head 1.2-1.8 mm high, composed of a basal cylinder 0.6-1.2 x 0.3-0.5 mm, only once seen with a basal ring 0.1 x 0.6 mm consisting of 10 acute spreading lobes (*Pereira* 4948 (MO)), a 5-lobed depressed globe 0.3-0.6 x 0.4-0.8 mm and a bilobed stigmoid apical part 0.2-0.3 x 0.2-0.3 mm. Ovules approximately 80-150 in each carpel. *Fruit* of 2 separate mericarps; mericarps bright red, green when immature, obliquely ellipsoid, 50-60 x 25-35 x 25-30 mm, recurved, rounded or apiculate at the apex, usually with a faint ridge at each side, densely covered with blunt excrescences

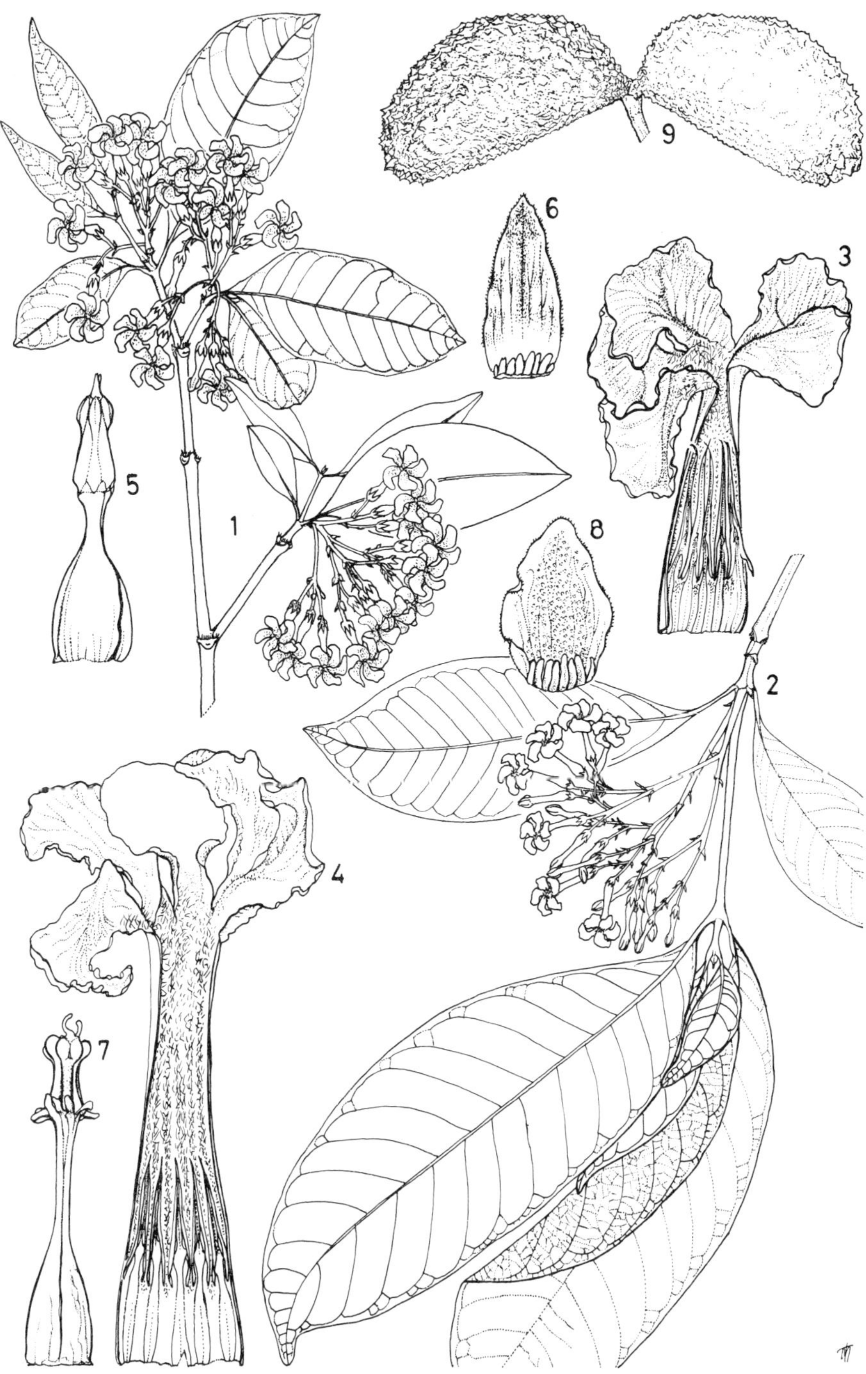

Fig. 80. *Tabernaemontana linkii.* **1-2,** habit (x 2/3); **3-4,** opened corollas (x 4); **5** and **7** pistils (x 10); **6** and **8** sepals inside (x 10); **9,** fruit (x 2/3). 1 from L.A.Mattos Silva et al. 695; 2-3 and 5-6 from Oldenburger 406; 4 and 7-8 from Pereira 4948; 9 from M.G.Silva & Bahia 3485.

mostly wider than long; wall 2-5 mm thick in dried fruits; aril orange, covering the hilar side only. *Seed* black, obliquely ellipsoid, 9-12 x 4-6 x 4-5 mm, with longitudinal grooves, dull, papillose; embryo 6.5-8 mm long; cotyledons ovate, 1.3-1.75 x as long as wide, 3.5-4 x 2-3 mm, obtuse at the apex, cordate or rounded at the base; rootlet 3.5-4.5 x 0.8 mm.

DISTRIBUTION: Suriname, French Guiana, Brazil, Peru and Bolivia.

ECOLOGY: Forest (Mata de capoeira de terra firme). Alt. 0-260 m. Flowering mainly from August to November. Fruiting season not clear.

Geographical selection of the approximately 100 specimens examined:

SURINAME. Sipaliwini Savanna, 5 km W of Morro Grande, on Brazilian border, *Oldenburger* et al. 406 (A, K, U, WAG).

FRENCH GUIANA. Tumac Humac, 2 km NW of Toukouchipan, on Brazilian border, *de Granville* 4488 (CAY, P).

BRAZIL. Amazonas: Maués, *N.T. Silva* 4513 (NY, WAG); Mun. Manicoré, near Santa Fe, *Krukoff* 6067 (A, BM, BR, G, K, MO, S, U, US); Mun. Humaitá, *Janssen* 194 (RB, WAG); Bôca do Acre, *Prance* et al. 2400 (GH, K, M, S, U, US, Z). Acre: km 50 Rio Branco-Placido de Castro Road, *Cid & Nelson* 2797 (MO, NY, US, WAG); km 14 Rio Branco-Brasiléia Road, *Lowrie* et al. 256 (GH, MO, US, WAG). Pará: Alto Tapajós, Rio Cururú, *Anderson* 10722 (C, F, K, MO, NY, US, Z); Tapajós National Park, *M.G. Silva & Rosário* 3910 (NY, WAG, Z); Serra de Parintins, *Ducke* RB 21802 (phot. of lost B sheet G, GH, MO, US, type of *T. stenantha*); Fordlandia, *Black* 48-2300 (MO); Santa Cruz, 14 km W of Belterra, *Bamps* 5275 (BR, WAG); near Santarém, *Ducke* 2931 (MG); ibid., *Maciel & Cordeiro* 207 (NY, U); ibid., *Spruce* 234 (BM, CGE, K, M, TCD, W, type of *P. ochracea*), 308 (K, P); ibid., near Taperinha, *Ginzberger* 336 (W, WU); near Obidos, *Spruce* 235 (BM, FI-W, G, M, TCD, W; phot. of lost B sheet GH, MO, US; phot. of G sheet GH, NY, US, type of *T. benthamiana* and *P. multiflora*) & 492 (K, P); Rio Curuá, Pacoval, *Jangoux & Ribeiro* 1642 (MG, NY); Rio Cumina-Mirim, *Ducke* 7964 (MG); BR 163, Cuiabá-Santarém, km 1131, near Igarapé Natal, *Prance* et al. 25428 (WAG); km 35 Palhão-Ramal, *M. Silva & Souza* 2331 (K, S, US, Z); km 28 Alenquer-Lauro Sodré Road, *Cid* 9401 (WAG); near Monte Alegre, *Trail* 516 (K); Sete Varas airstrip on Rio Curua, *Strudwick* et al. 4181 (K, MO, WAG); Mun. Altamira, Igarapé Ipixuna, tributary of Rio Xingu, *Balée* 2041 (WAG); Rio Xingu, Assurini, *Ballée* 2592 (WAG); Pombal, Black 52-15487 (MO); Mt Dourado, *M.R. Santos* 270 (INPA); Tucurui, *Lisboa* et al. 1372 (NY); ibid., Breu Branco, *M.G. Silva & Bahia* 3485 (NY); Marabá, *J.M. Pires* 4532 (US); Belém, *Burchell* 9390 (BR, K); ibid., *Martius* 2632 (M, paratype of *T. spixiana*); ibid., *J.M. Pires & Black* 590 (GH, MO, P); near Livramento, between Belém and Bragança, *Ducke* RB 21582 (US); Cacaual Grande, *Black & Ledoux* 50-10427 (MO); sin. loc., *Siber* in *Hoffmannsegg* s.n. (B-W 5190, BR, FI-W; phot of B-W sheet GH, MO, US, neotype). Amapa: Rio Jari, Serra da Arumanduba, *Egler & Irwin* 45980 (GH, K, NY, S, U, US, Z). Maranhão: 15 km SE of Fortuna, *J.U. Santos* et al. 652 (WAG). Rondônia: Fazenda São Antonio, *Pena* 29 (RB); near Porto Velho, *Maguire* et al. 56700 (F, K, Z); between Mutumparaná and Rio Madeira, *Prance* et al. 9008 (F, NY, U, US, Z, paratype of *Anacampta pendula*); km 158 Rondônia-Porto Velho Road, *Lleras* et al. P19446 (K, MO, NY, U, US, Z); Nova Vida, *Duarte* 7022 (RB); Mineraçao Taboca, 55 km from Ariquemec, *Vieira* et al. 333 (GH, MG, MO, NY, US, WAG); Serra do Balateiro, 7 km from Vila Campo Novo, *Cid* 8918 (WAG); Igarapé Rio Branco, tributary of Rio Jaci, *Cordeiro* 792 (Z); Road from Palheta to Bolivian border, *Cordeiro* 833 (Z). Sin. loc., *A.R. Fereira* 467 (K, LISU, P).

PERU. Madre de Dios: Maldonado, *A. Gentry & Young* 31771 (F, MO, P, USF); Tambopata Res., *Barbour* 5163 (LL, MO, USF); ibid., *Young* 166 (K).

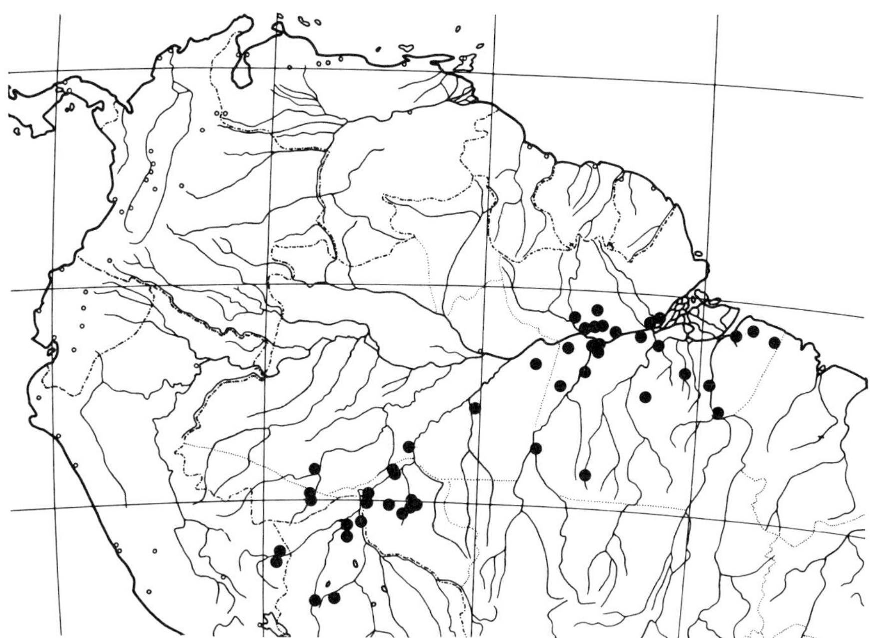

Map 42. *Tabernaemontana linkii.*

BOLIVIA. Pando: Junction of Rivers Beni and Madre de Dios, *Rusby* 2377 (BM, E, F, G, GH, K, MO, NY, P, US, W, WIS; phot. of US sheet WAG, type of *T. myriantha*). Beni: 5 km NW of Guayaramerin, *Anderson* 11835 (NY); Prov. Vaca Diez, Riberalta, *Solomon* 6149 (MO, WAG); Tumi Chucua, 30 km S of Riberalta, *Solomon* 7619 (U, WAG); near Chácobo, *Boom* 4100 (GH, HBG, MO, US, WAG, Z); Prov. Ballivian, Espiritu, *Beck* 5375 (Z).

CULT. Brazil, Pará, Belém, *Markgraf* 3798 (MO); ibid., Botanical Garden of Museu Goeldi, *Pereira* 4948 (B, F, MO, Z).

Notes. The type of *T. linkii* has very narrow corolla tubes and rather short lobes. The type of *T. benthamiana* has slightly wider corolla tubes and longer lobes. Many intermediates exist. As no other characters could be traced to distinguish between these two species, the name of the latter has been reduced to a synonym.

The fruits are not ridged, rounded and warty (± as in *T. cymosa*, in *Jangoux & Ribeiro* 1643 (NY), from Brazil).

T. linkii is easily confused with *T. laeta* in almost all characters. They differ from each other as follows:

Corolla white; tube 10-12 mm long, flask-shaped, stamens with apex 1.5-3.2 mm below mouth of corolla tube; pistil head with a ring of 10 acute lobes at the base **T. laeta**

Corolla yellow or orange (rarely white); tube 9-17 mm long, almost cylindrical; stamens with apex (2.8-)3.7-11.5 mm below mouth of corolla tube; pistil head not or occasionally (?) with a ring of 10 acute lobes at the base .... **T. linkii**

The secondary veins are mostly more clear in *T. linkii* than in *T. laeta*.

81. Tabernaemontana longipes Donn. Sm. in Bot. Gaz. 24: 397 (1897). – Type: Costa Rica, Limón, La Concepción, Llanuras de Santa Clara, *J.D. Smith* 6650 (holotype US; phot. WAG; isotypes GH, K). Fig. 81, p. 328; map 43, p. 329

Heterotypic synonyms:

T. costaricensis Mgf. in Notizbl. Bot. Gart. Berlin 14: 169, 183 (1938). *T. chrysocarpa* var. *costaricensis* (Mgf.) Allorge in Mém. Mus. natn. Hist. Nat. IIB, 30: 20 (1985). – Type: Costa Rica, San José, Río Candelaria, *Tonduz* 7936 (holotype B†; lectotype US, designated here; isotypes BR, G, P, WAG).

T. pendula Woods. in Ann. Miss. Bot. Gard. 27: 331 (1940), **syn. nov.** – Type: Panama, Coclé, N rim of El Valle de Antón, *Allen* 1734 (holotype MO; isotypes F, GH, US; phot. of US sheet WAG).

T. gentriana Allorge in Bull. Soc. Bot. Fr. 130, Lettres 4-5: 343, fig. 4 (1983) and in op. cit. 34, pl. 13, **syn. nov.** – Type: Panama, Panamá, Campo Tres, 5 km NE of Altos de Pacora, *Liesner* 553 (holotype MO).

Shrub or small tree 1-12 m high. Trunk up to at least 25 cm in diameter; bark pale brown. Branches pale brown, lenticellate or not; branchlets terete, glabrous. *Leaves* of a pair equal or unequal (larger up to 6 x as long as other and not comparatively narrower), petiolate; petiole glabrous, 5-25 mm long; (ocraea widened into intrapetiolar stipules or not); blade membranaceous when fresh, thinly papery and sometimes with slightly impressed venation above when dried, elliptic, narrowly elliptic or sometimes narrowly obovate, 2-4.5 x as long as wide, 1.5-26 x 0.6-9 cm, mostly acuminate or caudate at the apex, less often apiculate (acumen usually acute, 5-15 mm long), cuneate at the base, with a flat or slightly revolute margin, entire, glabrous on both sides, often with scattered black dots beneath, with 5-15 pairs of upcurved secondary veins forming an angle of 60-90° with the costa; tertiary venation reticulate, mostly conspicuous. *Inflorescence* very variable in size, pedunculate, 4-40 x 4-15 cm, many-flowered, lax or in last branchings often rather dense. Peduncle slender, glabrous, 1-30 cm long; pedicels glabrous, 3-18 mm long. Bracts sepal-like, 0.3-0.5 x as long as them. *Flowers* fragrant, open during the day. *Sepals* pale green, connate at the extreme base, erect, ovate or broadly ovate, 1.1-2 x as long as wide, (1.2-)2-3.5 x (1-)1.5-2.5 mm, obtuse or rounded, glabrous outside, not ciliate, glabrous inside and with 3-7 colleters in one row in the middle at the base; colleters 0.3-0.5 x 0.1-0.2 mm. *Corolla* white or yellow, sometimes with pinkish lobes, lobes sometimes turning brown, thin, 9-20 mm long in the mature bud and forming a comparatively large ovoid head 0.25-0.4 of the bud length (4-6 x 2-4 mm) with an acute or less often obtuse apex, glabrous outside, with a pubescent or pilose belt 2-4 mm wide inside just below the anthers, sometimes minutely pubescent from the insertion of the stamens to the base of the lobes (never both indumentums together), or entirely glabrous; tube 2.5-6 x as long as the calyx, 0.5-1.2 x as long as the lobes, (5-)7-19 mm long, almost cylindrical, (1.5-)2-3.5 mm wide above the base, narrowed below the insertion of the stamens to (1.2-)1.5-2.2 mm, again widened around the anthers to (1.5-)2-2.5 mm wide, not twisted, usually with 5 longitudinally bilobed thickenings in the throat mostly visible in the mouth; lobes obliquely elliptic and mostly with an acute or obtuse lateral lobe or sometimes dolabriform, 0.8-2 x as long as the tube, 1.6-4 x as long as wide, 8-19 x 3-7 mm, obtuse, slightly auriculate at the left side of the base, not undulate, spreading. *Stamens* with apex from 5 mm below to 1.5 mm above mouth of corolla tube, inserted 0.4-0.8 of the length of the corolla tube (at (3-)4.5-9 mm from the base); anthers sessile, narrowly triangular, 3-4 x as long as wide, 3-4.5 x 0.8-1.5 mm, apex acuminate, sterile for 0.3-0.4 mm, sagittate at the base (tails straight or slightly curved towards each other), glabrous. *Pistil* glabrous, (5-)6.5-10.5 mm long, with apex about halfway along anthers; ovary ovoid, 1.5-2.5 x 1-1.5 x 1-1.5 mm, of 2 separate carpels, gradually narrowed into the style or nearly so; style (2-)3-6.5 mm long; pistil head 1.5-1.7 mm high, composed of an undulate basal ring like a veil 0.2 x 1.5 mm, an apically 5-lobed stipitate depressed globe 0.6-0.8 x 1.1-1.3 mm, and a bilobed stigmoid apex 0.2 x 0.2 mm. Ovules approximately 50 in each carpel. *Fruit* of 2 separate mericarps; mericarps green, yellow or orange, creamy or orange inside, not dotted, smooth, sometimes warty, obliquely ellipsoid or reniform, 20-65 x 15-30 x 15-30 mm, rounded or apiculate at the apex (acumen 1-5 mm long), with or without one ridge at each side, often torulose when dried, approximately 10-40-seeded; wall about 1 mm thick in dried fruits; aril orange, enveloping the seed. *Seed* dark brown, obliquely ellipsoid or ovoid, 10-12 x 5-6 x 4-5 mm, with longitudinal grooves, papillose; embryo 7-8 mm long; cotyledons ovate or cordate, 1.1-1.4 x as long as wide, 3.5-4.5 x 2.5-4 mm, rounded at the apex, cordate at the base; rootlet 4-4.5 x 0.8 mm.

DISTRIBUTION: Nicaragua to Venezuela.
ECOLOGY: Forest understorey. Alt. 0-1500 m. Flowering throughout the year with a peak from April to June in Costa Rica. Fruiting throughout the year.

Geographical selection of the approximately 200 specimens examined:
NICARAGUA. Zelaya: Cãno Majagua, *Stevens* 6887 (MO, WAG); Mun. Siuna, Río Matis, *F. Ortiz* 143 (MO, WAG); Cãno Montecristo, *Moreno* 15165 (MO, WAG). Río San Juan: La Palma, *Stevens* 23440 (WAG).
COSTA RICA. Guanacaste: La Cruz de Abangares, *Haber & Bello* 2683 (WAG); Volcán Cacao, *J.F. Smith & Frost* 457 (USF); Rincón de la Vieja, *R. Baker* R 17 (BM); ibid., *Herrera* 743 (WAG). Puntarenas: Monte Verde, *Hammel* et al. 13787 (WAG); Carara Res., *Grayum* et al. 7604 (WAG); Rincón de Osa, Osa Peninsula, *Burger & J.L. Gentry* 9009 (BM, F, MO, U); Río Corozal, near Santo Domingo de Golfo Dulce, *Tonduz* 7083 (CR 9973) (BM, BR, G, K, US), 7084 (CR 9980) (BR, US); 6 km S of San Vito de Java, *Raven* 20889 (F, MO, U, UC); Cordillera de Talamanca, near Río Canasta, *Davidse* et al. 28412 (WAG). Alajuela: Pueblo Nuevo, Canton San Carlos, *A. Smith* 1920 (F, MO, U); San Luis, Canton Alfaro Ruiz, *A. Smith* P 2641 (F, MO, U); 12 km NNW of San Ramón, *Liesner & Judziewicz* 14929 (WAG, WIS); Río Naranjo, *Tonduz* 7608 (BR). San José: Río Candelaria, *Tonduz* 7936 (BR, G, P, US, WAG, type of *T. costaricensis*); San Lorenzo de Dota, *Pittier* 2274 (BR, G, US, Z); Santa Maria de Dota Road, *Tonduz* 7856 (BR, paratype of *T. costaricensis*); W of Jamaica Mts, *Grayum* et al. 5480 (WAG); Basin of El General, *Skutch* 4800 (A, F, MO, NA, US). Cartago: Turrialva, *Ørsted* 15547 (C); tributary of Quebrada Casa Blanca, Tapantí, *Grayum* 3605 (BM, USF, WAG). Limón: La Concepción, Llanuras de Santa Clara, *J.D. Smith* 6650 (GH, K, US; phot. of US sheet WAG, type); Tortuguero National Park, *Robles* 1618 (CR); Los Diamantes, Río Santa Clara, *Holm & Iltis* 381 (A, G, P); El Progreso, *Herrera & Chacón* 2707 (WAG).
PANAMA. Bocas del Toro: road to Chiriquí Grande, *McPherson* 7358 (MO); between Buena Vista coffee finca and Cerro Pilon, *Kirkbridge & Duke* 677 (MO); Cerro Colorado, *McPherson* 9542 (L, WAG). Chiriquí: 10 km N of Los Planes de Hornito, *Knapp & Vodicka* 5621 (MO, WAG); 50 km N of San Felix, *Mori & Dressler* 7816 (IJ, MO, U); Cerro Colorado, 28 km from Río San Felix, *Sullivan* 302 (USF, WAG). Veraguas: N of Santa Fe, *Mori & Kallunki* 2654 (AAU, GH, MO). Coclé: N of El Cope, *Folsom* 5844 (WAG); Cerro Pilon, *Lallathin* 20-1 (MO, NSW); N rim of Valle de Antón, *Allen* 1734 (F, GH, MO, US; phot. of US sheet WAG, type of *T. pendula*); El Valle de Antón, La Mesa, *Kennedy* et al. 3035 (C, F, MO, US, USF, WAG). Colón: Santa Rita Ridge, 19 km from highway, *McPherson* 10250 (WAG); S of Río Guanche, *Mori & Kallunki* 2028 (MO). Panamá: above El Valle on road to La

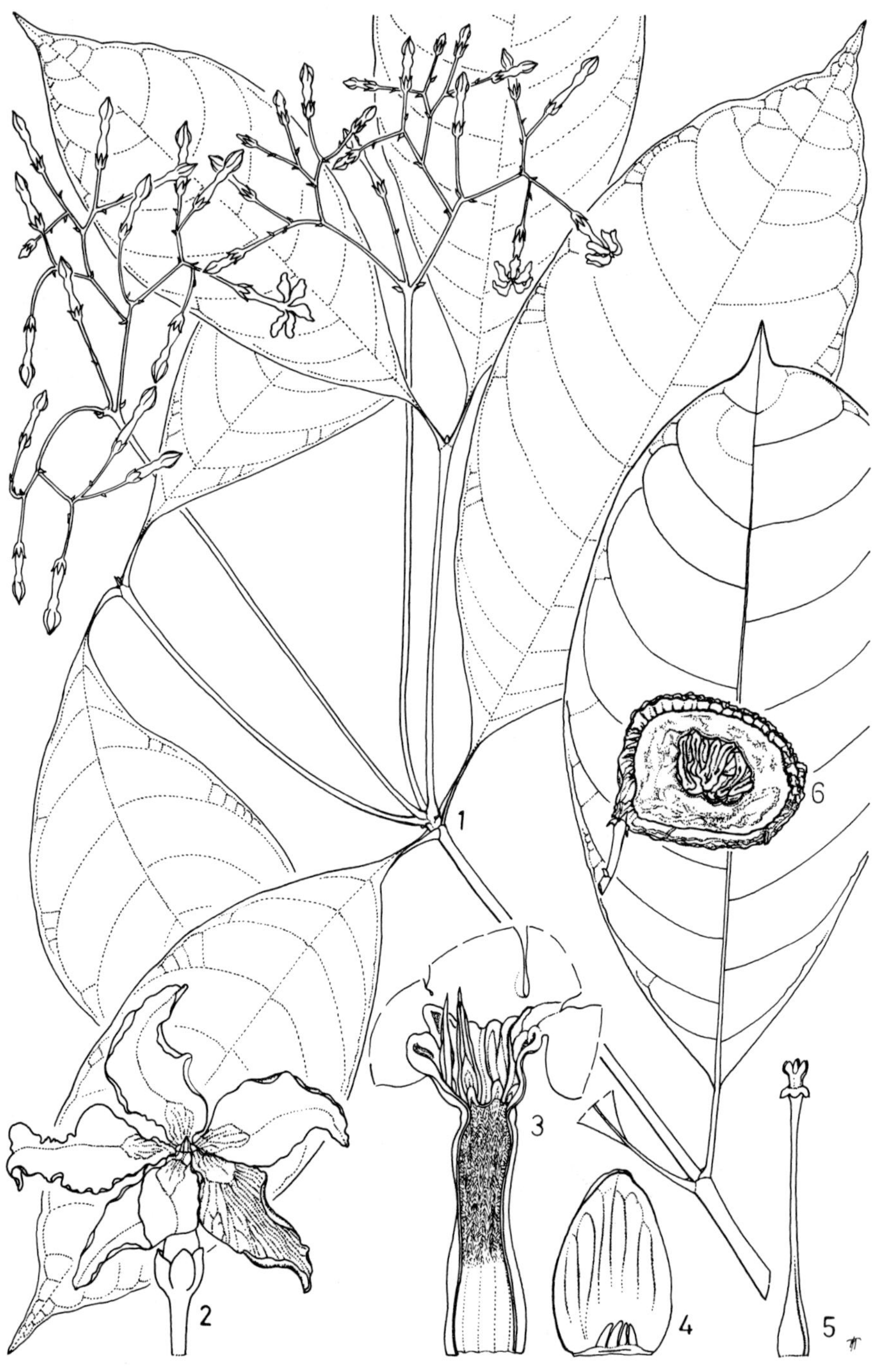

Fig. 81. *Tabernaemontana longipes.* **1,** habit (x 2/3); **2,** flower (x 1.6); **3,** opened corolla (x 4); **4,** sepal inside (x 8); **5,** pistil (x 4); **6,** section of fruit (x 2/3). 1 from Liesner et al. 14929; 2-5 from Hammel 13787; 6 from Folsom 5844.

Mesa, *Tyson* 6963 (NA); Cerro Jefe, *Dwyer* et al. 5039 (GH, MO); Campo Tres, 5 km NE of Altos de Pacora, *Liesner* 553 (MO, type of *T. gentriana*); 11 km N of Margarita, *Hammel* 6029 (MO, WAG); Majé, *Forster & Kennedy* 1985 (F, MO); headwaters of Río Piratí, *Knapp & Mallet* 5129 (MO, U, WAG); near Rancho Chorro, *Folsom* et al. 6999 (F, MO, USF). San Blas: Cangandi, *de Nevers* et al. 7626 (WAG); Río Chucunaque, *Duke* 8570 (GH, MO). Darién: Ensada del Guayabo, *Garwood* et al. 290 (BM); Río Ucurganti, *Bristan* 1134 (US). Río Chucunaque, Camp Ortiga, *Duke* 15525 (MO); Río Yape, *Bristan* 1393 (GH, MO).

COLOMBIA. Chocó: Mun. Acandí, Río Cuti, *Romero-Castañeda* 6454 (COL); Alto de Buey, *A. Gentry & Forero* 7316 (COL, MO). Antioquia: between San Jeronimo and Las Palmitas, *Soejarto* 2067 (COL); Turbo, *Brand & Cogollo* 66 (MO); Vijagual, 30 km S of Turbo, *Haught* 4561 (COL, MO, US); 2 km N of Carepa, *Zarucchi* et al. 5000 (GH, K, MO, WAG); 12 km N of Mutatá, *Callejas* et al. 5782 (WAG); Mun. Anorí, near Providencia, *Soejarto* 2805 (COL, GH, MO); 13 km N of Campamento, *Callejas* et al. 8206 (WAG); 4 km N of Amalfi, *Callejas* et al. 9189 (WAG). Sur de Santander: near Barranca Bermeja, *Haught* 1272 (COL, MO, US), 2017 (FI, MO, US). Valle: Río Yurumanguí, *Cuatrecasas* 16007 (COL, F, MO, US). Cauca: Chuare, *Haught* 5388 (COL, MO, US). Nariño: Río Telembí, *Idrobo & Weber* 1440 (COL); Barbacoas: *Lehmann* Bt.697 (K, NY); ibid., San Pablo, *J. Triana* 1890 (BM, FI); Mun. Espriella, Río Caunapí, *H. León* et al. 1489 (U).

VENEZUELA. D.F.: SE of Camuri Grande, *Morillo* et al. 3333 (Z).

Notes. *Tabernaemontana longipes* is extremely variable in peduncle length and also variable in height of insertion of stamens. However, the species is rather constant in leaves, calyces and fruits. A specimen with an extremely long peduncle is *W.C. Burger & C.J.L. Gentry Jr.* 9009 (F), 30 cm. The type has peduncles of 8-12 cm, and the type of *T. pendula* 7-13 cm, and those of the type of *T. costaricensis* are 1.5-4.5 cm long. The range of variation for this character is continuous in the material examined. Mme Allorge reduced *T. costaricensis* to a variety of *T. chrysocarpa* (here a synonym of *T. alba*), but it resembles *T. longipes* strikingly in the leaves and flowers and not *T. alba*.

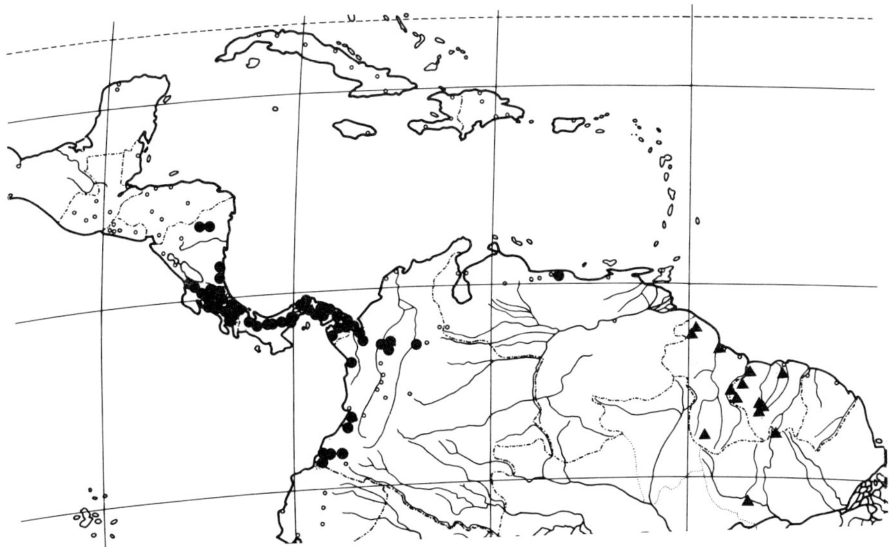

Map 43. ●. *Tabernaemontana longipes*, .▲ *T. lorifera*.

Finally, *T. gentriana* could not be maintained as a distinct species. Its type is exactly like the types of *T. longipes*, *T. pendula* and *T. costaricensis* as for the leaves and flowers, except for the height of insertion of the stamens. The stamens are 0.5 mm exserted in the types of *T. longipes* and *T. costaricensis* and with apex 2 mm below corolla mouth in that of *T. gentriana*. In this respect the extremely long-pedunculate inflorescences of *W.C. Burger & C.J.L. Gentry Jr.* 9009 has the stamens with apex 1 mm below mouth of corolla tube. Several more intermediates have been observed and therefore the names *T. costaricensis, T. gentriana* and *T. pendula* are considered synonyms of *T. longipes* here.

82. Tabernaemontana lorifera (Miers) Leeuwenberg, **comb. nov.** – Type: Guyana, sin. loc., *Robert Schomburgk* II 138 (= *Richard Schomburgk* 21) (lectotype BM, designated here; isolectotypes K, P). Fig. 82, p. 331; map 43, p. 329

Basionym and homotypic synonym:

Peschiera lorifera Miers, Apoc. S. Am. 47 (1878), partly, as for lectotype. *Bonafousia lorifera* (Miers) Boiteau & Allorge in Bull.Soc. Bot. Fr. 130, Lettres 4-5: 340 (1983); Allorge in Mém. Mus. natn. Hist. Nat. IIB, 30: 102, pl. 44 (1985).

Shrub or small tree, 0.5-3 m high. Trunk 0.5-4 cm in diameter; bark grey. Branches pale brown, lenticellate, with longitudinally fissured bark; branchlets terete, glabrous. *Leaves* petiolate; petiole glabrous, 1-5 mm long, but seemingly 5-10 mm long as blade decurrent into petiole; (ocreae not widened into intrapetiolar stipules); blade thinly coriaceous when fresh, papery when dried, elliptic or narrowly elliptic, 2.3-3.4 x as long as wide, 5-18 x 1.5-7.5 cm, acuminate or caudate at the apex, cuneate at the base and long-decurrent into the petiole, entire, glabrous on both sides, dull and with scattered black dots beneath, with 11-23 pairs of rather straight secondary veins forming an angle of 70-85° with the costa; tertiary venation reticulate, but inconspicuous. *Inflorescence* shortly pedunculate, 2-3 x 2-3 cm, 2-10-flowered, lax. Peduncle rather thin, glabrous, 1-5 mm long; pedicels thin, 5-14 mm long, glabrous. Bracts scale-like, about half as long as the sepals, often bracteoles at base of pedicels. *Flowers* fragrant, open during the day. *Sepals* pale green (?), connate at the base for 0.5 mm, erect, ovate, 1.3-1.8 x as long as wide, 1.5-2.5 x 1-1.5 mm, rounded, glabrous outside, ciliolate or not, glabrous inside and with 1-3 colleters in the middle 0.3 mm above the base; colleters forming an ovate patch 0.5 x 0.4 mm, if one emarginate or 3-lobed, if 2 or 3 respectively 0.2 or 0.13 mm wide. *Corolla* entirely white or white and with pale yellow throat, 14-20 mm long in the mature bud and forming a comparatively large ovoid head 0.4-0.5 of the bud length (5.5-10 x 3-6 mm) with a blunt apex, glabrous outside, pubescent inside in a belt 1-2 mm wide just below the insertion of the stamens, above it minutely glandular-puberulous or glabrous, otherwise glabrous; tube 5-6.5 x as long as the calyx, 0.5-0.9 x as long as the lobes, 10-13 mm long, almost cylindrical, 1.5-3 mm wide above the base, often slightly narrowed below and above the anthers to 1-2 mm wide, 2-2.7 mm wide around the anthers, 2-3 mm wide at the throat, not twisted; lobes obliquely obovate or elliptic, 1.1-2 x as long as the tube, 1.6-2 x as long as wide, 14-20 x 7-10 mm, rounded, neither auriculate, nor undulate, spreading. *Stamens* with apex 1.5-3 mm below mouth of the corolla tube, inserted 0.6 of the length of the corolla tube, (at 6-7.5 mm from the base), with tails 1 mm below; anthers sessile, narrowly triangular or nearly so, 3.2-3.5 x 1 mm, apex acuminate, sterile for 0.3 mm, sagittate at the base. *Pistil* glabrous, 7-8.5 mm long; ovary ovoid, 1.5 x 1 mm, with an undulate disk-like thickening, 1 mm high; style filiform, 4.5-6 mm long; pistil head 1-1.2 mm high, composed of an entire basal ring 0.3 x 0.6-0.8 mm, a stipitate 5-lobed depressed globe 0.3 x 0.5-0.6 mm and a stigmoid apical part

Fig. 82. *Tabernaemontana lorifera.* **1,** habit (x 2/3); **2,** flower (x 2); **3,** opened corolla (x 3); **4,** calyx with pistil (x 4); **5,** fruit (x 2/3); **6,** embryo (x 6). 1 from Dahlgren 201; 2-4 from Cavalcante 2532; 5-6 from Fanshawe F.D. 5476.

0.1-0.5 x 0.1-0.2 mm. Ovules approximately 50 in each carpel. *Fruit* of 2 separate mericarps; mericarps green, turning brown when ripening, not dotted, pod-like, 30-45 x 6-11 x 6-10 mm, recurved or not, acuminate sometimes with an upcurved apex, with a lateral ridge at each side and a raised line along the line of dehiscence, minutely papillose outside, approximately 5-10-seeded; wall thin; aril probably at hilar side only. *Seed* medium brown, obliquely ellipsoid, 8-10 x 3-4 x 2-3 mm, with longitudinal grooves, papillose.

DISTRIBUTION: Northern Brazil, Guyana and Suriname.

ECOLOGY: Forest understorey. Alt. 0-275 m. Flowering and fruiting at least from January to August.

Specimens examined:

GUYANA. NW Distr., Mabura Compound, *Archer* 2238 (K, US); Maburuma, Aruka R., *Fanshawe* F.D. 5094 (K, NY); near airstrip of Maburuma, *ter Welle* 3 (U); Waini R., *de la Cruz* 3761 (GH, MO, NY, UC, US); Biara Creek, Moruka, *Fanshawe* F.D. 5476 (K, NY, S, U); Rupununi Distr., Kuyuwini R., *Jansen-Jacobs* et al. 2857 (U, WAG); sin. loc., *Robert Schomburgk* II 138 = *Richard Schomburgk* 21 (BM, K, P, lectotype).

SURNAME. Kabalebo Dam project, Nickerie Distr., *Lindeman* et al. 183/80 (U); Corantyne R., Wonotobo Falls, *Stahel & Gonggrijp* BW 2858 (U); Upper Corantyne R., *Hulk* 93 (U), 97 (U); Maratakka R., near Snake Creek, *Maas* LBB 10776 (K, U); Upper Saramacca R., near Janbasigado, *Pulle* 138 (U); Wilhelmina Gebergte, Julianatop, *Schulz* LBB 10327 (A, U); 3 km S of Julianatop, *Maguire* et al. 54443 (C, F, K, MO, NY, U, US, W); Lucie R., *Irwin* et al. 55458 (NY); 9 km N of Lucie R., *Irwin* et al. 54525 (F, K, MO, NY, U, US); Oost R., *Maguire* et al. 54118 (C, F, K, MO, NY, U, US).

BRAZIL. Pará: Obidos, Rio Parú de Oeste, *Cavalcante* 799 (INPA, MG), 859 (INPA, MG), 2532 (K, NY, S, U, US, Z); Boa Vista, Rio Tapajós, *Dahlgren & Sella* 201 (MO, NY, US); Mun. Oriximiná, Rio Trombetas, 5 km below Porteira Fall, *Cid* et al. 1269 (NY).

Notes. *T. lorifera* resembles *T. albiflora* in the leaves, fruits and seeds. However, it has flowers almost the same as *T. rupicola*.
The three species can be distinguished as follows:

1. Corolla tube 20-27 mm long; mature bud often acute; stamens inserted at 0.4-0.5 of the length from the base of the corolla tube; mericarps acuminate, with lateral ridges, papillose; seed papillose; style not shed with the corolla .**T. albiflora**
   Corolla tube 8-14 mm long; mature bud obtuse or rounded; stamens inserted 0.54-0.7 of the length from the base of the corolla tube 2
2. Leaves subcordate or cuneate at the base, not decurrent into the petiole; mericarps rounded, without lateral ridges, glabrous; seed with spongy testa; style shed with the corolla...**T. rupicola**
   Leaves decurrent into the petiole; mericarps acuminate, with lateral ridges, papillose; seeds papillose; style not shed with the corolla ....**T. lorifera**

83. Tabernaemontana macrocalyx Muell. Arg. in Linnaea 30: 403 (1860), non Pitard (1933). – Type: French Guiana, Acarouany, *Sagot* 394 (holotype G; phot. F, MO, NY, US; isotypes FI, P). Fig. 83, p. 335; map 45, p. 338

Homotypic synonyms:

Codonemma macrocalyx (Muell. Arg.) Miers, Apoc. S. Am. 73 (1878). *Anacampta macrocalyx* Muell. Arg.) Mgf. in Notizbl. Bot. Gart. Berlin 14: 163 (1938). *Bonafousia macrocalyx* (Muell. Arg.) Boiteau & Allorge in Mém. Mus. natn. Hist. Nat. IIB, 30: 97, pl. 42 (1985).

Heterotypic synonyms:

T. benthamiana Muell. Arg. in Martius Fl. Bras. 6, 1: 70 (1860), not p. 80. *T. muelleriana* Mart. ex Muell. Arg. in op. cit. 181 (1865), **syn. nov.** *C. calycinum* Miers, l.c. *Anacampta muelleriana* (Mart. ex Muell. Arg.) Mgf., l.c. *B. muelleriana* (Mart. ex Muell. Arg.) Boiteau & Allorge in Bull. Soc. Bot. Fr. 130, lettres 4-5: 340 (1983) and in op. cit. 83. pl. 34. – Type: Brazil, Amazonas, near São Gabriel da Cachoeira, *Spruce* 2110 (lectotype G, designated here; isolectotypes BM, BR, CGE, K, OXF, P, TCD; phot. of lost B sheet GH, MO, NY, US). The BM sheet is the holotype of *C. calycinum.*

Quadricasaea inaequilateralis Woods. in Ann. Miss. Bot. Gard. 28: 271 (1941). *T. inaequilateralis* (Woods.) Pichon in Mém. Mus. natn. Hist. Nat. II, Bot. 1: 148 (1950). – Type: Colombia, Caquetá, Florencia, *Cuatrecasas* 8814 (holotype US; isotype MO).

Q. caquetensis Woods. in op. cit. 272. – Type: Colombia, Caquetá, Sucre, *Cuatrecasas* 9062 (holotype US).

Bonafousia morettii Allorge in Bull. Soc. Bot. Fr. 130, Lettres 4-5: 341, fig. 1 (1983); in Mém. Mus. natn. Hist. Nat. IIB, 30: 75, pl. 30 (1985), **syn. nov.** – Type: French Guiana, Upper Maroni R., Litani R., *Moretti* 711 (holotype P, not seen; isotype CAY).

Shrub or small tree 1-10 m high. Branches pale or dark brown, lenticellate; branchlets triangular in section when fresh, glabrous or sometimes pubescent. *Leaves* of a pair equal or subequal, petiolate; petiole 2-15 mm long, glabrous or sometimes pubescent; (ocreae not or slightly widened into intrapetiolar stipules); blade coriaceous or subcoriaceous when fresh, more or less papery or coriaceous when dried, elliptic, narrowly elliptic or less often ovate, 1.6-5 x as long as wide, 8-40 x 3-14 cm, acuminate or sometimes apiculate at the apex, cuneate or rounded at the base, entire or sometimes undulate, sometimes bullate, with scattered black dots on both sides, occasionally pubescent on the costa above, often pale green and dull and sometimes pubescent beneath, with 10-25 pairs of rather straight or upcurved secondary veins forming an angle of 60-80° with the costa; tertiary venation reticulate, rather conspicuous. *Inflorescence* shortly pedunculate, 3-8 x 3-7 cm, 3-12-flowered. Peduncle robust, glabrous or sometimes puberulous, 5-15 mm long; pedicels glabrous or sometimes puberulous, 3-10 mm long. Bracts very small, 0.1-0.3 x as long as the sepals and resembling them. *Flowers* open during the day. *Calyx* white, pinkish, greenish-white or white with pink, campanulate, (5-)9-25 mm long; sepals united into a tube for 0.2-0.67 of their length; lobes mostly unequal, (3-)5-15 x 3-7 mm, the longer lobes mostly 1-3 mm longer than the shorther ones, ovate to obovate, 1-2.7 x as long as wide, the shorter lobes often wider than the longer ones, rounded or occasionally obtuse, entire, glabrous or sometimes puberulous outside, not ciliate, glabrous inside and with 10-40 colleters within united parts in 1-4 curved or straight rows or scattered over the basal 4-5 mm; colleters 0.2-1 x 0.1-0.5 mm. *Corolla* pale yellow, white, pinkish or with tube pinkish and limb yellow or orange, 27-39 mm long in the mature bud and forming a comparatively large broadly ovoid head 0.13-0.26 of the bud length (4-10 x 4-8 mm) with an obtuse apex, glabrous or puberulous outside on the part of the lobes not covered in bud, not ciliate, with a pilose or villose belt 2-8 mm wide inside from 1.5-8 mm below the anthers to the insertion of the stamens or up to the level of the apices of the anthers, sometimes puberulous on the base of the lobes; tube 1.3-3.6 x as long as the calyx, 1.25-2 x as long as the lobes, 23-45 mm long, almost cylindrical, mostly narrow, 1.5-3 mm wide, often wider above the base and up to 5 mm wide, not or obscurely widened around the anthers, not twisted; lobes obliquely and often narrowly obovate or elliptic, (0.5-)1.5-3 x as long as wide, 12-28 x 6-18 mm, rounded, undulate, spreading and recurved later. *Stamens* with apex 3.5-17.5 mm below mouth of corolla tube, inserted 0.4-0.7 of the length of the corolla tube (at 10-20 mm from the base); anthers sessile, narrowly triangular, 4-5.5 x as long as wide, 5-7 x 1.2-1.5 mm, apex acuminate, sterile for 0.4-0.8 mm, sagittate at the base, glabrous. *Pistil* glabrous, 17-22 mm long, with apex 0-0.6 of their length below the apices of the anthers; ovary ovoid, 2-3 x 1.5-2 x 1.5-2 mm, acuminate, of two carpels connate for 0.5-1 mm, often with a disk-like thickening at the base; style filiform, 12-17 mm long, persistent; pistil head composed of an entire or undulate ring 0.3-0.5 x 1.2-1.5 mm, a broadly ovoid central part 0.6-1 x 0.8-1 mm and a bilobed apex 0.2-0.7 x 0.2-0.3 mm. Ovules approximately 70-200 in each carpel. *Fruit* of 2 separate mericarps; mericarps dark green, white inside, pod-like, 32-60 x 12-18 x 10-15 mm, straight or recurved, acuminate or acute at the apex, with 2 rather faint lateral ridges, not torulose, obscurely and minutely tuberculate when dried, approximately 20-30-seeded; wall about 3 mm thick; aril white (?). *Seed* medium brown, obliquely ellipsoid, 7 x 3 x 3 mm, with deep longitudinal grooves, papillose.

DISTRIBUTION: Northern South America.

ECOLOGY: Forest understorey. Alt. 0-1000 m. Flowering in Colombia from July to October, in Venezuela from July to December, in Guyana at least in September and November, in Suriname in July and August, in French Guiana from February to November with peaks in March and July-August, in Brazil March to July and October to December, in Ecuador almost exclusively in August, and in Peru mainly in July and August. A fruiting season could not be deduced from the collections in existence.

Geographical selection of the approximately 280 specimens examined:

COLOMBIA. Vaupés: Cerro La Campana Río Ajajú, *Schultes* 5552 (US); Mitú, *Schultes* et al. 24224 (ECON); Villa Fatima, *Romero-Castañeda* 3434 (COL, NY); Río Apaporis, Jirijirimo Fall, *Schultes & Cabrera* 12926 (COL, GH, MO, US). Caquetá: Sucre, *Cuatrecasas* 9062 (US, type of *Quadricasaea caquetensis*); Florencia, *Cuatrecasas* 8814 (MO, US, type of Q. *inaequilateralis*). Putumayo: 15 km NW of Puerto Asís, *King & Guevara* 6239 (NY, US). Amazonas: Puerto Porvenir, *Cuatrecasas* 10657 (COL); W of Leticia, *Plowman & R. Martin* 142 (ECON).

VENEZUELA. Amazonas: Oromana, Alto Ventuari, *Lister & Colchester* 495 (K); near San Carlos de Río Negro, *Spruce* s.n. (K, P); ibid., *Stergios & Aymard* 4263 (PORT, US); Piedra de Cucuí, *Daly* et al. 5528 (WAG); Río Baria, *J.S. Miller* 1785 (MO); SE of Cerro de la Neblina, *Liesner* 16595 (MO); Dept. Atabapo, near Culebra, *Liesner* 17869 (MO); Cerro Duida, *Steyermark* 57952 (MO); Río Ocamo, *Guanchez* 2207 (MO); Río Siapa, *Liesner & Delascio* 21929 (MO, WAG); Río Yatua, *Maguire* et al. 36784 (MO, NY, US); Sierra Parima, *O. Huber & Colchester* 8397 (WAG); Ugueto, *Croizat* 807 (MO, NY). Bolívar: Mun. Raul Leoní, 3 km from base of Cerro Camarón, *Aymard & Fernández* 7154 (MO); Mun. Aripao, El Cacaro, *Marin* 326 (MO); Río Caura, Guiñuna, *Stergios & Delgado* 12496 (MO); Mun. Sucre, near Santa Maria de Erebato, *Sanoja* 2535 (MO); Río Canaracuni, *Stergios* 11827 (MO); Mun. Foráno Aripao, *Aymard & Delgado* 6860 (MO).

GUYANA. Cuyuni-Mazaruni Region, Ayanganna Plateau, *Pipoly* 10905 (P); Upper Essequibo R., *J.G. Meyers* 5765 (K); Gunn's, *Jansen-Jacobs* 1513 (U, WAG); Konashen area, *Jansen-Jacobs* 1794 (U, WAG); Corantyne R., W of New R., F.D. 7330 (NY).

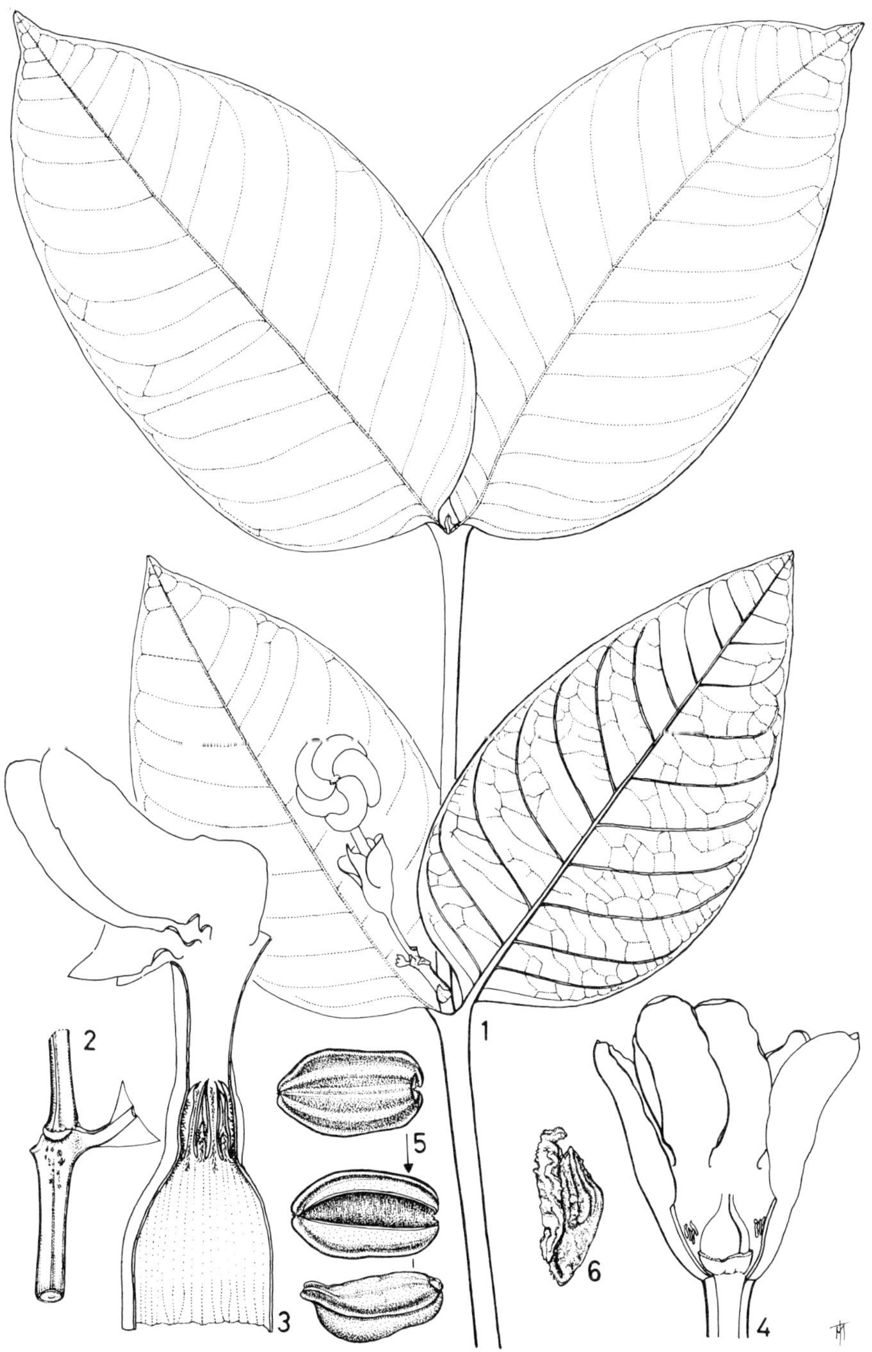

Fig. 83. *Tabernaemontana macrocalyx.* **1,** habit (x 2/3); **2,** portion of branchlet (x 2/3); **3,** opened corolla (x 2); **4,** calyx with ovary (x 2); **5,** dehiscing fruit (x 2/3); **6,** seed (x 2). 1 and 3-4 from Prévost 368; 2 and 5-6 from Leeuwenberg 11648.

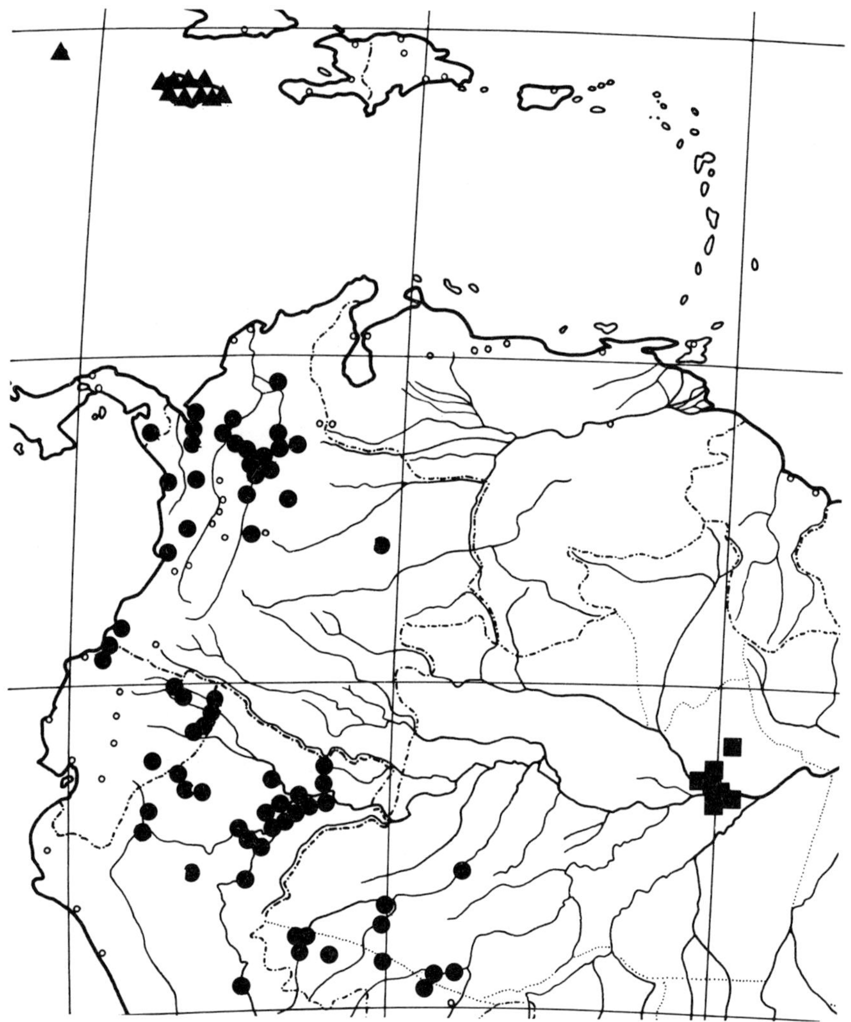

Map 44. ▲. *Tabernaemontana laurifolia*, ●. *T. markgrafiana*, ■. *T. muricata.*

SURINAME. Upper Suriname R., near Goddo, *Tresling* 176 (U); Brokopondo Distr., near Gran Dam, van *Donselaar* 3806 (U); Saramacca R., Tafelberg Creek, *Maguire* 24102 (NY); Wilhelmina Gebergte, 2-5 km SW of Julianatop, *Irwin* et al. 54902 (C, F, K, MO, NY, U, US); Lucie R., *Maguire* et al. 54165 (F, G, GH, K, MO, NY, S, U, UC, US); Paloemeu and Tapanahoni Rs. confluence, *Wessles Boer* 1286 (NY, U); Upper Maroni R., Mt Domeke, *Sastre & Moretti* 3883 (CAY, K, P, WAG); near S headwater of Sipaliwini R., *Oldenburger* et al. 1117 (U).

FRENCH GUIANA. S of St. Laurent du Maroni, *Sastre* 293 (CAY, P, U); Acarouany (= Karouany), *Sagot* 393 (K, P), 394 (FI, G, P; phot. of G sheet F, MO, NY, US, type); Iracoubo R., *Oldeman* B 597 p.p. (P); SW of Sinnamary, Piste St. Elie, km 13, *Leeuwenberg* 11648 (WAG); km 14, Prévost 368 (CAY, P, WAG); Mt Française, *Sastre & Moretti* 4093 (CAY, P); Mana R., Dallas Fall, *de Granville* 4608 (CAY, P);

Upper Baboune Creek, ca 15 km from its confluence with Mana R., *de Granville* 4717 (CAY, P, WAG); Approuague R., Arataye, Pararé Fall, *Sabatier* 929 (CAY, U); ibid., *Sastre* 5694 (CAY, P, WAG); Cayenne, *von Rohr* s.n. (BM); km 52 Cayenne-Regina, *Prévost* 502 (CAY, P, WAG); Pic des Trois Pitons, *Moretti* 1159 (CAY, P); Oyapock R., Maripa Road, *Oldeman* T 856 (CAY, P, U); Pedra Alice, *Irwin* et al. 47563 (B, GH, K, M, NY, S, U, UC, US, Z); Notaye Creek, *Oldeman* T 363 (CAY, P, U); Approuague R., Couata Fall, *Oldeman* B 1940 (CAY, P); Petite Ouaqui R., Baille Nom Fall, *de Granville* 1834 (CAY, K, P, WAG); 2 km SW of Saül, *de Granville* 1567 (CAY, G, HBG, P, WAG); near Saül, *Prévost* 1821 (BR, CAY); Upper Maroni R., Litani R., *Moretti* 711 (CAY, type of *Bonafousia morettii*); Upper Oyapock R., Camp Poivre Creek, *Sastre* 4544 (P, WAG); Trois Sauts, *Grenand* 1001 (CAY); Yaloupi R., *de Granville* 513 (CAY, MO, P, U).

BRAZIL. Roraima: near Auaris, *Prance* et al. 9729 (NY, US, Z); Rio Uraricoera, *J.M. Pires* et al. 16880 (F, INPA, MG, Z); between Surucucu and Uaicá, *Prance* et al. 10857 (NY). Amazonas: near mouth of Rio Embira (= Envira), tributary of Rio Tarauca, *Krukoff* 4939 (A, G, K, MO, NY, S, U, US); São Paulo de Olivença, *Krukoff* 9051 (NY); Upper Rio Negro, Rio Içana, *Doyle* 356 (MG); Taraquá, *J.S. Rodrigues* 183 (NY, Z); Rio Vaupés, *Fróes* 28861 (MO); near Panurè, *Spruce* 2872 (A, CGE, K, P, W; phot. of W sheet US); near São Gabriel da Cachoeira, *Spruce* 2110 (BM, BR, CGE, G, K, OXF, P, TCD; phot. of lost B sheet GH, MO, NY, US, type of *T. muelleriana* and *Codonemma calycinum*); ibid., *Nascimento* 695 (F, MO, NY, WAG); Mun. São Gabriel da Cachoeira, BR 307, km 48 from S.G., *W.A. Rodrigues* et al. 10687 (NY); Serra do Curicuari, *Nascimento* 728 (F, MO, NY, WAG); near Tapurucuara, *N.T. Silva* 4463 (NY, WAG); Mun. Maraã, Rio Japurá, *Cid & Lima* 3421 (NY); Mun. Manaus, ca 80 km NNE of Manaus, Fazenda Esteio, *Nee* 42851 (WAG); km 129 Manaus-Carcarai Highway, *Nelson & Lima* P21117 (K, MO, NY, US, Z); Mun. Novo Aripuanã, *Cid* et al. 5957 (K, MG, US, WAG), Mun. Axinim, lower Rio Paca, *Zarucchi* 2940 (WAG); Mun. Maués, Rio Urupadi, *Zarucchi* et al. 3108 (WAG). Acre: Mun. Mancio Lima, Serra do Divisor, *Cid* et al. 10100 (NY); Mun. Cruzeiro do Sul, Rio Bagé, *Daly* et al. 7329 (WAG). Pará: Rio São Manoel, Igarapé Preto, *J.M. Pires* 3774 (US); between Porto Trombetas and Viveiro, *Soares* 115 (INPA); Upper Rio Tapajós, Rio Cururú, *Anderson* 10572 (C, F, K, MO, NY, U, US, W, Z); Rio Trombetas, Porteira Fall, *N.T. Silva & M.R. Santos* 4754 (NY, WAG); Tapagem, *Cid* et al. 1421 (NY, WAG); Mun. Oriximiná, Rio Caxipacora, *Davidson & Martinelli* 10658 (NY, WAG); Rio Paru do Oeste, *Cid* et al. 2190 (GH, NY, US, WAG), 2345 (MO, NY, US, WAG); Rio Tapajós, *Markgraf* RB 17469 (RB); ibid., Boã, Vista *Ducke* 17469 (K, S, U, US); Serra dos Carajás, *A.S. Silva* et al. 64 (NY); ibid., AMZA camp 3-Alfa, *Sperling* et al. 5949 (F, K, MO, US, WAG), 6084 (GH, WAG); Marabá, *Secco* et al. 172 (MG, WAG); ibid., Serra Norte-Carajás, *M.F.F. da Silva* et al. 1515 (MG); Belém, *A. Silva* 232 (P, US). Amapa: Rio Oiapoque, Clevelandia, *Egler* 1408 (NY); Rio Oiapoque, 1-3 km N of Fall Tres Saltos, *Irwin* et al. 48202 (NY).

ECUADOR. Napo: Sendero La Hormiga, *Jaramillo & Romoleroux* 8482 (QCA); Lagunas de Cyabeno, *Brandbyge* et al. 36101 (AAU, F, MA, U); between San Carlos and Guamayacu, *Brandbyge & Asanza* 30337C (AAU); 5 km N of Coca, El Chuncho F.R., *Neill* et al. 7254 (MO); Res. Biol. Jatun Sacha, 8 km from Puerto de Misahualli, Palacios 1655 (MO). Pastaza: Río Curaray, *Jaramillo & Coelho* 3562 (AAU, QCA); ibid., Río Querano mouth, *Neill & Palacios* 6795 (MO, WAG); 20 km S of Curaray, *Zak & Espinoza* 5198 (WAG); Montalvo, *Brandbyge* et al. 35493 (AAU, WAG); Kapawí, Amuntai, *Lewis* et al. 13838 (MO, WAG); ca 10 km E of Puerto Sarayacu, *Lugo* 4016 (GB, USF); Pacayacu, *Lugo* 5268 (USF). Morona-Santiago: Taisha, *Cazalet & Pennington* 7761 (B, K); Mendez-Morona Road, *Hirtz* 4387 (MO).

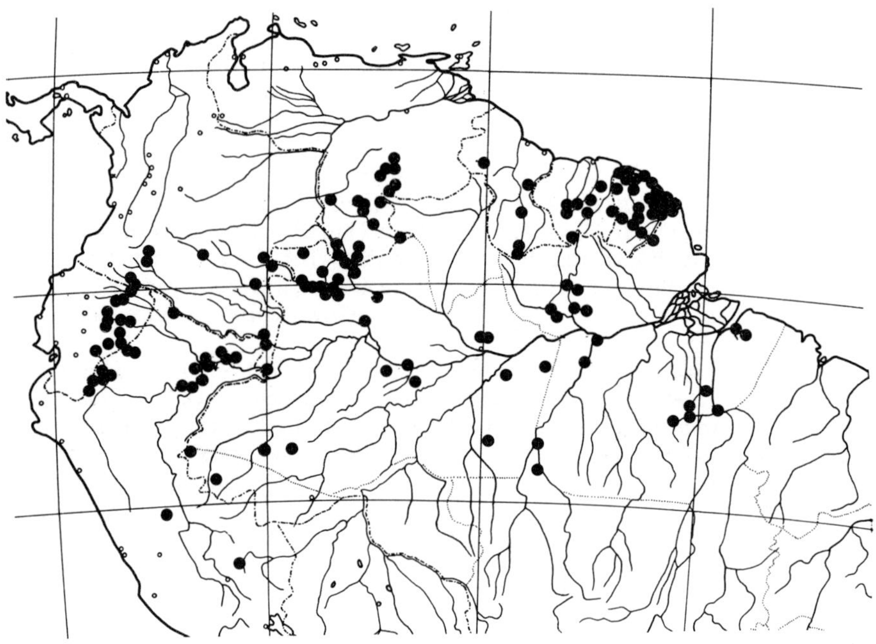

Map 45. *Tabernaemontana macrocalyx.*

PERU. Amazonas: Río Santiago, Quebrada Caterpiza, *Huashikat* 670 (MO); 3 km from La Poza, *Berlin* 3523 (F, MO); near Quebrada Shimpunts, Upper Río Cenapa, *Berlin* 872 (Z); N of Cenapa, *Berlin* 1886 (MO, Z); near Huampami, *Kujikat* 43 (AAU, MO, NY); Huambisa, near Caterpiza, *Tunqui* 365 (MO). Loreto: Capihuari, *A. Gentry & Díaz* 28210 (MO); Nueva Jerusalem, *Lewis* et al. 10996 (MO); Río Yubineto, *Barrier* 510 (P); Nauta-Iquitos Road, *Vásquez* 2246 (MO, WAG); Quebrada Tahuayo, above Tamishiyaco, *Croat* 19695 (MO, Z); Maynas Prov., Puerto Almendras, *Vásquez* & *Jaramillo* 4610 (MO, WAG); Mishuyacu, near Iquitos, *Killip* & *A.C.Smith* 29985 (MO, NY, US); Mazán, *Croat* 19453 (L, MO, NY, WAG); near Sucusari, *Pipoly* et al. 13147 (WAG); Quebrada de Yanamono, *Vásquez* et al. 14112 (WAG); Brillo Nuevo, Río Yaguasyacu, *Treacy* & *Alcorn* 392 (F); between Pucaurquillo and Pebas, *Ayala* et al. 2709 (GH, MO, NY, WAG); Pebas, *Fosberg* 29156 (US). Pasco: Oxapampa, 2 km SW of Puerto Bermudez, *Forster* 8542 (MO, P). Madre de Dios: Prov. Manu, downriver from Atalaya, *Forster* & *Baledon* 12840 (WAG).

Notes. The types of *Tabernaemontana macrocalyx* and *T. muelleriana* are remarkably similar in all characters. The type of *Bonafousia morettii* is slightly hairy as are several other specimens collected in French Guiana and Suriname. The most hairy specimen is *Anderson* 10572 collected in Brazil, which is in all other characters exactly like the glabrous *Prévost* 368, doubtlessly belonging to *T. macrocalyx*. Therefore, *T. muelleriana* and *B. morettii* are reduced to synonyms here.

Phot. 1. *Tabernaemontana cerea*. Guyana. Photo P.J.M. Maas. *Polak & Maas* 507.

Phot. 2. *Tabernaemontana citrifolia*. Nevis. Photo C.D. Adams.

Phot. 3. *Tabernaemontana citrifolia*. Nevis. Photo C.D. Adams.

Phot. 4. *Tabernaemontana cymosa*. Trinidad. Photo C.D. Adams.

Phot. 5. *Tabernaemontana cymosa.* Venezuela. Photo F.J. Breteler. *Breteler* 3904.

Phot. 6. *Tabernaemontana disticha.* French Guyana. Rorata Lake. Photo M.F. Prévost. *Prévost* 166.

Phot. 7. *Tabernaemontana heterophylla.* French Guyana. Photo A. Raynal. *Leeuwenberg* 11611.

Phot. 8. *Tabernaemontana heterophylla.* French Guyana. Photo A. Raynal. *Leeuwenberg* 11611.

Phot. 9. *Tabernaemontana heterophylla*. French Guyana. Photo A. Raynal. *Leeuwenberg* 11611.

Phot. 10. *Tabernaemontana undulata*. Guyana, Mabura Hills. Photo P.J.M. Maas.

Phot. 11. *Tabernaemontana undulata*. French Guyana, piste St. Elie. Photo M.F. Prévost. *Prévost* 212.

Phot. 12. *Stemmadenia pauli.* Costa Rica. Photo P.J.M. Maas. *Maas* 7850.

Phot. 13. *Stemmadenia pauli.* Costa Rica. Photo P.J.M. Maas. *Maas* 7850.

84. Tabernaemontana markgrafiana Macbride in Publ. Field Mus. Bot. Ser. 13, 5: 406 (1959). – Type: Brazil, Amazonas, Rio Jurua, Marari, *Ule* 5179 (holotype B†; lectotype K, designated here; isotypes G, HBG, L, MG).

Fig. 84, p. 340; map 44, p. 336

Basionym:

Bonafousia longituba Mgf. in Notizbl. Bot. Gart. Berlin 14: 166, 180 (1938), non *T. longituba* Pichon (1948).

Shrub or small tree, 1.3-6 m high. Trunk up to at least 4 cm in diameter. Branches with fissured bark, pale or dark brown, lenticellate; branchlets probably terete or elliptic in section when fresh, glabrous. *Leaves* large, shortly petiolate; petiole glabrous, 5-15 mm long; (ocreae not or slightly widened into intrapetiolar stipules); blade coriaceous or subcoriaceous when dried, elliptic or narrowly elliptic, 1.5-3.5 x as long as wide, (14-)22-57 x (7-)9-27 cm, acuminate at the apex, cuneate or rounded at the base, entire or undulate, glabrous and with scattered black dots on both sides, with 12-20 pairs of rather straight secondary veins forming an angle of 50-70° with the costa; tertiary venation reticulate. *Inflorescence* shortly pedunculate, 4-9 x 3-7 cm, few- or many-flowered, dense, 1-3 x furcate or subumbellate; branches ending thyrsoid and densely covered with flowers. Peduncle robust, glabrous or puberulous, 7-40 mm long; pedicels puberulous or sometimes glabrous, 1-2 mm long. Bracts numerous, 0.5-0.6 x as long as the sepals, triangular, obtuse, deciduous, with flowers in axils. *Flowers* open during the day. *Sepals* pale green (?), subequal, connate at the base for 0.3-1 mm, erect, circular, ovate or elliptic, 1-2 x as long as wide, 2.5-6 x 2.5-3 mm, rounded, entire, puberulous to subglabrous outside, ciliate or ciliolate, glabrous inside and with 4-8 colleters in 1 row in the middle of the base; colleters 0.3-0.7 x 0.2 mm. *Corolla* white or greenish-white, 22-27 mm long in the mature bud and forming a comparatively small subglobose head 0.11-0.15 of the bud length (3-4 x 3-4 mm) with a rounded apex, puberulous or pubescent at the apex of the tube and on the part of the lobes not covered in bud, or sometimes entirely glabrous outside, mostly ciliate on the same part of the lobes, with a pilose belt 4-6 mm wide inside from 0.2 mm above the insertion of the stamens downwards, especially on the filament ridges and mostly puberulous at the base of the lobes; tube 6-9 x as long as the calyx, 1.8-4 x as long as the lobes, 22-27 mm long, almost cylindrical, 2-4.5 mm wide, not or obscurely widened around the anthers, twisted 0-0.4 turn from the base to just below the insertion of the stamens to the left and from there 0.5-0.7 turn to the right around the anthers; lobes obliquely and narrowly obovate or elliptic, 0.25-0.55 x as long as the tube, 1.3-2.3 x as long as wide, 7-15 x 3-8 mm, rounded, undulate, spreading, recurved later. *Stamens* with apex 1.5-3 mm below mouth of corolla tube, inserted 0.6-0.76 of the length of the corolla tube (at 15.5-19 mm from the base); anthers sessile, narrowly triangular, 3.5-4 x as long as wide, 5.2-6 x 1.2-1.5 mm, apex acuminate, sterile for 0.3 mm, sagittate at the base, glabrous. *Pistil* glabrous, 17-20 mm long, with apex 0.3-0.4 way along anthers; ovary ovoid, 2-2.5 x 1.5-2 x 1.5-2 mm, acuminate, of two carpels, connate for 0.3-0.5 mm, sometimes with a disk-like thickening at the base; style filiform, 13-16.5 mm long, persistent; pistil head composed of an entire or undulate ring 0.4-0.5 x 1.2-1.5 mm, a broadly obovoid apically 5-lobed central part, 0.7-0.9 x 0.9-1 mm and a short bilobed apex, 0.2-0.3 x 0.2-0.3 mm. Ovules approximately 50-100 in each carpel. *Fruit* of 2 separate mericarps; mericarps yellow or orange, subglobose, obliquely ellipsoid or ovoid or pod-like, 30-55 x 15-30 x 15-20(-30?) mm, with a longitudinal ridge at each side, smooth or somewhat warty when fresh, wrinkled or wrinkled-verrucose when dried, approximately 15-40-seeded; wall about 2-3 mm thick in dried fruits; aril white, enveloping the seed. *Seed* pale

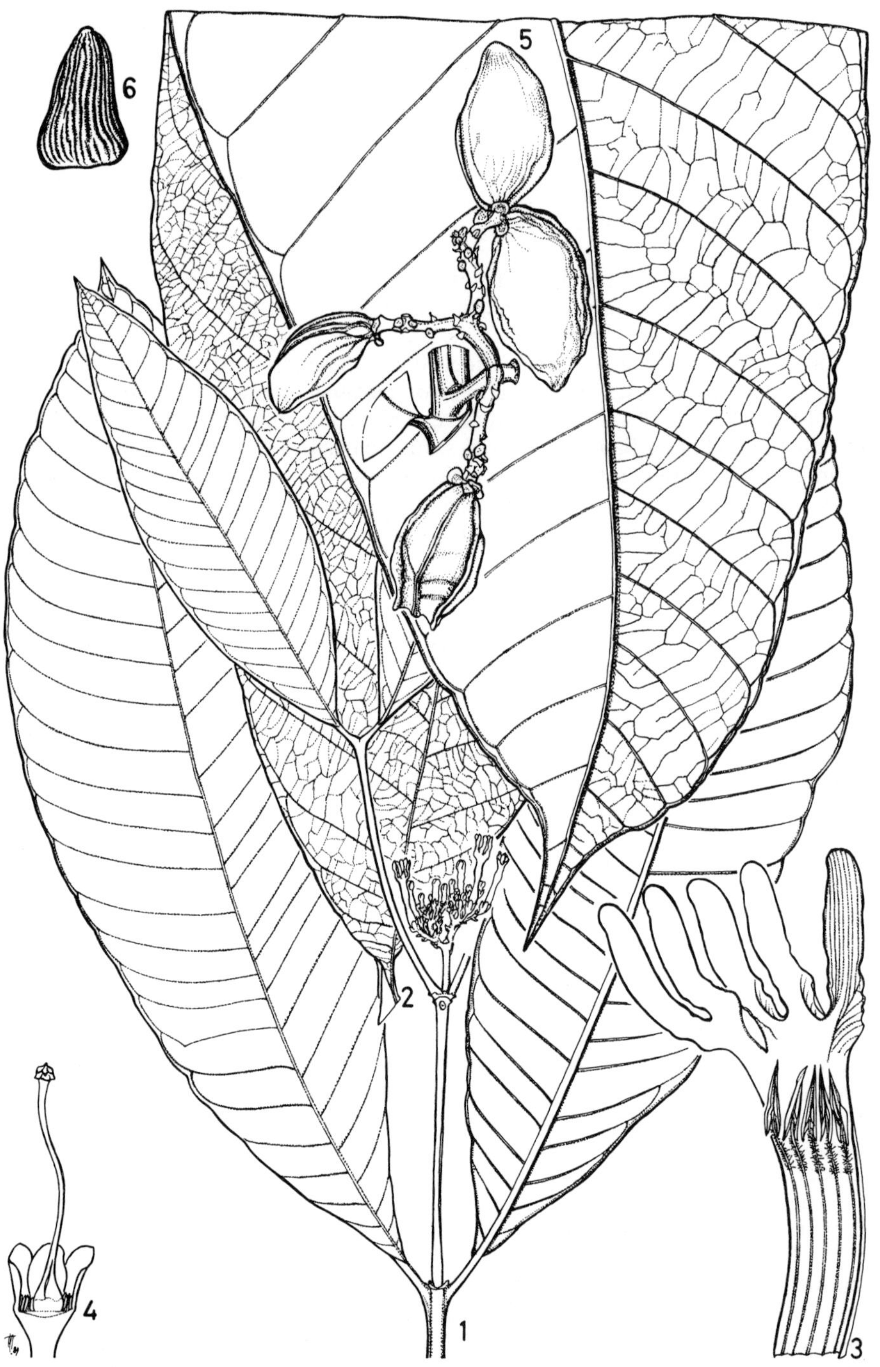

Fig. 84. *Tabernaemontana markgrafiana.* **1,** habit (x 1/3); **2,** leaf (x 2/3); **3,** opened corolla (x 2); **4,** calyx with pistil (x 2); **5,** fruits (x 2/3); **6,** seed (x 2). 1 and 3-4 from Haught 1749; 2 and 6 from A.Gentry & Juncosa 41097; 5 from Rentería et al. 733.

brown, obliquely ellipsoid, 6-10 x 3-5 x 3-4 mm, with deep longitudinal grooves; embryo 6-9 mm long; cotyledons ovate or cordate, 1-1.2 x as long as wide, 2.5-3 x 2.5-3 mm, obtuse or rounded at the apex, cordate or subcordate at the base; rootlet 1.5-2 x as long as the cotyledons, 4-6 x 0.6-0.8 mm.

DISTRIBUTION: Panama and northwestern South America.

ECOLOGY: On river banks, periodically inundated or on non-inundated places in forest understorey (tierra firma). Alt. 0-1200 m. Flowering mainly from August to October. A fruiting season could not be delimited from the data in extant collections.

Geographical selection of the approximately 120 specimens examined:

PANAMA. Darién: Río Congo, *Bristan* 637 (MO).

COLOMBIA. Choco: ca 1.5 km NE of Camp Curiche, *Duke & Idrobo* 11337 (MO, US); between Tutunendo and Ichó, *Juncosa* 1212 (MO). Antioquia: Turbo, *Brand* 1210 (COL, MO); Mun. Chigorodó, Malagón Path, *Rentería* et al. 4539 (MO); 12 km N of Mutatá, *Callejas* et al. 5754 (WAG); Mun. Caucasia, road to Nechí, *Brant* 1176 (MO, WAG); 10-15 km SE of Cáceras, *Callejas* et al. 5376 (WAG); near Planta Providencia, *Shepherd* 922 (COL, MO, WIS); 9-18 km SW of Remedios, *Callejas* et al. 8080 (NY); Peñas Blancas, *Bossé* 8717 (LE). Bolívar: San Martin de Loba, *Curran* 200 (S, US). Santander: Campo Capote, 30 km E of Carare, *A. Gentry* et al. 19940 (COL, MO); near Barranca Bermeja, *Haught* 2093 (COL, MO, US); 3 km NE of Cimitarra, *Romero-Castañeda* 4323 (COL); near Puerto Berrio, *Haught* 1749 (US); Puerto Wilches Region, road to Pedral, *Rentería* et al. 2168 (COL); Sitio los Cayenos, *Rentería* et al. 733 (COL, MO, NY); Río Carare, *Hodge* 6514 (GH). Boyacá: Puerto Boyacá, *Avellaneda* 39 (COL). Caldas: 13 km N of La Dorada, *A. Gentry* et al. 18210 (COL, MO, NY). Cundinamarca: between Nilo and Quebrada de Agua de Diosito, *Murillo* et al. 284 (COL). Valle: Río Calima, Quebrada La Brea, *Schultes & Villarreal* 7378 (F). Nariño: Tumaco, *Romero-Castañeda* 5144 (COL). Amazonas: Quebrada Arara, 2 hours N of Leticia, *Croat* 7555 (MO).

BRAZIL. Amazonas: Rio Jurua, Marari, *Ule* 5179 (G, HBG, K, L, MG, type); Matupiri, *Krukoff* 4595 (A, G, K, NY, S, U, US); near mouth of Rio Embira (= Envira), tributary of Rio Tarauca, *Krukoff* 5132 (A, G, K, MO, NY, S, U, US); near Rio Iaco, *Cid & Nelson* 2695 (NY); Porto Alegre, Upper Rio Purus, *J. Huber* 4470 (RB). Acre: Cruzeiro do Sul, *Prance* et al. 2895 (NY, US, Z); ibid., near Carlota, *Maas* et al. P12912 (NY, Z); near Rio Juruá Mirim, *Ule* 5847 (HBG); near Tarauacá, *Prance* et al. 7469 (G, K, NY, S, US, Z); near mouth of Rio Maucahan, *Krukoff* 5432 (A, G, K, NY, S, U, US).

ECUADOR. Esmeraldas: San Lorenzo, *Palacios* 2612 (MO, WAG); Zapallo Grande, *Kvist & Asanza* 40750 (QCA). Napo: Cantón Lago Agrio, Dureno, *Cerón* 405 (WAG); San Pablo de los Secoyas, *Asanza* 32938 (AAU); Río Napo, near confluence with Río Aguarico, *Harling* et al. 7376 (GB, USF); Rocafuerte, *Heinrichs* 482 (G, M, Z). Pastaza: Lorocachi, *Jaramillo* et al. 31729 (AAU, WAG); Ceilán, *Brandbyge & Asanza* 31651 (AAU). Morona-Santiago: NW of Pumpuentza, Brandbyge & Asanza 32312 (AAU).

PERU. Amazonas: Río Santiago, 1 km downstream from La Poza, *Leveau* 241 (F, MO, Z); Prov. Bagua, Río Marañon, *Wurdack* 2266 (MO, US). Loreto: Pongo de Manseriche, *Tessmann* 4260 (NY, S); Andoas, *Ayala* 2106 (MO, WAG); Pampa Hermosa, *Lewis* et al. 9943a (MO, WAG); Puranchim, *Lewis* et al. 12155 (MO); Yurimaguas, *Schunke* 6353 (F, K, MO, US); San José de Parinari, *Vásques* et al. 2303 (MO, WAG); Río Samiria, *Ayala & Criollo* 3987 (MO, WAG); Prov. Requena, near Jenero, *Peters* 118 (WAG); Negro Urco, *R.T. Martin & Lau-Cam* 1325 (ECON);

Nauta, *Vásquez & Jaramillo* 8623 (MO); Santa Maria de Nanay, *Vásquez & Jaramillo* 12268 (WAG); Río Nanay mouth, *Revilla* 1708 (MO); near Río Napo mouth, *Croat* 20206 (AAU, MO, NY); Sanangal, *Vásquez* et al. 351 (MO); Quebrada Tamshiyacu, *Ayala* et al. 3582 (MO); Yarina, *Vásquez & Espiritu* 4862 (MO, WAG); Yanomono, *A. Gentry* et al. 37158 (MO, P); Brillo Nuevo, *Balick* et al. 976 (MO); ibid., *Treacy & Alcorn* 341 (F, WIS); Distr. Pebas, *Río Ampiyacu,* Revilla 837 (F, MO, Z). Ucayali: Bosque Nacional de Iparia *Schunke* 2805 (F, G, NY, US).

CULT. Ecuador, Napo, WSW of San Pablo de los Secoyas, *Brandbyge* et al. 33309 (AAU).

85. Tabernaemontana maxima Mgf. in Notizbl. Bot. Gart. Berlin 10: 1036 (1930). – Type: Brazil, Amazonas, Manaus, Cachoeira do Mindú, *Ducke* 21605 (holotype B†; lectotype US; phot. of lost holotype GH, MO, NY, US; phot. of lectotype, designated here, WAG). Fig. 85, p. 343; map. 40, p. 318

Homotypic synonyms:

Anacampta maxima (Mgf.) Mgf. in op. cit. 14: 162 (1938). *Bonafousia maxima* (Mgf.) Boiteau & Allorge in Bull. Soc. Bot. Fr. 130, Lettres 4-5: 340 (1983); in Mém. Mus. natn. Hist. Nat. IIB, 30: 95, pl. 40 (1985).

Shrub or small tree, 2-10 m high. Branches thick, pale brown, with few lenticels; branchlets terete or nearly so, glabrous. *Leaves* sessile, or subsessile; petiole glabrous, up to 3 mm long; (ocreae not or obscurely widened into intrapetiolar stipules); blade coriaceous both when fresh and when dried, narrowly elliptic or narrowly obovate, 1.8-4 x as long as wide, 16-50 x (5-)8-23 cm, acuminate or apiculate at the apex, cordate or subcordate at the base, entire or undulate, glabrous and with scattered black dots on both sides, with 15-25 pairs of rather straight secondary veins forming an angle of 50-70° with the costa; tertiary venation reticulate. *Inflorescence* shortly pedunculate, 7-9 x 6-7 cm, 15-50-flowered, corymbose, dense, Peduncle robust, glabrous, 10-20 mm long; branches covered with pedicel and bract scars with age; pedicels glabrous, 4-8 mm long. Bracts numerous, almost sepal-like and 0.2-0.4 x as long, obtuse, deciduous, with flowers in axils or not. *Flowers* fragrant, open during the day. *Calyx* campanulate, 1.5-2 x as long as wide; sepals white or green, subequal, connate at the base for 0.25-0.55 of their length (2-5 mm), clasping the corolla base when fresh, elliptic, 2-3 x as long as wide, 6-12 x 3-4 mm, rounded, entire, glabrous outside, not ciliate, glabrous inside and with 6-7 colleters in 1 row about 1 mm above the base; colleters 0.3-0.5 x 0.1-0.3 mm. *Corolla* white, yellow in the throat, 20-25 mm long in the mature bud and forming a comparatively rather large broadly ovoid head 0.2-0.25 of the bud length (4-5 x 4 mm) with an obtuse apex, puberulous or, in immature buds, glabrous outside on the part of the lobes not covered in bud, with a pubescent belt 5 mm wide inside from 1.5 mm below to 3.5 mm above the insertion of the stamens and puberulous on the entire lobes, otherwise glabrous; tube 2-3.5 x as long as the calyx, 2-2.5 x as long as the lobes, 18-30 mm long, almost cylindrical, 3-4 mm wide above the base, narrowed below the insertion of the stamens to 2 mm wide, widened again around the anthers to 3 mm wide and from there more or less gradually widened to the mouth and 4-5 mm wide, not twisted; lobes 0.4-0.5 x as long as the tube, 1.2-2.2 x as long as wide, obliquely ovate, 10-14 x 6-8 mm, strongly curved to the right, rounded, auriculate at the left side of the base, not undulate, upcurved. *Stamens* with apex 3-5 mm below mouth of corolla tube, inserted 0.52-0.6 of the length of the corolla tube (at 12-16.5 mm from the base); anthers sessile, with tails 0.5-1 mm below insertion, narrowly triangular, 4-5.5 x as long as wide, 5.5-6 x 1-1.5 mm, apex acuminate, sterile for 0.5 mm, sagittate at the base with straight tails.

Fig. 85. *Tabernaemontana maxima.* **1,** apex branchlet (x 2/3); **2,** inflorescence (x 2/3); **3,** opened corolla (x 2); **4,** corolla lobe (x 4); **5,** pistil head (x 14); **6,** sepals inside (x 2.4); **7,** fruit (x 2/3); **8,** dehiscing fruit (x 2/3); **9,** seeds in mericarp (x 2/3); **10,** seeds (x 2). 1-2 from Stergios et al. 4317; 3-6 from Zarucchi et al. 1870; 7-8 from Romero-Castañeda 10115; 9-10 from Romero-Castañeda 4348.

Pistil glabrous, 13-19 mm long, with apex about halfway along anthers; ovary subglobose, 3 x 3 x 3 mm, with a disk-like thickening 1 mm high; style filiform, 10-15 mm long; pistil head 1.2 mm high, composed of a veil-like basal ring 0.2 x 1-1.2 mm, a stipitate 5-lobed depressed globe 0.3 x 0.8-1 mm and a stigmoid apical part 0.2 x 0.2-0.3 mm. Ovules approximately 100 in each carpel. *Fruit* of 2 separate mericarps; mericarps white, obliquely ellipsoid, 30-50 x 17-20 x 15-17 mm, obtuse, immature often apiculate, with a lateral ridge at each side, tuberculate when dried, approximately 20-seeded; wall about 3 mm thick in dried fruits; aril only at the hilar side. *Seed* dark brown, obliquely ellipsoid, 8-9 x 4-5 x 3-4 mm, with longitudinal grooves, papillose; embryo 7-7.5 mm long; cotyledons elliptic, 1.3-1.9 x as long as wide, 2.8-4 x 1.5-3 mm, rounded at the apex, cordate at the base; rootlet 4.5 x 0.6-0.9 mm.

DISTRIBUTION: Northern South America.

ECOLOGY: Forest understorey, on terra firme. Alt. low. Flowering probably mainly from August to October. Fruits known from December and January.

Geographical selection of the approximately 40 specimens examined:

COLOMBIA. Antioquia: Mun. San Luís de Cocorná, *Romero-Castañeda* 10115 (AAU). Santander: Barranca Bermeja, *Uribe* 3055 (COL); between Guayabito and Oponcito, *Romero-Castañeda* 4348 (AAU, COL). Vaupés: Mitú, *Zarucchi* et al. 1870 (COL, GH, K, MO, US, USF, WAG), 2230 (COL, ECON, GH, INPA, K, MO, US, USF, WAG); Santa Teresita, *Romero-Castañeda* 3730 (COL); Río Piraparana, near San Miguel, *Davis* 131 (COL, GH, K, U). Guainia: San Felipe, *Schultes* et al. 18055 (GH, US).

VENEZUELA. Amazonas: San Carlos de Río Negro, road to Solano, *Ramos* 38 (MY); ibid., *Stergios & Aymard* 4317 (PORT); ibid., *Klinge* 3 Apr. 1978 (MO, VEN); Cerro Yapacana, *Maguire* et al. 30647 (NY).

BRAZIL. Amazonas: Mun. São Gabriel de Cachoeira, Vista Alegre, *Farney* et al. 1760 (WAG); Rio Içana, *Fróes* 28049 (MG, Z); near Serra Araçá, *Rosa & Cordeiro* 1659 (F, MG, MO, NY); ibid., Rio Jauari, *Cordeiro* 208 (NY, WAG); Rio Demeni, *Fróes* 28947 (MO); Manaus, Igarapé de Santa Maria, *W. Rodrigues* INPA 2225 (MG, NY); Igarapé da Bolivia, BR 17, km 19 from Manaus, *L.F. Coelho* INPA 3213 (INPA, MG); Manaus, *Ducke* 1302 (A, NY, RB, US); ibid., Mindú Fall, *Ducke* 21605 (US; phot. WAG; phot. of lost B sheet GH, MO, NY, US, type); ibid., *Maas* 487 (NY, U); Tarumã Açu, *Webber* et al. 1253 (WAG); Rio Tarumã, *Fróes* 26560 (Z).

86. Tabernaemontana muricata Link ex Roem. & Schult., Syst. 4: 431 (1819), not Willd. ex Roem. & Schult., op. cit. 797; Leeuwenberg in Agric. Univ. Wageningen Pap. 87, 5: 15 (1988). – Types: Brazil, sin. loc., herb. *Link* s.n. (holotype B†). Brazil, Amazonas, near Barra do Rio Negro (= Manaus), *Spruce* 1470 (neotype BM, designated by Leeuwenberg in 1988; isoneotypes BR, CGE, F, G, GH, GOET, K, LD, NY, OXF, P, TCD, W; phot. of lost B sheet F, GH, MO, NY, US).

Fig. 86, p. 346; map 44, p. 336

Homotypic synonyms:

Peschiera muricata (Link ex Roem. & Schult.) A. DC, Prod. 8: 361 (1844).
Bonafousia muricata (Link ex Roem. & Schult.) Mgf. in Notizbl. Bot. Gart. Berlin 14: 166 (1938).

Heterotypic synonyms:

T. macrophylla Muell. Arg. in Martius, Fl. Bras. 6, 1: 75 (1860), non Lam. (1792). *Phrissocarpus rigidus* Miers, Apoc. S. Am. 72, pl. 9, A (1878). *Anacampta rigida* (Link ex Roem. & Schult.) Mgf. in op. cit. 163. *T. rigida* (Miers) Leeuwenberg in Journ. Ethnopharmacology 10: 16 (1984). Type: The lost B sheet of *Spruce* 1470 was the holotype *T. macrophylla* Muell. Arg. The BM sheet of it is the holotype of *Phrissocarpus rigidus*.

Shrub or small tree, 1-7 m high. Trunk 1-8 cm in diameter. Branches medium brown, lenticellate, often with 3 ridges; branchlets triangular in section, glabrous. *Leaves* shortly petiolate or sessile; petiole up to 6 mm long, glabrous; (ocreae widened into intrapetiolar stipules); blade coriaceous when dried, narrowly oblong or elliptic, 1.9-4 x as long as wide, (10-)15-50 x (4.5-)8-16 cm, apiculate or subacuminate at the apex with a blunt acumen, cuneate or rounded at the base, entire or sinuate, sometimes slightly bullate, glabrous on both sides. mostly much paler and dull beneath, with scattered or more or less regularly arranged often black dots or not, with (10-)15-20 pairs of rather straight secondary veins forming an angle of 60-70° with the costa; tertiary venation not conspicuous, reticulate. *Inflorescence* shortly pedunculate, 3-7 x 3-5 cm, 5-30-flowered, congested. Peduncle robust, glabrous or puberulous, 5-25 mm long; pedicels 0-4 mm long, glabrous or puberulous. Bracts numerous, sepal-like, about half as long as the sepals. *Flowers* open during the day. *Calyx* often subtended by one bracteole (often also bracteoles on pedicel); sepals green (?), connate at the base for 0.3-0.5 of their length (1-1.5 mm), erect, ovate or circular, 1-1.6 x as long as wide, 3-5 x 2-3 mm, rounded, glabrous or puberulous outside, ciliate, glabrous inside and with 2-7 colleters in 1 row in the middle at the base; colleters 0.8-1 x 0.3-0.5 mm. *Corolla* white, with often pinkish tube, 20-30 mm long in the mature bud and forming a comparatively small broadly ovoid or subglobose head 0.12-0.16 of the bud length and 1-1.2 x as wide as the upper part of the tube, (3-5 x 3-4 mm) with a blunt or rounded apex, puberulous outside on the base of the lobes and sometimes at the apex of the tube, puberulous inside at the base of the lobes with a pubescent belt 2 mm wide just below the anthers; tube 6-9 x as long as the calyx, 1.6-2.1 x as long as the lobes, 22-28 mm long, almost cylindrical and 2.5-3 mm wide, slightly narrowed below the insertion of the stamens to 2 mm wide, not twisted; lobes obliquely elliptic, 0.45-0.6 x as long as the tube, 2.4-2.8 x as long as wide, 13-17 x 5-7 mm, rounded, not undulate, spreading. *Stamens* with apex 3-4 mm below mouth of corolla tube, inserted 0.7-0.8 of the length of the corolla tube, (at 15-22 mm from the base); anthers sessile, narrowly triangular, ca 4 x as long as wide, 5 x 1.2-1.3 mm, apex acuminate, sterile for 0.4 mm, sagittate at the base. *Pistil* glabrous, 16-22 mm long; ovary ovoid, 2 x 1.5 x 1.5 mm, of two separate carpels, connate at the base by a disk-like ring 1 mm high, gradually narrowed into the style; style filiform, 13.5-18.5 mm long; pistil head 1.5 mm high, composed of a basal veil-like ring 0.2 x 1 mm, a stipitate 5-lobed depressed globe 0.3 x 0.7 mm and a cleft stigmoid apical part 0.5 x 0.2 mm. Ovules approximately 50 in each carpel. *Fruit* of 2 separate mericarps; mericarps green, obliquely ellipsoid, with acute partly fused triangular prickles, 25-35 x 15-20 x 15-20 mm, often recurved, rounded or nearly so at the apex, without ridges, approximately 20-30-seeded; wall thin; aril very small, only partly covering the hilar side of the seed. *Seed* dark brown, obliquely ellipsoid, 9-10 x 3-3.5 x 2.5 mm, with clear longitudinal grooves, with honeycomb-structure; embryo comparatively small, 4-4.2 mm long; cotyledons oblong, 2 x 0.6-0.7 mm, obtuse or rounded at the apex, truncate at the base; rootlet 2-2.2 x 0.4 mm.

Distribution: Brazil (Amazonas, near Manaus).

Ecology: Non-inundated forest understorey. Alt. low. Flowering and fruiting probably from February to August.

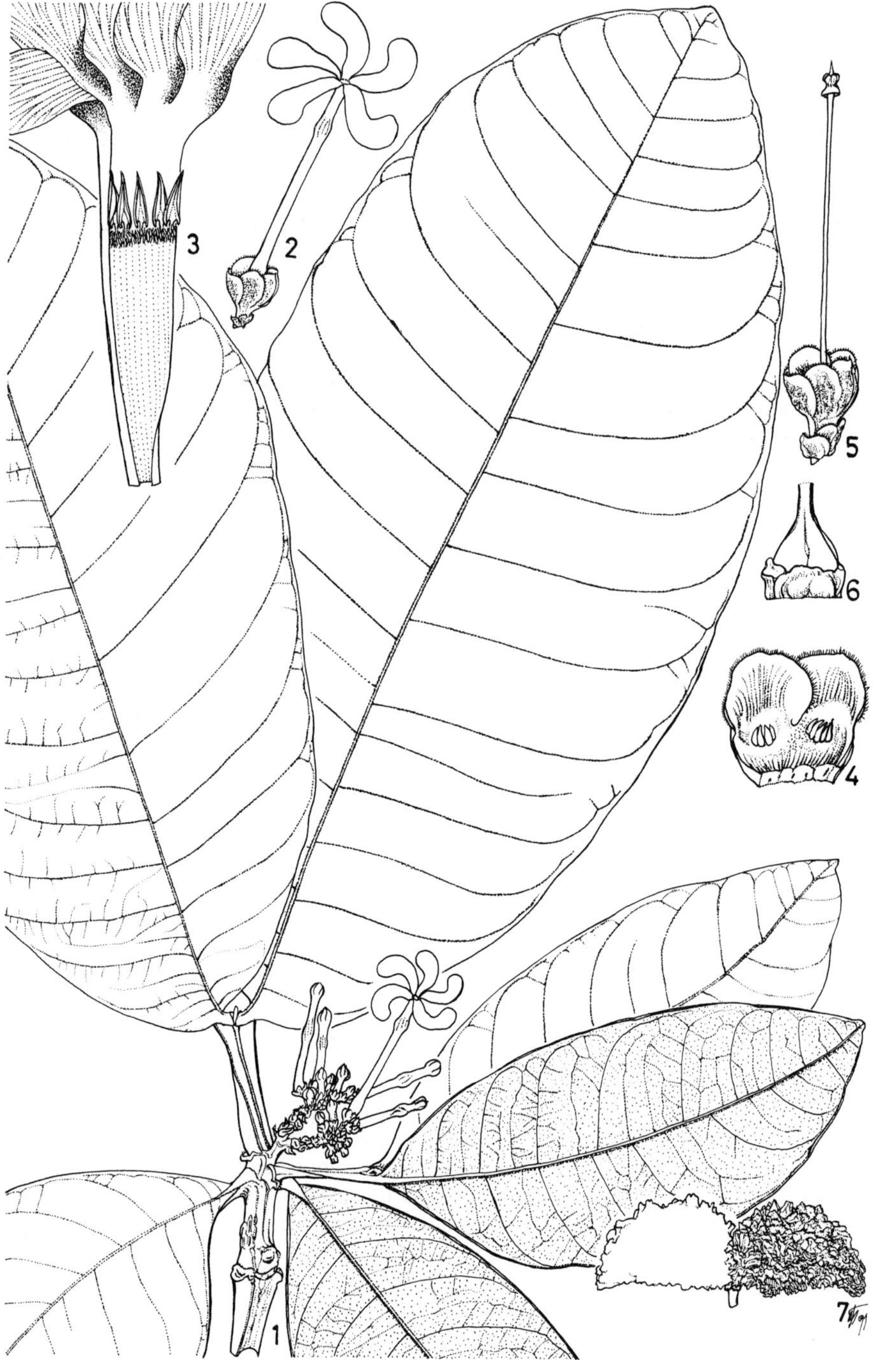

Fig. 86. *Tabernaemontana muricata.* **1,** habit (x 2/3); **2,** flower (x 1); **3,** opened corolla (x 2); **4,** sepals inside (x 4); **5,** calyx with pistil (x 2); **6,** ovary (x 6); **7,** fruit (x 2/3). 1 from Rodrigues et al. 2769; 2-6 from Ducke 186; 7 from Ducke 984.

Geographical selection of the 26 specimens examined:

BRAZIL. Amazonas: Lago de Jaradá, Rio Cueiras, *Mori* et al. 20345 (WAG); near Rio Uatumã, *Cid* et al. 934 (NY); Rio Cueiras, just below mouth of Rio Bracinho, *Prance* et al. 14192 (K, NY, US, Z); Barra do Rio Negro (= Manaus) *Spruce* 1470 (neotype, see above); ibid., *Ule* 5181 (G, HBG, K, L); ibid., *Prance* et al. 18768 (K, NY, U, US, Z); km 26 Manaus-Itacoatiara Highway, *Prance* et al. 2195 (GH, K, NY, S, U, US, Z); km 80 of same highway, Rio Prêta da Eva, *Prance* et al. 23382 (S, Z); Manaus, Cachoeira Grande, *Ducke* 186 (A, K, MG, MO, NY, S, US), 984 (K, MG, MO, NY, US); Rio Tarumã, *Miranda* 235 (INPA); Cachoeira Alta do Tarumã, *W.Rodrigues* et al. 2769 (ECON, INPA); Manaus, BR 17, km 8, *W. Rodrigues & Chagas* 2124 (F, INPA).

CULT. Brazil, Ponta Negra Road, Zoological Garden of COSAC, *Albuquerque* INPA 50111 (INPA).

Notes. Three species are easily confused when flowering. They differ as follows:

1. Branchlets triangular in cross section when fresh 2
   Branchlets terete in cross section when fresh .... **T. coriacea**
2. Fruit muricate; corolla in the mature bud forming a comparatively small broadly ovoid or subglobose head 0.12-0.16 of the bud length, 1-1.2 x as wide as the upper part of the tube, puberulous outside on the base of the lobes and sometimes at the apex of the tube .... **T. muricata**
   Fruit smooth; corolla in the mature bud with a broadly ovoid head 0.25-0.35 of the bud length and twice as wide as the tube, puberulous outside on the base of the lobes and on the tube from 5 mm below the mouth .... **T. angulata**

87. Tabernaemontana ochroleuca Urb., Symb. Antill. 6: 34 (1909). Type: Jamaica. lower slopes of Dolphin Head, *Harris* 10297 (holotype BM; isotypes C, F, GH, K, NY, P, US; phot. of US sheet in WAG). Fig. 87; p. 348; map 46, p. 349

Shrub or small tree, 2-10 m high. Branches pale grey-brown, lenticellate; branchlets terete, glabrous. *Leaves* petiolate; petiole glabrous, 2-12 mm long; (ocreae widened into intrapetiolar stipules); blade thinly papery when dried, elliptic or narrowly elliptic, 2-4 x as long as wide, 2-14 x 0.6-4.7 cm, acuminate at the apex, cuneate at the base and even decurrent into the petiole, glabrous on both sides, dotted beneath, with 5-10 pairs of rather straight secondary veins forming an angle of 50-60° with the costa, inconspicuous beneath, tertiary venation inconspicuous. *Inflorescence* pedunculate, 3-7 x 3-7 cm, 5-15-flowered, lax. Peduncle slender, 10-25 mm long, glabrous; pedicels thin, 8-14 mm long, glabrous. Bracts and bracteoles sepal-like 0.2-0.3 x as long, deciduous. *Flowers* open during the day. *Sepals* pale green (?), free, erect, ovate or broadly ovate, 1-1.3 x as long as wide, 1.5-2 x 1.5 mm, acute or obtuse, glabrous outside, not ciliate, glabrous inside and with 3-8 colleters in 2 rows in the middle at the base; colleters 0.3-0.5 x 0.1-0.3 mm. *Corolla* pale yellow, or tube yellow and lobes white, 12-13 mm long in the mature bud and forming a comparatively small ovoid head 0.23-0.25 of the bud length (3 x 2.5 mm) acute at the apex, glabrous on both sides; tube 5.5-8 x as long as the calyx, 1.15-1.5 x as long as the lobes, 11-12 mm long, almost cylindrical, rather slender, 2-2.5 mm wide at the base, narrowed below and above the stamens to 1.2-1.8 mm wide, widened around them to 2-2.3 mm wide, not twisted; lobes obliquely oblong, slightly falcate, 0.67-0.86 x as long as the tube, 2.4-3.3 x as long as wide, 9-9.5 x 3-3.5 mm, rounded, with an acute lobe at the left side (seen from outside), spreading. *Stamens* barely included, inserted 0.7-0.8 of the length of the the corolla tube, (at 8.5 mm from the base); anthers sessile, with tails 0.5 mm below insertion, narrowly triangular, 3 x 1 mm, apex acuminate, sterile for 0.3 mm, sagittate at the base, with tails curved towards each

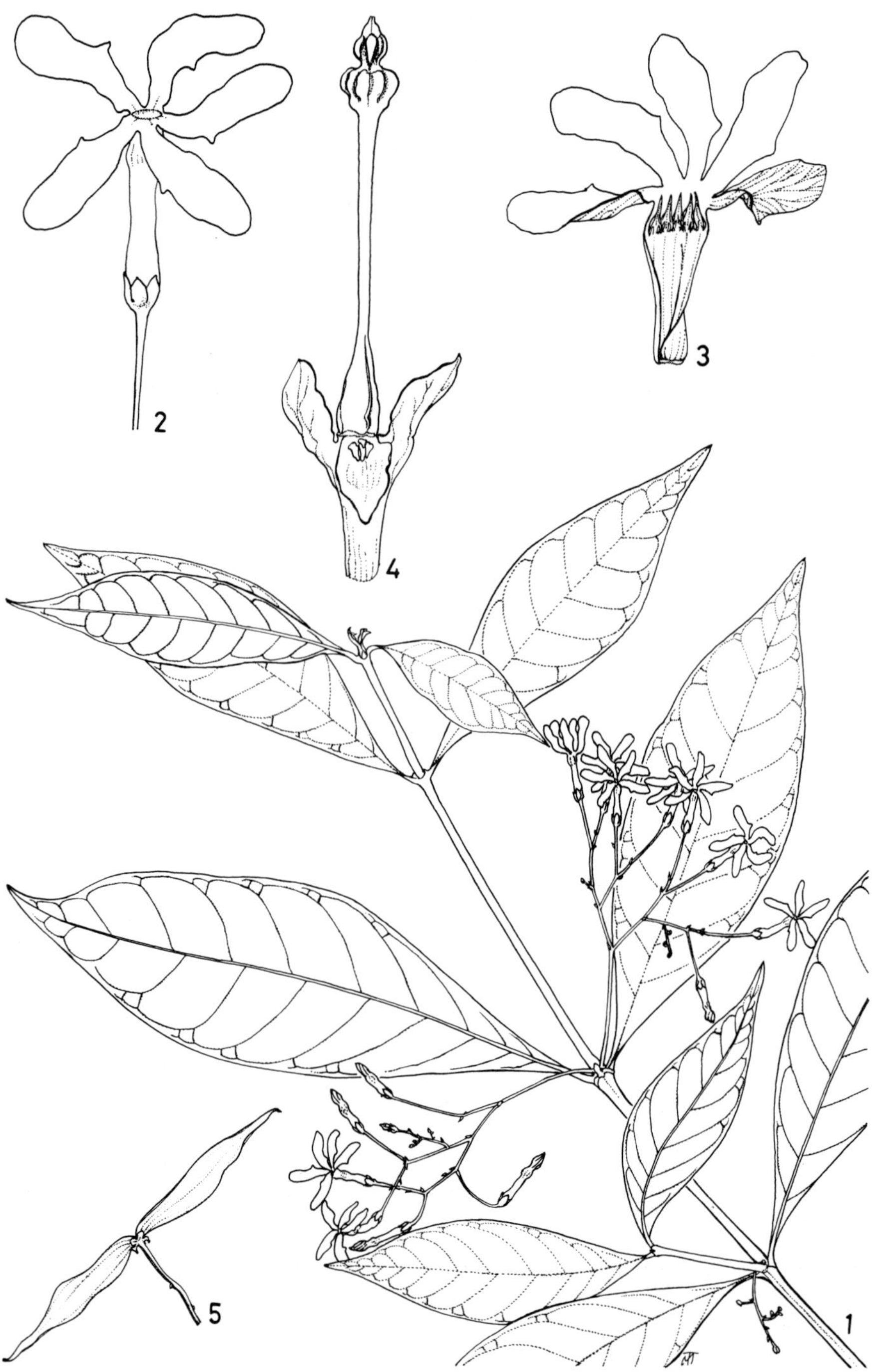

Fig. 87. *Tabernaemontana ochroleuca.* **1,** habit (x 2/3); **2,** flower (x 2); **3,** opened corolla (x 2); **4,** calyx with pistil (x 6); **5,** immature fruit (x 2/3). 1 and 5 from Harris 10297; 2-4 from Webster & K.A.Wilson 5081.

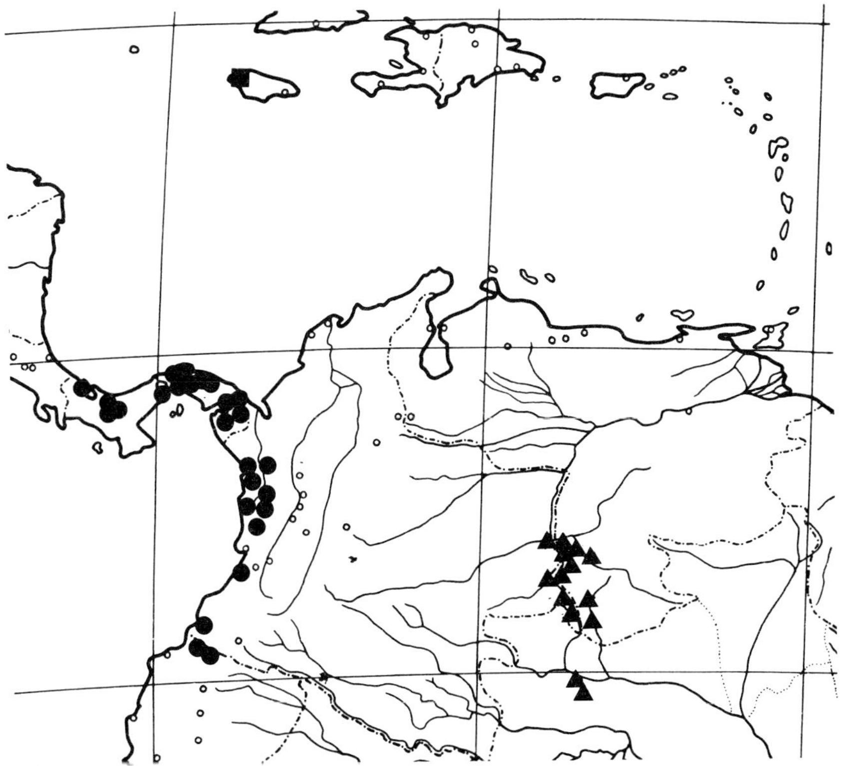

Map 46. ■. *Tabernaemontana ochroleuca, T. ovalifolia,* ▲ *T. palustris,* ●, *T. panamensis.*

other. *Pistil* glabrous, 8.6-9.3 mm long; ovary ovoid, 2-2.5 x 1-1.2 x 1-1.2 mm; style slender, 4.6-4.8 mm long; pistil head comparatively large, 2 mm high, composed of a basal ring 1 x 1.5 mm, consisting of 10 rounded lobes, a stipitate 5-lobed depressed globe 0.5 x 0.7 mm and a stigmoid apical part 0.3 x 0.3 mm. Ovules approximately 30 in each carpel. *Fruit* of 2 separate mericarps; mericarps only immature seen, pod-like, 40 x 7 x 5(?) mm, long-acuminate at the apex, few-seeded, smooth, with a faint ridge at each side.

DISTRIBUTION: Jamaica (Hanover Parish, Dolphin Head).

ECOLOGY: Forest on hills. Alt. 300-500 m. Probably flowering and fruiting the whole year around. Flowering from November to March as far as is known.

Other specimens examined:

JAMAICA. Hanover: Dolphin Head, *Adams* 8591 (BM, M, MO); Britton 2293 (NY), 2311 (NY); *Proctor* 11307 (IJ); *Purdie* anno 1844 (K); *Webster & K.A. Wilson* 5081 (A, BM, IJ, US).

88. Tabernaemontana oppositifolia (Spreng.) Urb., Symb. Antill. 4: 493 (1910). – Type: Puerto Rico, sin. loc., *Berter* Apr. 1820 (holotype G-DC; phot. F, GH, MO, US). Fig. 88, p. 351; map 47, p. 352

Basionym and homotypic synonym:

Rauvolfia oppositifolia Spreng., Neue Entd. 3: 33 (1822). *T. berteri* A. DC., Prod. 8: 367 (1844), as *berterii*.

Shrub or small tree, 2-7.5 m high. Trunk up to at least 13 cm in diameter; bark grey, smooth or slightly fissured. Branches pale brown, lenticellate; branchlets terete, glabrous. *Leaves* of a pair equal or unequal, petiolate; petiole glabrous, 5-30 mm long; (ocreae widened into intrapetiolar stipules); blade subcoriaceous when dried, elliptic or obovate, 2-3 x as long as wide, 4-21 x 1.5-8.5 cm, subacute, apiculate or acuminate at the apex, with a blunt acumen, cuneate at the base, entire, with a mostly slightly revolute margin, glabrous on both sides, with scattered black dots beneath, with 8-14 pairs of upcurved secondary veins forming an angle of 70-90° with the costa; tertiary venation reticulate and often impressed above, not conspicuous beneath. *Inflorescence* pedunculate, 5-9 x 4-9 cm, 2-20-flowered, lax. Peduncle slender, glabrous, 6-30 mm long; pedicels glabrous, 10-16 mm long. Bracts sepal-like, about 0.3 x as long as the sepals. *Flowers* fragrant, open during the day. *Sepals* pale green (?), connate at the extreme base, erect, ovate or broadly ovate 1-2 x as long as wide, 3-4 x 1.5-3 mm, obtuse or rounded, glabrous outside, not ciliate, glabrous inside and with 4-8 colleters in 1 row in the middle at the base; colleters 0.3-0.8 x 0.2-0.3 mm. *Corolla* white or cream, 15-20 mm long in the mature bud and forming a comparatively large ovoid head 0.35-0.4 of the bud length (6-8 x 3-5 mm) with an obtuse or subacute apex, glabrous on both sides; tube 3.2-5.3 x as long as the calyx, 0.6-1 x as long as the lobes, 12-16 mm long, almost cylindrical, 2.5-3 mm wide above the base, narrowed at the insertion of the stamens to 1.5-2 mm wide, again widened around the anthers to 2.5-3 mm wide, not twisted, with 5 longitudinally bilobed thickenings in the throat not reaching the mouth; lobes obliquely elliptic, mostly slightly falcate, 1-1.5 x as long as the tube, 2.1-2.8 x as long as wide, 13-22 x 5.5-8 mm, rounded, with a short acute lateral lobe, slightly auriculate at the left side of the base, spreading. *Stamens* with apex 0.5-2 mm below mouth of corolla tube, inserted 0.6-0.7 of the length of the corolla tube (at 7.5-10 mm from the base); anthers sessile, narrowly triangular, 2.7-4 x as long as wide, 3-4 x 0.8-1.5 mm, apex acuminate, sterile for 0.3-0.5 mm, sagittate at the base, glabrous. *Pistil* glabrous, 9-13 mm long, with apex about halfway along anthers; ovary ovoid, or narrowly so, 3-4.5 x 1.5 x 1.5 mm, of two separate carpels, gradually narrowed into the style; style 4-7.5 mm long; pistil head 1-1.5 mm high, composed of an undulate basal ring like a veil 0.2 x 1-1.5 mm, a stipitate apically 5-lobed depressed globe 0.5-0.7 x 0.6-1.1 mm, and a bilobed stigmoid apex, 0.2 x 0.2 mm. Ovules approximately 100 in each carpel. *Fruit* of 2 separate mericarps; mericarps yellow, not dotted, obliquely pod-like, 45-50 x 10 x 10 mm, acuminate with 2 ridges (one at each side of the line of dehiscence); approximately 20-seeded; wall 1 mm thick in dried fruits; aril enveloping the seed. *Seed* medium or dark brown, obliquely ellipsoid, 6 x 3 x 3 mm.

DISTRIBUTION: Endemic to Puerto Rico.

ECOLOGY: Limestone hill forest. 0-800 m. Flowering throughout the year. Fruiting season unknown.

Geographical selection of the approximately 35 specimens examined:

PUERTO RICO. Serpentine Hill, N of San German, *N.L. & E.G. Britton* 9921 (BH, MO, NY); near Mayaguez, *E.G. Britton & Marble* 670 (NY, US); Maricao For., *Wagner* 1843 (A, U); ibid., *Little* 21701 (BM, NY, US); ibid., *Proctor & Padrón* 40475 (IJ); near Yabacoa, *Wagner* 1312 (A, U); Adjuntas, *Sintenis* 4434 (GOET, US); near Toa Baja, *Little* 21687 (BM, NY, US); near Bayamon, Bro. *Alain Liogier* 10315 (F, GH, NY, US); Sierra de Loquillo, *Eggers* 1069 (BR, FI, G, M, P, TL, UPS, WU); ibid., *Sintenis* 1632 (BM, F, FI, G, GH, HBG, K, L, LD, M, MO, NY, P, S, US, W); ibid., *Howard* 16569 (A); El Yunque, *Howard & Nevling* 15452 (A, U); Yabucoa, *Sintenis* 5073 (G, GH, LD, M, NY, S); sin. loc., *Berter* Apr. 1820 (G-DC; phot. F, GH, MO, US, type).

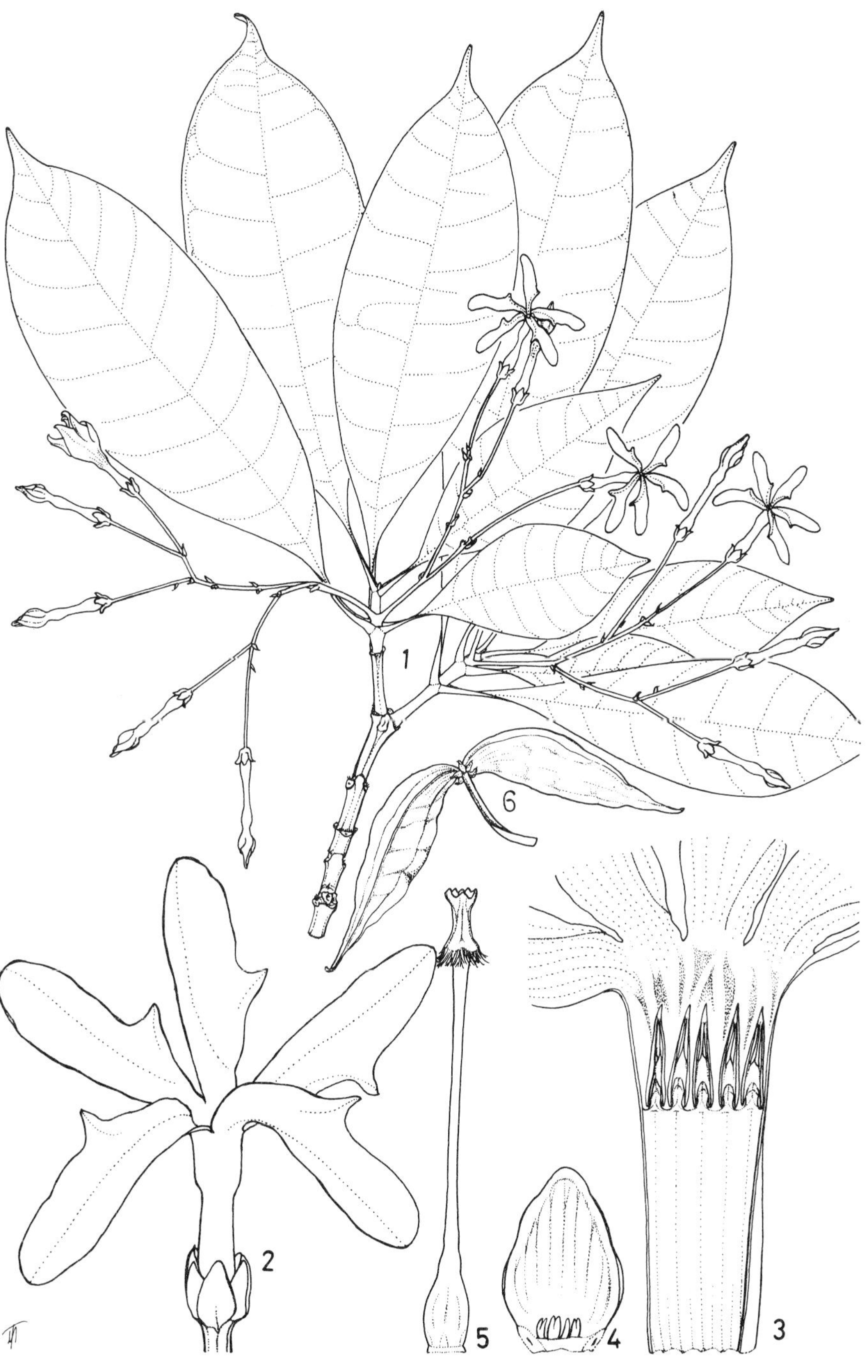

Fig. 88. *Tabernaemontana oppositifolia.* **1,** habit (x 2/3); **2,** flower (x 2); **3,** opened corolla (x 4); **4,** sepal inside (x 8); **5,** pistil (x 6); **6,** fruit (x 2/3). 1-5 from Howard 16569; 6 from Britton et al. 9921.

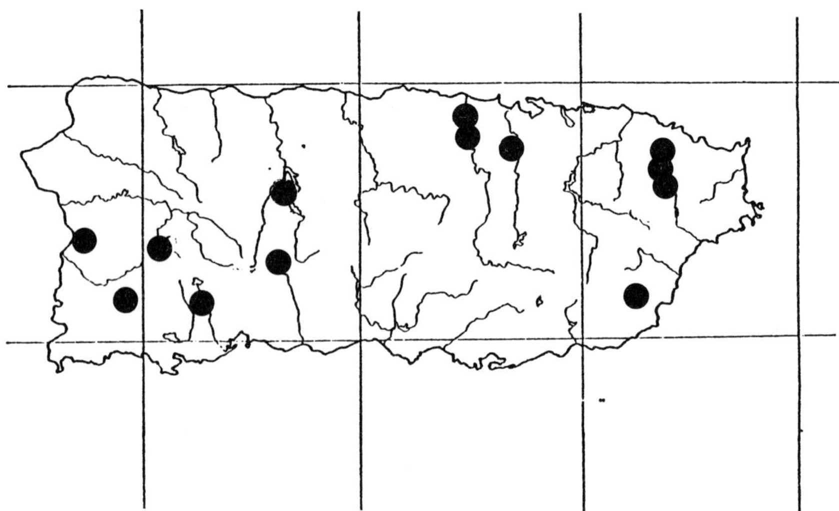

Map 47. *Tabernaemontana oppositifolia.*

CULT. Puerto Rico, Rio Piedras Exp. Stat., *Wagner* 651 (BM, S, U).

Notes. *T. oppositifolia* is closely allied to *T. citrifolia*, which is not known from Puerto Rico, but from islands nearby. Flower and fruit are quite similar. The two species can be distinguished as follows:

Corolla tube 8.5-12.5 mm long; stamens with apex (0-)0.5-1 mm above mouth of corolla tube; thickenings in corolla throat visible in the mouth ....**T. citrifolia**

Corolla tube 12-16 mm long; stamens with apex 0.5-2 mm below mouth of corolla tube; thickenings in corolla throat not visible in the mouth ......**T. oppositifolia**

89. Tabernaemontana ovalifolia Urb., Symb. Antill. 5: 462 (1908), non Glaziou (1910); Stearn in Journ. Arn. Arb. 52: 617 (1971). – Type: Jamaica, Hanover, between Askenish and Dolphin Head, *Harris* 9239 (holotype BM; isotypes: A, F, K, US; phot. of US sheet in WAG). Fig. 89, p. 353; map 46, p. 349

Small tree, 7 m high. Branches pale to dark brown, not lenticellate; branchlets terete, glabrous. *Leaves* long-petiolate; petiole 10-30 mm long; (ocreae widened into intrapetiolar stipules); blade subcoriaceous when dried, elliptic or narrowly elliptic, 1.7-3 x as long as wide, 4-21 x 2-8 cm, shortly acuminate to obtuse at the apex, cuneate or almost rounded at the base, entire, with a revolute margin, glabrous on both sides, with or without scattered black dots, with 5-14 pairs of rather straight secondary veins forming an angle of 60-70° with the costa; tertiary venation reticulate. *Inflorescence* pedunculate, 5-9 x 5-10 cm, 5-15-flowered, lax. Peduncle rather robust, 12-45 mm long, glabrous; pedicels glabrous, 5-15 mm long. Bracts sepal- or scale-like and 0.2-0.5 as long as the sepals. *Flowers* fragrant, open during the day. *Sepals* pale green (?), connate at the base for 0.5 mm, erect, ovate, 1.2-1.3 x as long as wide, 3.5-4 x 3 mm, rounded, glabrous outside, not ciliate, glabrous inside and with 5-6 colleters in 1 row in the middle at the base; colleters about 0.4 x 0.2 mm. *Corolla* white, 13-17 mm long in the mature bud and forming a comparatively large ovoid head 0.4-0.53 of the bud length (5-8 x 3-5 mm) with an acute or obtuse apex, glabrous on both sides; tube 3-4 x as long as the calyx, 0.6-1 x as long as the lobes, 12-14 mm long, almost cylindrical, 3-4 mm wide above the base, slightly widened around the anthers

Fig. 89. *Tabernaemontana ovalifolia.* **1,** habit (x 2/3); **2,** sepal inside (x 4); **3,** flower (x 2); **4,** opened corolla (x 6); **5,** pistil (x 8). 1 from Harris 9239; 2-5 from Harris 10276.

to 2.5-3 mm wide, again narrowed at the throat to 2-3 mm wide, not twisted; lobes obliquely oblong, 1-1.6 x as long as the tube, 1.9-3.2 x as long as wide, 11-17 x 4-8 mm, obtuse, with an acute lateral lobe about halfway along the length and auriculate at the base at the left side, not undulate, with recurved margin, spreading. *Stamens* with apex 1.5-2 mm below mouth of corolla tube, inserted 0.5-0.53 of the length of the corolla tube, (at 6-7.5 mm from the base); anthers sessile, with tails 0.5 mm below insertion, narrowly triangular, 5 x 1.3 mm, apex acuminate, sterile for 0.4 mm, sagittate at the base with straight tails, glabrous. *Pistil* glabrous, 7.5-9 mm long, with apex almost halfway along anthers; ovary ovoid, 3 x 2 x 2 mm; style 2.5-4 mm long; pistil head 2 mm high, composed of a basal ring 1 x 1.7 mm, consisting of 10 rounded suberect lobes, a stipitate 5-lobed depressed globe 0.5 x 1 mm and a stigmoid apex 0.2 x 0.2 mm. Ovules approximately 80 in each carpel. *Fruit* unknown.

DISTRIBUTION: Jamaica, only known from the Dolphin Head and its surroundings.
ECOLOGY: Wooded hill. Alt. 300-400 m. Flowering found in March and May.

Specimens examined:

JAMAICA: Hanover, Dolphin Head and vicinity, *Britton* 2300 (NY); between Askenish and Dolphin Head, *Harris* 9239 (A, BM, F, K, US, type), 10276 (BM, NY, US).

90. Tabernaemontana palustris Mgf. in Notizbl. Bot. Gart. Berlin 12: 297 (1935). – Type: Brazil, Amazonas, Rio Negro mouth, Curicuriari, *Ducke* 23955 (holotype B†; lectotype US, designated here; phot. WAG; isotype K).

Fig. 90. p. 356; map 46, p. 349

Homotypic synonyms:

Anacampta palustris (Mgf.) Mgf. in op. cit. 14: 163 (1938); *Bonafousia palustris* (Mgf.) Boiteau & Allorge in Bull. Soc. Bot. Fr. 130, Lettres 4-5: 340 (1983); Allorge in Mém. Mus. natn. Hist. Nat. IIB, 30: 90, pl. 38 (1985).

Heterotypic synonyms:

T. rimulosa Woods. in Bot. Mus. Leafl. Harvard Univ. 18: 179, 317 (1958), **syn. nov.** *B. rimulosa* (Woods.) Boiteau & Allorge, l.c.; Allorge in op. cit. 90, pl. 37. – Type: Colombia, Guainia, near San Felipe, *Schultes* et al. 17983 (holotype MO; isotypes A, BM, GH, NY, U, US; phot. of MO sheet NY, US; phot. of US sheet WAG).

T. tenuis Monachino in Mem. N.Y. Bot. Gard. 10, 4: 64, t. 44, H-O (1961). – Type: Venezuela, Amazonas, Río Atabapo, 15 km above Guarinuma, *Wurdack & Adderley* 42965 (holotype NY).

Shrub or small tree 0.30-4 m high. Branches 1-3 together with 1 inflorescence, pale or dark brown, with few lenticels, often thick and with bark more or less peeling off, often more or less ridged when dried; branchlets probably terete when fresh, neither ridged nor angular, glabrous. *Leaves* of a pair equal or subequal, petiolate; petiole glabrous, 3-15 mm long; (ocreae not widened into intrapetiolar stipules); blade coriaceous, also when fresh, variable in shape and size, elliptic or narrowly elliptic, 2-7 x as long as wide, 4-21 x 0.8-9.5 cm, obtuse or rounded at the apex, cuneate or rounded at the base or decurrent into the petiole, entire or undulate, with a revolute margin, glabrous and with or without scattered or more or less regularly arranged black dots on both sides, with 12-15 pairs of rather straight conspicuous or obscure sencondary veins forming an angle of 70-90° with the costa; tertiary venation conspicuous and reticulate or inconspicuous. *Inflorescence* shortly pedunculate, 2-4 x 2-3 cm, 3-17-flowered, dense, racemose or racemoid. Peduncle glabrous, 1-2 mm long, robust; pedicels glabrous, 3-7 mm long. Bracts 0.2-0.5 x as long as the sepals and more or less similarly shaped, obtuse or acute. *Flowers* open during the day. *Sepals* pale green, subequal, connate at the base for 0.5 mm, erect, ovate, 1.4-2 x as long as wide, 2-3.5 x 1-2.5 mm, obtuse or rounded, entire, glabrous or sometimes minutely puberulous outside, not ciliate, glabrous inside and with 3-7 colleters in 1 row in the middle of the base; colleters 0.5-0.6 x 0.1-0.2 mm. *Corolla* white, often pink-flushed on the tube, with yellow throat, 14-23 mm long in the mature bud and forming a comparatively rather large ovoid head 0.36-0.38 of the bud length (5-7 x 3-5 mm) with an obtuse apex, glabrous or on the part of the lobes not covered in bud and the apex of the tube puberulous outside, not ciliate, with an interrupted pubescent belt 4-5 mm wide inside from 2 mm below to 2-3 mm above the insertion of the stamens and (if hairy outside) puberulous at the base of the lobes; tube 4-7 x as long as the calyx, 0.7-1.1 x as long as the lobes, 10-18 mm long, almost cylindrical, slightly widened above the base and 2-3.5 mm wide, narrowed below the insertion of the stamens to 1.5-2.5 mm wide, widened again around the anthers to 2-3 mm wide, not twisted; lobes obliquely obovate, 0.9-1.5 x as long as the tube, 1.3-2 x as long as wide, 14-17 x 6-12 mm, rounded, undulate, spreading. *Stamens* with apex 0-4 mm below mouth of corolla tube, inserted 0.5-0.6 of the length of the corolla tube (at 6-10 mm from the base); anthers sessile, narrowly triangular, 3-5 x as long as wide, 4-5 x 1-1.2 mm, apex acuminate, sterile for 0.3 mm, sagittate at the base, glabrous. *Pistil* glabrous, 8-12 mm long, with apex about halfway along anthers; ovary ovoid, 2-3 x 1-2 x 1-2 mm, gradually narrowed into the style, of 2 carpels connate basally by a 0.5-1 mm high annular disk-like thickening; style filiform, 5-8 mm long; pistil head composed of an entire or undulate ring 0.3-0.4 x 0.8-1.2 mm, a broadly obovoid apically 5-lobed central part 0.5-0.8 x 0.6-0.8 mm and a stigmoid bilobed apex about 0.1 x 0.1 mm. Ovules approximately 50 in each carpel. *Fruit* of 2 separate mericarps; mericarps green (?), obliquely ellipsoid or ovoid, 30 x 15 mm, acute or acuminate at the apex, often slightly recurved, not ridged, smooth, about 3-10-seeded (?); wall about 2-3 mm thick (?). *Seed* unknown.

DISTRIBUTION: Southern Venezuela and adjacent Colombia and Brazil.
ECOLOGY: Periodically inundated riverine forest. Alt. 100-200 m. Flowering and fruiting throughout the year.

Geographical selection of the 27 specimens examined:
COLOMBIA. Guiania: near Coitara, ca 7 km S of San Fernando de Atabapo (Venezuela), *Davidse* 16828 (MO); Río Guainia, N of Puerto Colombia, *Maguire* et al. 41841 (NY); near San Felipe, *A. Gentry* & *Stein* 46453 (MO); ibid., *Schultes* et al. 17983 (A, BM, GH, MO, NY, U, US; phot. of MO sheet NY, US; phot. of US sheet WAG, type of *T. rimulosa*).
VENEZUELA. Amazonas: Río Atabapo, above San Fernando de Atabapo, *A. Gentry* 10970 (MO, VEN); ca 15 km SW of S.F., *Stergios* et al 11653 (MO); same r., 15 km above Guarinuma, *Wurdack & Adderley* 42965 (NY, type of *T. tenuis*); same r., caño Viejta, *Melgueiro* 127 (MO); Cerro Yapacana, *Holt & Blake* 760 (US); Cucurital de Yagua, *Davidse* et al. 17350 (MO); near Patacame, *O. Huber* 2631 (MO, Z); Caño Temi, just above Yavita, *Wurdack & Adderley* 42936 (NY, paratype of *T. tenuis*); 0.5 km E of Maroa, *Maguire* et al. 41720 (F, G, GH, K, NY, S, U, US, W); E of Maroa, *Maguire* et al 41735 (NY, UC); 8 km above mouth of Río Siapa, *Maguire & Wurdack* 37612 (NY); Dept. Río Negro, lower part of Río Baria, *Davidse* 27609 (MO).
BRAZIL. Amazonas: between Manaus and São Gabriel, along Rio Marié, Marauná, *Alencar* et al. 469 (NY, US, WAG); Rio Curicuriari, *Ducke* 23955 (K, US; phot. US sheet WAG, type); ibid., *Poole* 1999 (NY).

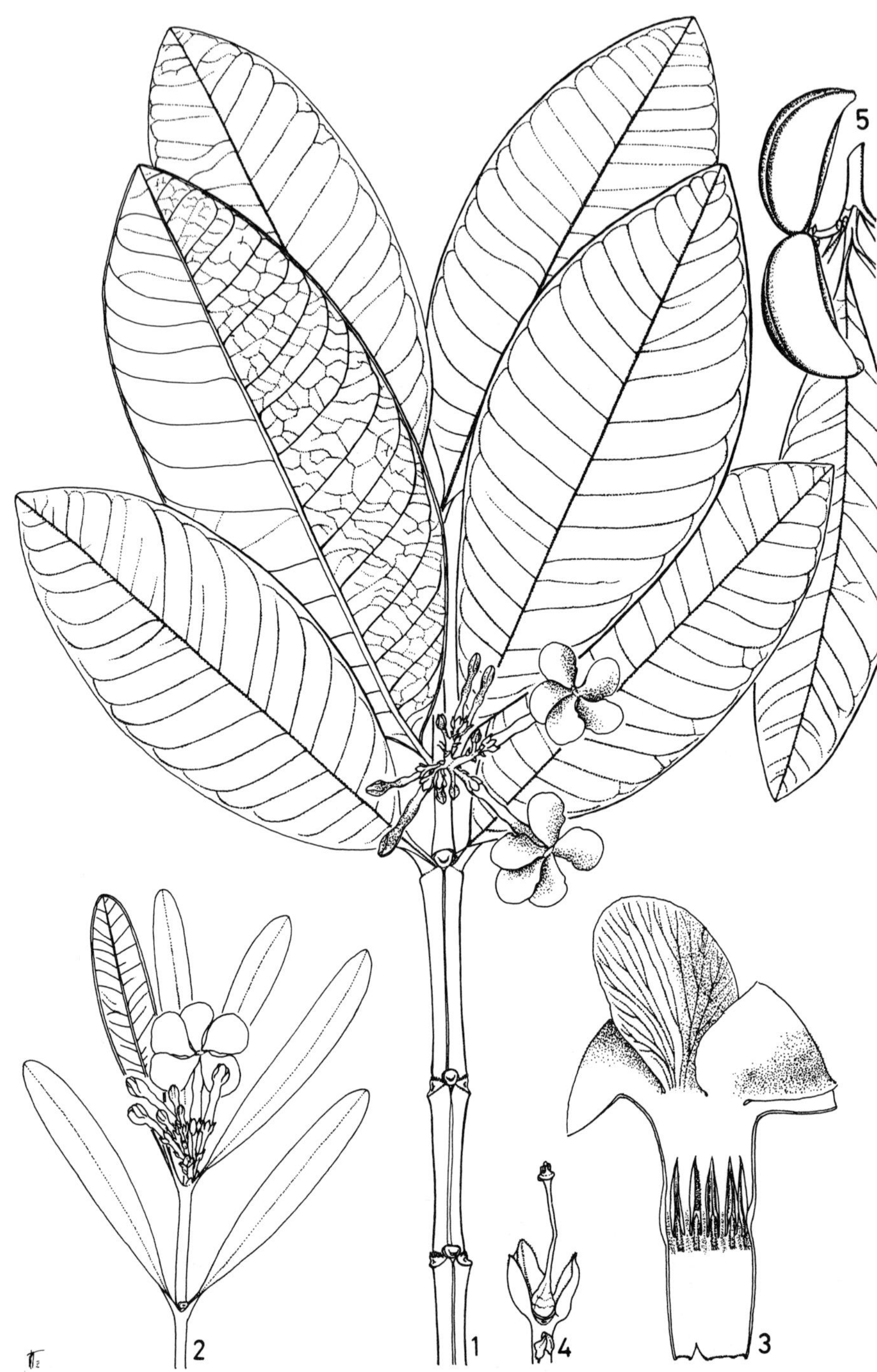

Fig. 90. *Tabernaemontana palustris.* **1-2,** habit (x 2/3); **3,** opened corolla (x 2); **4,** calyx with pistil (x 2); **5,** fruit on branchlet (x 2/3). 1 and 3-5 from Maguire et al. 41720; 2 from Wurdack & Adderley 42965.

Notes. *Tabernaemontana rimulosa* and *T. palustris* are distinguished by Allorge (1985) as follows:

Branches thick, with bark peeling off. Corolla lobes pubescent, shorter than tube .... *T. rimulosa*

Branches slender, with persistent bark. Corolla lobes glabrous, as long as tube .... *T. palustris*

Intermediates have been found and therefore *T. rimulosa* has not been maintained as a distinct species:

Slender branches with bark peeling off and corolla lobes glabrous and shorter than tube are in *A. Gentry* 10970 and *Melgueiro* 127.

Thick branches with bark peeling off and corolla lobes glabrous, shorter than tube are in *Maguire* et al. 41735 (UC) and lobes pubescent in NY sheet of the same number.

Slender branches with persistent bark and corolla lobes glabrous and shorter than tube are in *Stergios* et al. 11653.

Rather thick branches with persistent bark and corolla lobes glabrous outside and longer than tube have been observed in *Wurdack & Adderley* 42936.

91. Tabernaemontana panamensis (Mgf., Boiteau & Allorge) Leeuwenberg in Meded. Landbouwh. Wageningen 83, 7: 60 (1984). – Type: Panama, Coclé, N of El Valle, *A.Gentry* 6847 (holotype Z; isotypes F, MO). Fig. 91, p. 358; map 46, p. 349

Basionym:

Bonafousia panamemsis Mgf., Boiteau & Allorge in Ann. Miss. Bot. Gard. 68: 677, fig.1 (1981); Allorge in Mém. Mus. natn. Hist. Nat. IIB, 30: 114, pl. 50 (1985).

Shrub or small tree 1.50-7.50 m high. Trunk 2-5 cm in diameter or more. Branches pale grey-brown, lenticellate, branchlets terete, glabrous. *Leaves* shortly petiolate; petiole glabrous, 6-22 mm long; (ocreae slightly widened into intrapetiolar stipules); blade coriaceous when dried, narrowly to very narrowly elliptic, 2.8-5 x as long as wide, 12-45 x 4-19 cm, acuminate at the apex, cuneate at the base, glabrous on both sides, entire, with more or less regularly arranged slightly darker dots beneath, with 10-25 pairs of rather straight secondary veins forming an angle of 70-80° with the costa; tertiary venation reticulate, but not darker, prominent beneath. *Inflorescence* much shorter than the leaves, shortly pedunculate, 3-5 x 3-5 cm, 10-20-flowered, dense. Peduncle robust, glabrous, 3-15 mm long; pedicels glabrous, 3-10 mm long. Bracts sepal-like and approximately 0.25 x as long as them. *Flowers* open during the day. *Calyx* not subtented by bracteoles, persistent, even under the fruit; sepals green, connate at the extreme base, erect, subequal, oblong or ovate, 1.4-2.2 x as long as wide, 6-12(-15) 3.5-7.5 mm, rounded, glabrous outside, mostly ciliolate at the apex, glabrous inside and with 10-20 colleters in 2-3 dense rows at the base; colleters 0.3-1 x 0.1-0.2 mm. *Corolla* yellow or orange, often with a darker throat and a paler, sometimes white, tube, 17-30 mm long in the mature bud and forming a comparatively small broadly ovoid head 0.17-0.2 of the bud length (3-5 x 3-5 mm) with an obtuse apex, glabrous outside, pubescent inside in longitudinal stripes 3-4 mm long from 1.5-2 mm below to 1.5-2 mm above the insertion of the stamens; tube 1.7-2.5 x as long as the calyx, 1.2-1.9 x as long as the lobes, 15-30 mm long, almost cylindrical, 3-5.5 mm wide above the base, slightly narrowed to 2.5-4 mm wide below the insertion of the stamens, widened again to 3-4.5 mm wide around the anthers and above, twisted 0.4-0.5 turn around the anthers; lobes obliquely obovate or oblong, 0.55-0.8 x as long as the tube, 1.6-2.2 x as long as wide, 8-15 x 4-9 mm, rounded, not undulate, recurved. *Stamens* with apex 0-3.5 mm below mouth of corolla tube, insert-

Fig. 91. *Tabernaemontana panamensis.* **1,** habit (x 2/3); **2,** corolla (x 2); **3,** opened corolla with style and pistil head (x 4); **4,** calyx outside (x 2); **5,** calyx with ovary (x 2); **6,** fruit (x 2/3); **7,** open fruit (x 2/3). 1 from Hammel 4792; 2-5 from Knapp 4751; 6 from A.Gentry 5784; 7 from A.Gentry 4951.

ed 0.6-0.7 of the length of the corolla tube, (at 10-14 mm from the base); anthers sessile, narrowly triangular, 3.2-3.3 x as long as wide, 5-5.5 x 1.5-1.7 mm, apex acuminate and sterile for 0.3-0.7 mm, sagittate at the base, glabrous. *Pistil* glabrous, 12-16 mm long, with apex about halfway along anthers; ovary broadly ovoid, 2-2.5 x 2-2.5 x 1.5-2.5 mm, of 2 separate carpels, with an annular disk-like thickening at the base, 0.8-1 mm high, gradually narrowed into the style; style 8-12 mm long; pistil head 1.4-1.5 x 1.4-1.5 mm, composed of a veil-like ring 0.5-0.8 x 1.4-1.5 mm, a stipitate apically 5-lobed depressed globe 0.5 x 1-1.2 mm and a bilobed stigmoid apex 0.2 x 0.2 mm. Ovules approximately 100 in each carpel. *Fruit* of 2 separate mericarps; mericarps orange or yellow, green when immature, like small oranges, subglobose, 30-45 x 20-45 x 20-45 mm, not dotted, rugulose when dried, rounded or bluntly short-acuminate at the apex, with a faint ridge at each side, with approximately 20-80 seeds; wall 3-5 mm thick in dried fruits; aril orange or white, not enveloping the seed. *Seed* dark brown or black, obliquely polyhedral, 9-11 x 4-6 x 3-4.5 mm, with longitudinal grooves; embryo 8-10 mm long; cotyledons broadly ovate, 1.1-1.2 x as long as wide, 3.5-4.5 x 3-4 mm, rounded at the apex, cordate at the base; rootlet 1.3-1.7 x as long as the cotyledons, 5.3-6 x 0.8-1 mm.

DISTRIBUTION: Panama, Colombia and Ecuador.
ECOLOGY: Wet forest understorey; older leaves often covered with epiphyllous mosses. Alt. 200-1600 m. Flowering and fruiting throughout the year.

Geographical selection of the approximately 90 specimens examined:
PANAMA. Bocas del Toro: Almirante Region, *Cooper* 610 (F, NY, US). Veraguas: NW of Santa Fe, *Mori & Kallunki* 5348 (MO); near Río Colovebora, *Croat & Folsom* 34103 (MO, NY). Coclé: N of El Cope, *McPherson* 12422 (MO); N of El Valle, *A. Gentry* 6847 (F, MO, Z, type); 3 km E of El Valle, *Hammel* 4792 (MO, Z); Cerro Pilon, *Croat* 22919 (F, GH, MO, NA, NY); Cerro Gaital Caracoral, *Dwyer* 8862 (F, MO). Panamá: 16 km from Pan-American Highway, on the El Llano-Cartí Road, *Knapp* et al. 4751 (MO, WAG); same road, 8.5 km from highway, *Mori & Kallunki* 4553 (AAU, NY, U). Colón: Santa Rita Ridge, 6-8 km from Pan-American Highway, *A. Gentry* 6100 (MO, Z); Cerro Brewster, *McPherson* 7571 (MO, WAG); Cerro Campana, *A. Gentry* 4951 (MO), 5784 (MO, Z); Cerro Jefe, *Sytsma* et al. 2891 (MO, WAG); near La Eneida, *Luteyn* 1125 (F, MO, NY). San Blas: Cerro Habú, *Sytsma* et al. 2709 (MO, WAG); km 19 El Llano-Cartí Road, *de Nevers & Herrera* 5137 (MO, WAG); near Puerto Obaldía, *Pittier* 4354 (F, MO, US). Darién: near Upper Río Membrilla, *Duke* 10876 (MO); Cerro Pirre, *Folsom* 4512 (MO); Cerro Tacarcuna, *A. Gentry & Mori* 14054 (MO).
COLOMBIA. Chocó: ca 10 km E of Mecana, *A. Gentry & Juncosa* 41097 (MO, P); Alto de Buey, *A. Gentry & Forero* 7340 (COL, MO); ca 14 km NE of Quibdó, *A. Gentry & Rentería* 24502 (MO, P); Lower Río Baudó, *Fuchs* et al. 21892 (COL, US); between Condoto and Andagoya, *Idrobo* 1951 (COL); near Noanamá, *Forero* et al. 4509 (COL, MO, P). Antioquia: Las Orquídeas National Park, *Callejas* et al. 2753 (NY), 2945 (NY). Valle: Río Maya, upriver from Puerto Merizalde, *A. Gentry & Juncosa* 40677 (MO). Nariño: Mun. Tumaco, La Gayacana, *Romero-Castañeda* 2920 (COL, MO). Putumayo: near La Tagua, *Romero-Castañeda* 4213 (COL).
ECUADOR. Carchi: San Marcos, *Barford* 41632 (QCA); 25 km NW of El Chical, *Rubio* et al. 1026 (WAG); near Río Gualpi Chico, *Hoover* et al. 2519 (MO, QCA).

92. Tabernaemontana rupicola Benth. in Hooker, Journ. Bot. 3: 243 (1841). – Type: Brazil, Amazonas, Pedrera, Rio Negro, W of Manaus, *Robert Schomburgk* I 898 (holotype K; isotypes BM, CGE, E, F, FI, FI-W, G, L, NY, P, SING, TCD, US, W; phot. of G sheet GH, MO, NY, US). Fig. 92, p. 361; map 48, p. 363

Homotypic synonyms:

Bonafousia rupicola (Benth.) Miers, Apoc. S. Am. 52 (1878). *Anacampta rupicola* (Benth.) Mgf. in Pulle, fl. Surinam 4, 1: 452 (1937).

Heterotypic synonyms:

T. rupicola var. *poeppigii* Muell. Arg. in Martius, Fl. Bras. 6, 1: 74 (1860). – Type: Brazil, Amazonas, Lago Ega (=Tefé), *Poeppig* 2504 p.p. (holotype G; isotypes BM, F, G-DC, GH, GOET, MO, P, W; phot. of G sheet MO, NY, US; phot. of W sheet MO, NY, US).

T. rupicola var. *oblongifolia* Muell. Arg. in op. cit. 75. *B. rupicola* var. *oblongifolia* (Muell. Arg.) Allorge in Mém. Mus. natn. Hist. Nat. IIB, 30: 106 (1985). – Type: Brazil, Amazonas, Rio Negro, sin. loc., *Riedel* s.n. (holotype P; isotype LE).

T. rupicola var. sprucei Muell. Arg, l.c. *Bonafousia polyneura* Miers, op. cit. 53. – Type: Brazil, Amazonas, near Barra do Rio Negro (= Manaus), *Spruce* 1758 (lectotype G, designated here; isolectotypes AWH, BM, BR, CGE, K, P, TCD, W). The BM sheet is the holotype of *B. polyneura*.

B. rariflora Miers, l.c. – Type: Brazil, Amazonas, Barra do Rio Negro (= Manaus) *Spruce* 1005 (holotype BM; isotypes CGE, FI-W, K, M, TCD, W; phot. of M sheet GH, MO, paratype of *T. r. var. sprucei*).

T. versicolor Woods. in Bull. Torr. Bot. Cl. 75: 559 1948), **syn. nov.**, non Muell.Arg. ex Miers, op. cit. 61. *B. rupicola* var. *versicolor* (Woods.) Allorge in Mém. Mus. natn. Hist. Nat. IIB, 30: 104 (1985). – Type: Guyana, Potaro R., below Tukeit, *Maguire & Fanshawe* 23482 (holotype NY; isotypes A, K, MO).

Shrub or small tree, 0.50-8 m high. Branches pale brown, lenticellate; branchlets terete (?) when fresh, glabrous. *Leaves* of a pair equal or unequal (larger up to 2 x as long as other and similarly shaped), sessile or with petiole up to 2 mm long; petiole glabrous; (ocreae not widened into intrapetiolar stipules); blade coriaceous or subcoriaceous when dried, elliptic, ovate or narrowly so, (1.8-)2-5 x as long as wide, 2-15 x 1-7 cm, acuminate at the apex, cuneate, rounded or cordate at the base, with base often unequal-sided, with often revolute margin, sometimes bullate, glabrous and with scattered black dots on both sides, with 15-40 pairs of straight secondary veins forming an angle of 70-90° with the costa, anastomosing into submarginal veins; tertiary venation conspicuous and reticulate or inconspicuous. *Inflorescence* shortly pedunculate, 2-5 x 2-4 cm, 1-11-flowered, rather dense. Peduncle glabrous or puberulous, 2-12 mm long; pedicels glabrous or puberulous, slender, 3-20 mm long. Bracts sepal-like, 0.3-0.7 x as long as them. *Flowers* fragrant, open during the day. *Sepals* pale green or sometimes pink, subequal, connate at the base for 0.3-0.7 mm, erect, ovate or broadly ovate, 1-1.6 x as long as wide, 1.5-3.5 x 1-2.5 mm, rounded or obtuse, glabrous or pubescent outside, mostly ciliate, glabrous inside and with 2-8 colleters in 1 row just above the middle of the base; colleters 0.3-0.7 x 0.1-0.2 mm. *Corolla* white, pink or more frequently with pink tube and white limb, 10-14 mm long in the mature bud and forming a comparatively large ovoid or subglobose head 0.25-0.5 of the bud length (3-6 x 3-4.5 mm) with a rounded or obtuse apex, glabrous or on the lobes (sparsely) puberulous outside, often ciliate, with a pilose or pubescent interrupted belt 2-4 mm wide inside from 1-2.5 mm below to 1-2 mm above the insertion of the stamens and often minutely puberulous on the base of the lobes; tube 4-6.7 x as long as the calyx, 0.44-1 x as long as the lobes, 8-14 mm long, almost cylindrical, widened above the base to 2-4 mm wide, mostly narrowed at the throat to 2-3 mm wide, not twisted; lobes obliquely obovate or elliptic, 1-2.3 x as long as the tube, 1.7-2.5 x as long as wide, 8-25 x 3-15 mm, rounded, undulate, spreading and recurved later. *Stamens* with apex 0-2 mm below mouth of corolla tube, inserted 0.54-0.7 of the

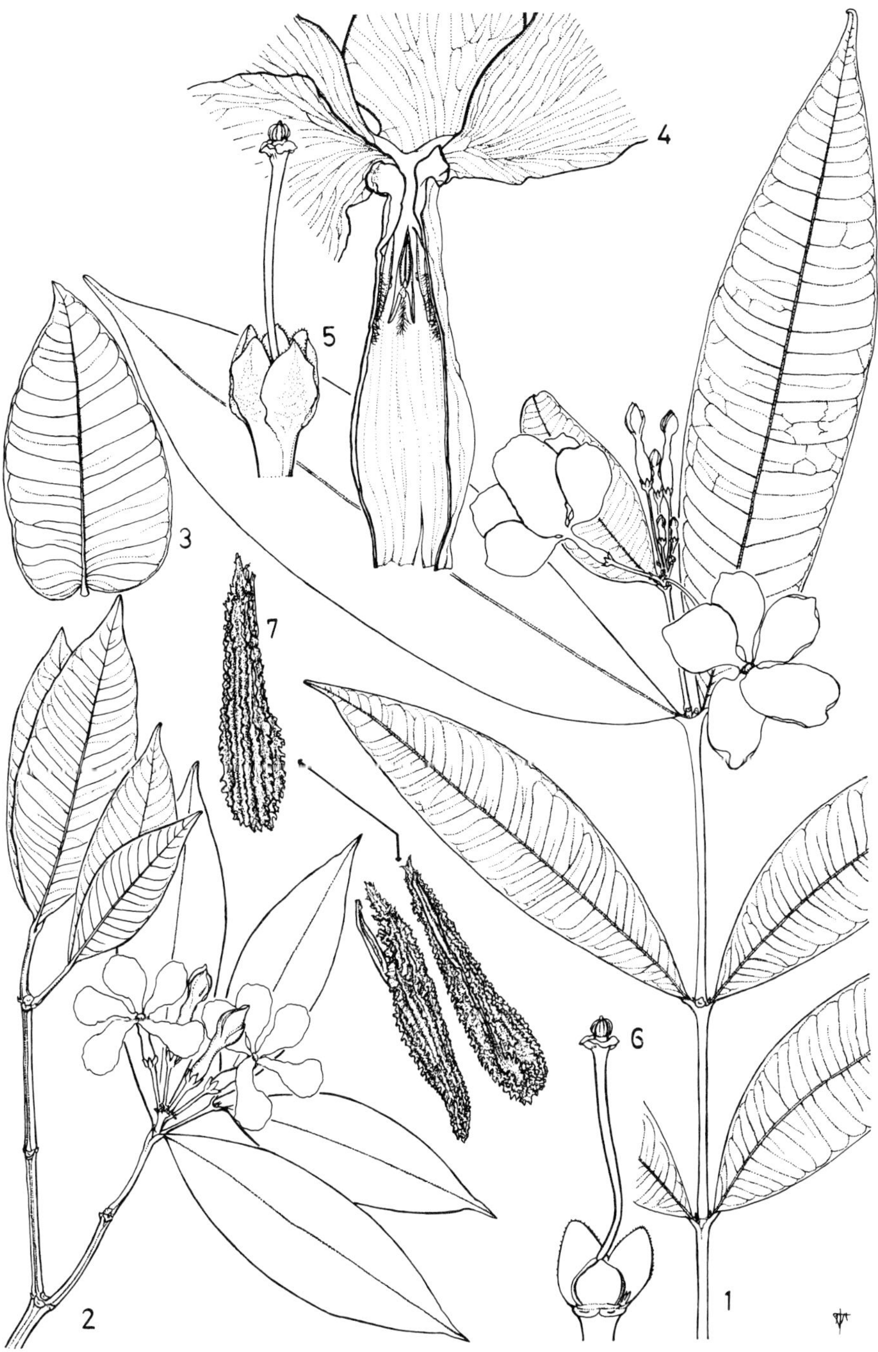

Fig. 92. *Tabernaemontana rupicola.* **1-2,** habit (x 2/3); **3,** leaf (x 2/3); **4,** opened corolla (x 4); **5-6,** calyx with pistil (x 4); **7,** seeds (x 4). 1 and 4-5 from Alencar 231; 2 and 6 from Alencar 125; 3 from Cid et al. 7673; 7 from Zarucchi et al. 2889.

length of the corolla tube, (at 4-7.5 mm from the base); anthers sessile, narrowly triangular, 3-4 x as long as wide, 3-5 x 0.8-1.4 mm, apex acuminate, sterile for 0.2-0.4 mm, sagittate at the base, glabrous. *Pistil* glabrous, 6-11 mm long, with apex 0.5-0.7 way along anthers, with an abscission layer at the apex of the ovary and therefore style and pistil head shed with the corolla; ovary subglobose or ovoid, 1.5-2 x 1-2 x 1-2 mm, rounded or obtuse, of two carpels connate basally by a 0.5-1.2 mm high disk-like thickening, style cylindrical or filiform, 4-8.5 mm long; pistil head composed of an entire or undulate ring 0.3-0.4 x 0.6-1.4 mm, a broadly obovoid apically 5-lobed central part 0.4-0.8 x 0.5-0.8 mm and a short bilobed apex 0.1-0.3 x 0.2-0.3 mm. Ovules approximately 100-150 in each carpel. *Fruit* of 2 separate mericarps; mericarps green, obliquely ellipsoid, 25-35 x 12-17 x 12-17 mm, straight, rounded or shortly and bluntly acuminate when dried, with or without 2 faint angles, smooth and wrinkled when dried, about 10-20-seeded; wall 1-2 mm thick in dried fruits; aril white (?), partly covering the seed. *Seed* ochraceous, obliquely and narrowly ellipsoid, 10 x 2 x 2 mm, with deep longitudinal grooves in a spongy testa with long papillae.

DISTRIBUTION: Northern South America (Venezuela, the three Guianas and Brazil).

ECOLOGY: Periodically inundated riverine forests. Alt. 0-250 m. Flowering and fruiting throughout the year.

Geographical selection of the approximately 150 specimens examined:

COLOMBIA. Guainia: near San Felipe, *A. Gentry & Stein* 46423 (MO).

VENEZUELA. Amazonas: San Carlos de Río Negro, *Liesner* 9052 (MO, VEN); ibid., *Spruce* Oct. 1851 (K, P).

GUYANA. Mazaruni R., Jenman 5494 (K, NY, US); Bartica, Essequibo R., For. Dept. 6886 (K); Potaro R., between Kaieteur Falls and Tukeit, *Kvist* et al. 185 (P); below Tukeit, *Maguire & Fanshawe* 23482 (A, K, MO, NY, type of *T. versicolor);* Kamoa R., *Jansen-Jacobs* et al. 1674 (U, WAG); ibid., Toucan Mt, *Jansen-Jacobs* et al. 1612 (U); sin. loc., *Robert Schomburgk* II 178 (= *Richard Schomburgk* 117) (BM, G, K, P, W).

SURINAME. Kabalebo R., *im Thurn* Sept. 1879 (K); Nat. Res. Raleigh Falls, *Heyde* 656 (K, MO, U, US); ibid., *Mennega* 138 (U); Bakhuis Mts, *Florschütz & Maas* 2781 (K, NY, U, US); Wilhelmina Gebergte, Lucie R., *Stahel* 307 in BW 7104 (U); ibid., *Maguire* et al. 54142 (NY); Oelemari R., *Wessels Boer* 1072 (U).

FRENCH GUIANA. Cayenne, *Martin* s.n. (K), in coll. *Rudge* s.n. (BM); Petite Ouaqui R., Baille-Nom Fall, *de Granville* B 4974 (CAY, MO, P); Grand Inini R., near Etats-Unis Fall, *de Granville* C 22 (CAY, P, U); Carbet Brûlé, *de Granville* 1933 (CAY, GH, MO, P); ca 22 km from Dégrad Claude on Emerillons Road, *de Granville & Oldeman* 2271 (CAY, P, U); Petit Tamouri, *Lescure* 211 (CAY, MO, P).

BRAZIL. Roraima: Mun. Caracaraí, *J.L. dos Santos & J.A. Côelho* 691 (K, WAG); between Caracaraí and Rio Branco, *Coradin & Cordeiro* 1053 (NY, Z); Ajarani, *Kuhlmann* 3004 (S, U, US); Rio Anauá, *J.M. Pires* et al. IPEAN 14430 (Z); Mun. São Luiz do Anauá, near Equador, *Cid* 9075 (WAG); near Vila São José de Biaçu, *Cid* 9108 (WAG); Rio Xeriuini, *J.M. Pires* et al. IPEAN 13965 (U, Z). Amazonas: Rio Negro, upper Rio Padauiri, *Fróes* 22613 (MO, US); Sierra Aracá, *W.A. Rodrigues* et al. 10530 (USF); Rio Tiquié, *Fróes* 12542/236 (A); near Panurè, Rio Vaupès, *Spruce* 2559 (BM, BR, CGE, K, NY, P, TCD, paratype of *T. rupicola* var. *sprucei*); Rio Negro, between Manaus and São Gabriel, Tapereira, *Alencar* 231 (NY, US, WAG); São Luiz, *Alencar* 125 (GH, NY, US, WAG); Barcelos, *Fróes* 22059 (NY); Lago Maraá, *Amaral* et al. 182 (MO, WAG); Rio Negro, 45 km above mouth of Rio Branco, *Madison* et al. 6073 (F, NY, U); near junction of Rio Negro and Rio Branco, *Poole* 1632 (NY, WAG); Rio Jaú mouth, *Trail* 512 (K); Rio Auati Paraná, *Byron* 324 (MO); Rio Negro, Santo Antonio, *Mori* et al. 21294 (WAG); Est. Ecol. Anavilhanas,

Knob et al. 266 (WAG); basin of Rio Negro-Rio Cuieras, just above mouth of Rio Bracinho, *Prance* et al. 14818 (F, K, M, NY, P, S, U, US, Z), 14835 (NY, U, Z); Lago do Castanho-Mirim, *Byron* et al. 739 (F); Rio Negro, sin. loc., *Riedel* s.n. (LE, P, type of *T. rupicola* var. *oblongifolia*); between Ilha Jacaré and Airao, *Prance* et al. 15073 (F, GH, K, NY, P, S, U, US, Z); Pedrera, W of Manaus, *Robert Schomburgk* I 898 (BM, CGE, E, F, FI, FI-W, G, K, L, NY, P, SING, TCD, US, W; phot. of G sheet GH, MO, NY, US, type); Rio Cuieras, 50 km upstream, *Ongley & Ramos* P21777 (NY, Z); Est. Biol. de INPA, Res. Campina, km 45 Manaus-Caracaraí Road, *Plowman* et al. 12656 (K, WAG); 2 km S of Maués, *Campbell* et al. P22137 (K, NY, U, US, Z); Mun. Maués, between Laranjal and Villa Darcy, *Zarucchi* et al. 3064 (WAG); Rio Urubu, *Prance* et al. 4901 (K, S, U, US, Z); lower Rio Pacoval, *Zarucchi* et al. 3182 (WAG); ca 5 km S of Vila de Canuma, *Zarucchi* et al. 2889 (WAG); near Barra do Rio Negro (= Manaus), *Spruce* 1005 (BM, CGE, FI-W, K, M, TCD, W; phot. of M sheet GH, MO, type of *B. rariflora* and paratype of *T. rupicola* var. *sprucei*) & 1463 (AWH, BM, BP, BR, C, CGE, F, G, GH, GOET, K, LD, MPU, NY, P, TCD, W, paratype of *T. r.* var. *sprucei*) & 1758 (AWH, BM, BR, CGE, G, K, P, TCD, W, lectotype of *T. r.* var. *sprucei*); ibid.(?), *Spruce* 1841 (CGE, K, P, W); ibid., *Ule* 5335 (G, HBG, K, L); Rio Tarumá, *Ducke* 252 (A, K, MO, NY, S, UC, US); ibid., *Prance* et al. 11756 (F, K, M, NY, U, US, Z); Lago de Tefé, *Cid & Lima* 3150 (K, WAG); ibid., *Poeppig* 2504 p.p. (BM, F, G, G-DC, GH, GOET, MO, P, W; phot. of G sheet MO, NY, US; phot. of W sheet MO, NY, US, type of *T. r.* var. *poeppigii*); ibid., *Prance* et al. 24453 (MO, NY, S, Z); Rio Tupuna, *Campbell* et al. P20830 (K, MO, NY, U, US, Z); Pauini, *Miranda & Souza* 526 (GUA); Rio Curuquetê, near Santo Antonio Fall, *Prance* et al. 14185 (F, K, NY, S, U, US, Z); km 150 Humaitá-Jacarencanga Road, *Teixeira* et al. 1295 (K, US, WAG); Rio Ipixuna, *Prance* et al. 3353 (K, S, US, Z). Pará: Tapajós, Rio Cururu, *Egler & Raimundo* 1007 (NY). Rondônia: Mun. Santa Barbara, *Teixeira* et al. 641 (WAG); Rio Curuquetè, *Prance* et al. 14588 (F, GH, K, M, NY, P, S, U, US, Z). Pará: Belém, *Martius* s.n. (L, paratype of *T. rupicola* var. *sprucei*), s.n. (BP, M); Tapajós National Park, Ilha do Pacú, *M.G. Silva & Rosário* 4028 (NY, USF); Mun. Oriximiná, Rio Mapuera, *Cid* et al. 7673 (K, WAG).

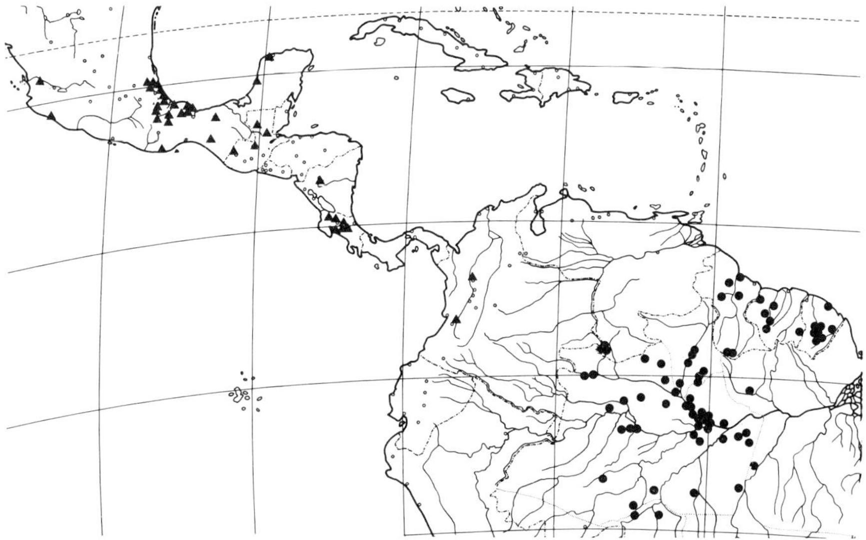

Map 48. ●. *Tabernaemontana rupicola,* ▲. *Stemmadenia litoralis.*

Note. *T. rupicola* is easily confused with *T. lorifera* (q.v.).

93. Tabernaemontana salzmannii A. DC., Prod. 8: 362 (1844). – Type: Brazil, Bahia, sin. loc., *Salzmann* 13 (holotype G-DC; phot. MO, NY, US).
Fig. 93, p. 365; map 49, p. 366

Homotypic synonym:

Peschiera salzmannii (A. DC.) Miers, Apoc. S. Am. 40 (1878).

Heterotypic synonyms:

T. rauwolfiae A. DC., op. cit. 364. *T. salzmannii* var, *lanceolata* Muell. Arg. in Martius, Fl. Bras. 6, 1: 78 (1860). – Type: Brazil, Bahia, sin. loc., *Blanchet* 713 (holotype G-DC; phot. GH, US; isotype G).
T. salzmannii var. *longifolia* Muell. Arg., l.c. – Type: Northern Brazil, sin. loc., *Sellow* 1641 (lectotype NY, designated here; isolectotypes BR, CGE, K, P).

Shrub or small tree 2-12 m high. Trunk 5-20 cm in diameter or less. Branches pale to dark grey-brown, lenticellate; branchlets terete, glabrous. *Leaves* petiolate; petiole glabrous, 3-30 mm long; (ocreae not widened into intrapetiolar stipules); blade subcoriaceous or in fruiting specimens even coriaceous and shining above when dried, elliptic, narrowly elliptic or narrowly obovate, 2-3.2 x as long as wide, 3-22 x 1-7 cm, obtuse or obscurely and obtusely acuminate at the apex, cuneate at the base, entire, glabrous on both sides, with scattered black dots beneath or not, costa prominent above or not, but not impressed, in dried leaves; 6-25 pairs of rather straight secondary veins forming an angle of 70-90° with the costa, mostly anastomosing near the margin; tertiary venation invisible. *Inflorescence* pedunculate, 2-7 x 2-5 cm, 1-12-flowered, lax. Peduncle short, glabrous, 2-17 mm long; pedicels 2-10 mm long, glabrous. Bracts scale-like, 0.3-0.5 x as long as the sepals. *Flowers* fragrant, open during the day. *Sepals* green, connate at the base for about 0.5 mm, mostly recurved, sometimes only slightly spreading, oblong, ovate or broadly ovate, mostly unequal and often variously shaped in a single flower, 1-4 x as long as wide, 3-7 x 1.5-3 mm, obtuse or rounded, glabrous outside, sometimes ciliate, glabrous inside and with 3-9 colleters in one row in the middle at the base; colleters very variable in shape and size, 0.2-0.5 x 0.1-0.4 mm, when broad often lobed. *Corolla* yellow, cream or white, 12-16 mm long in the mature bud and forming a comparatively large ovoid head 0.33-0.47 of the bud length (4-7 x 2-4 mm) at least twice as wide as narrow part of corolla tube with a blunt apex, glabrous outside, pilose inside in stripes from 0.5-1.5 mm below the anthers tails to the insertion of the stamens and from there in a belt up to the base or less often the middle of the lobes; tube 3.3-6.5 x as long as the calyx, when sepals recurved, 1.7-3 x, when sepals straightened out, 0.8-1.7 x as long as the lobes, 10-13 mm long, flask-shaped, 2.5-3.5 mm wide above the base, narrowed around the anthers and even more above to 1.5-2.5 mm wide, widened again at the mouth to 2.5-3.5 mm wide, not twisted; lobes obliquely obovate to almost dolabriform, 0.6-1.3 x as long as the tube, 1.1-2.1 x as long as wide, 7-13 x 4-8 mm, rounded, neither auriculate, nor undulate, spreading. *Stamens* with apex 2-5 mm below mouth or corolla tube, inserted 0.3-0.4 of the length of the corolla tube, (at 3.5-4.5 mm from the base); anthers sessile, narrowly triangular, 3-3.7 x as long as wide, 4.3-5 x 1.2-1.5 mm, apex acuminate, sterile for 0.4-0.8 mm, sagittate at the base. *Pistil* glabrous, with apex almost halfway along anthers, 5-5.5 mm long; ovary ovoid, 1.8-2.5 x 1-1.5 x 1-1.5 mm; style short, 1.2-2.5 mm long, thickened at apex; pistil head, 1.2-1.8 mm high, composed of a basal ring 0.2-0.3 x 1-1.2 mm consisting of 10 upcurved acute lobes, a stipitate 5-lobed depressed globe 0.4-0.5 x 0.7-0.8 mm and a stigmoid apical part 0.2-0.3 x 0.2-

Fig. 93. *Tabernaemontana salzmannii.* **1,** habit (x 2/3); **2,** opened corolla (x 2); **3,** pistil (x 8); **4,** sepal inside (x 2); **5,** fruit (x 2/3). 1-4 from Duarte 14152; 5 from Mori et al. 9734.

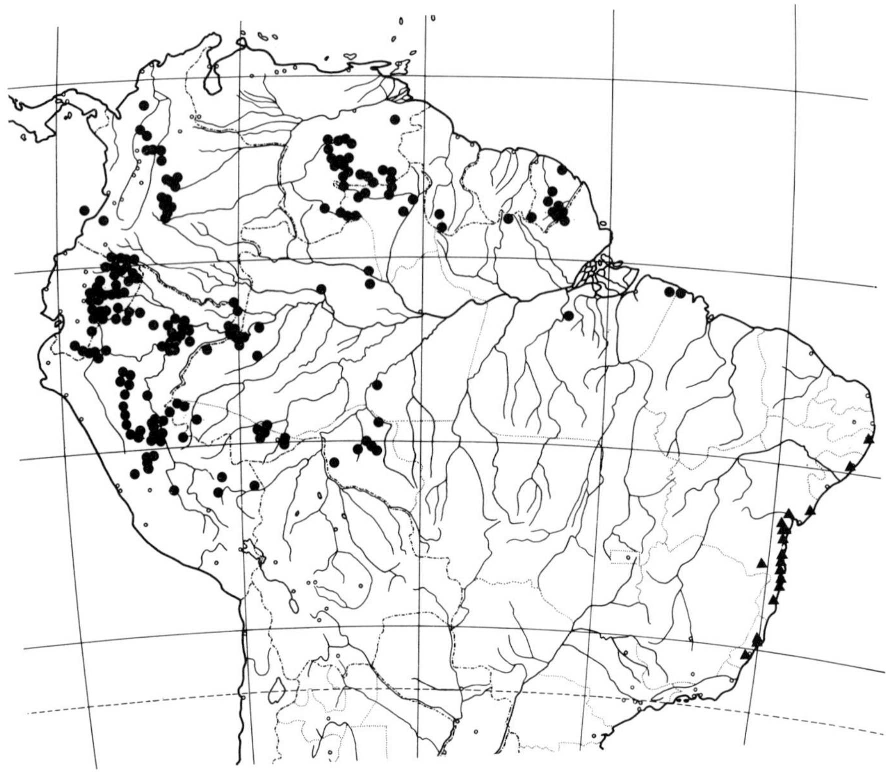

Map 49. ▲. *Tabernaemontana salzmannii,* ●. *T. sananho.*

0.3 mm. Ovules approximately 100 in each carpel. *Fruit* of 2 separate mericarps; mericarps green, obliquely ellipsoid, 30-45 x 20-30 x 20-30 mm, recurved, rounded at the apex, densely covered with short blunt excrescences; wall rather thick. *Seed* brown, obliquely ellipsoid, about 7 x 4 x 3 mm, with longitudinal grooves, papillose.

DISTRIBUTION: Eastern Brazil.

ECOLOGY: Dry forest, mainly near the coast. Alt. 0-70 m. Flowering from September to December with a peak in October. Fruiting March to May.

Geographical selection of the approximately 45 specimens examined:

BRAZIL. Pernambuco: Recife, Dois Irmãos For., *Soares* in herb. *Tavares* 2193 (US). Algoas: São Miguel dos Campos, *Paiva* 3321 (US). Bahia: near Porto Seguro, *Duarte* 6828 (Z); between Amarosa and Santo Amaro, *Duarte* 10591 (RB, WAG, Z); Represa For., near Castro, Duarte 14152 (K, M, NY); Ilha de Tinharé, *L.A. Mattos Silva* & *T.S. dos Santos* 1914 (WAG); km 13 Taperoã-Valença, *Hage* et al. 433 (WAG); between Maraú and Ubaitaba, *T.S. dos Santos* 3543 (WAG, Z); Itacaré, *Almeida* & *T.B. Santos* 155 (NY, P, Z); km 7.3 Serra Grande-Itacaré, *de Carvalho* et al. 3671 (WAG), 3641 (WAG); Ilhéus, *Luschnath* anno 1859 (BR); Pontal dos Ilhéus, *Belém* 3582 (K, S, U, US, Z); Mun. Una, *Mori* & *Thompson* 11010 (K, NY, P); Una, *T.S. Santos* 485 (P, Z); km 40 Una-Santa Luzia Road, *E.B. Santos* & *M.C. Alves* 100 (WAG); ca km 26 Belmonte-Itapebi Road, *Harley* et al. 17407 (K, MO, NY, U, US, Z); Est. Ecol. Pau-brasil, *Euponino* 356 (NY, P, WAG); Mun. Prado, 61 km N of Alcobaça, *Mori* et al.

9734 (Z); sin. loc., *Blanchet* 367 (BM, G), 713 (G, G-DC; phot. of G-DC sheet GH, US, type of *T. rauwolfiae*), 1556 (BM, G); *Salzmann* 13 (G-DC; phot. MO, NY, US, type). Espirito Santo: CVRD F.R., Linhares, *I.A. Silva* 259 (MO, WAG); Porto Seguro F.R., *Folli* 980 (WAG); Barra de Jucú, *Prince Maxim Vidensis* comm. anno 1827 (BR). Sin. loc., *Sellow* 116 (K), 501 (LISU), 1641 (BR, CGE, K, NY, P, lectotype of *T. salzmannii* var. *longifolia*).

CULT. Brazil, D.F., Horto Floresta da Gouvêa, *P .Rosa* 294 (RB, WAG).

94. Tabernaemontana sananho Ruiz & Pav., Fl. Peruv. 2: 22, t.144 (1799). – Type: Peru, San Antonio de Playa (not localized), *Ruiz & Pavón* s.n. (holotype MA; isotypes BR, F, G, G-DC, US; phot. of G-DC sheet F, MO, NY, US; phot. of US sheet WAG). Fig. 94, p. 369; map 49, p. 366

Homotypic synonyms:

Merizadenia sananho (Ruiz & Pav.) Miers, Apoc. S. Am. 78 (1878). *Bonafousia sananho* (Ruiz & Pav.) Mgf. in Notizbl. Bot. Gart. Berlin 14: 166 (1938).

Heterotypic synonyms:

T. poeppigii Muell. Arg. in Linnaea 30: 405 (1860). *Taberna poeppigii* (Muell. Arg.) Miers, op. cit. 63. Type: Peru, San Martín, Tocache, *Poeppig* 1923 (holotype W; phot. A, GH, MO, US).

Shrub or tree 1-15 m high. Trunk 1-15 cm in diameter or more; bark smooth, pale grey-brown, with (white) lenticels, wood white. Branches pale brown, with few lenticels; branchlets elliptic in section when fresh, often sharply angular when dried, glabrous. *Leaves* of a pair equal or unequal (larger up to twice as long as the other and comparatively narrower), shortly petiolate; petiole glabrous, 3-20 mm long; (ocreae not or slightly widened into intrapetiolar stipules); blade coriaceous when fresh, papery or chartaceous when dried, elliptic, (1.5-)2-3 x as long as wide, 6-30 x 3-20 cm, acuminate or apiculate at the apex, cuneate or rounded at the base, entire or less often undulate, glabrous and with scattered black dots on both sides, with 6-17 pairs of rather straight or upcurved secondary veins forming an angle of 50-70° with the costa, tertiary venation reticulate. *Inflorescence* shortly pedunculate, 3-9 x 3-6 cm, few- or many-flowered, rather dense, 0-3 x furcate or subumbellate; branches each ending thyrsoid (mostly less rich than in *T. siphilitica*). Peduncle glabrous, 5-40 mm long; pedicels glabrous, 3-7 mm long. Bracts numerous, about half as long as the sepals, triangular or nearly so, obtuse, deciduous, with flowers in the axils or not. *Flowers* fragrant, open during the day. *Sepals* pale green (?), subequal, connate at the base for about 0.5 mm, erect, ovate or broadly ovate, 1-2 x as long as wide, 3-6 x 2-4 mm, rounded, entire, glabrous outside, ciliate, glabrous inside and with 2-6 colleters in 1 row in the middle at the base; colleters 0.3-0.7 x 0.15-0.2 mm. *Corolla* white or creamy, yellow in the throat, 13-27 mm long bud in the mature and forming a comparatively rather large ovoid head 0.25-0.33 of the bud length (4-8 x 3-6 mm) with an obtuse or acute apex, glabrous or sometimes pubescent on base of lobes outside, ciliate at the edge of the part of the lobes not covered in bud, with a 2.5-4 mm wide pilose or lanate belt inside just below the insertion of the stamens or for 1 mm above it, especially on the filament ridges; tube 3.6-5 x as long as the calyx, 0.8-1.2 x as long as the lobes, 14-25 mm long, almost cylindrical, 2.5-4.5 mm wide, often widest above base, not widened around anthers, twisted 0-0.2 turn to the left just below the insertion of the stamens and from there 0.4-0.7 turn to the right around the anthers; lobes obliquely and often narrowly obovate or elliptic, 0.8-1.3 x as long as the tube, 2-2.4 x as long as wide, 13-26 x 6-10 mm, rounded, undulate, recurved. *Stamens* with apex

2-4 mm below mouth of corolla tube, inserted 0.6-0.65 of the length of the corolla tube, (at 8-16 mm from the base); anthers sessile, narrowly triangular, 3-4 x as long as wide, 4-5 x 1.2-1.5 mm, apex acuminate, sterile for 0.3-0.5 mm, sagittate at the base, glabrous. *Pistil* glabrous, 10-18 mm long, with apex 0.4-0.5 way along anthers, persistent; ovary subglobose or ovoid, 1.5-2 x 1-1.5 x 1-1.5 mm, acuminate, of two basally hardly connate carpels, sometimes with a disk-like thickening at the base; style filiform, 7-14.5 mm long; pistil head composed of an entire or undulate ring 0.5 x 1.2 mm, a broadly ovoid apically 5-lobed central part 0.7 x 0.8-1 mm and a stigmoid apex about 0.2 x 0.2 mm. Ovules approximately 50-100 in each carpel. *Fruit* of 2 separate mericarps; mericarps yellow or orange, obliquely subglobose, 35-50 x 35-50 x 30-50 mm, rounded or sometimes mucronate at the apex, with 2-3 faint ridges, especially near the base, smooth, not dotted, immature broadly ellipsoid, about 20-40-seeded; wall 2-3 mm thick; aril white, enveloping the seed only at the hilar side. *Seed* pale brown when dried, obliquely and irregularly ellipsoid, 8-12 x 4-6 x 4-5 mm, with clear longitudinal grooves, papillose; embryo 6-9 mm long; cotyledons broadly ovate or ovate, 1-1.6 x as long as wide, 3-4 x 2.5-3.5 mm, obtuse at the apex, cordate at the base; rootlet 1-1.6 x as long as the cotyledons, 4-6 x 0.5-0.8 mm.

DISTRIBUTION: Northern and western South America.

ECOLOGY: Forest understorey, not along rivers. Alt. 100-1200 m. Flowering in Colombia from August to March, in Venezuela found in flower only once in May, and once in both October and November, in French Guiana in August, in Brazil from July to December, in Ecuador almost throughout the year with peaks February-March and August-October, in Peru February-March and June to November with a peak in August. This species has been collected mainly with immature fruits. A fruiting season could not be deduced from the extant collections.

Geographical selection of the approximately 420 specimens examined:

COLOMBIA. Antioquia: Puerto Valdivia, Bro. *Daniel* 3381 (US); 8-27 km NE of Amalfi, *Callejas* et al. 9089 (WAG); San Luis, *Hernandez* et al. 595 (COL); Mun. San Luís, Quebrada La Cristalina, *Ramírez & Cárdenas* 756 (COL). Boyaca: Mt Chapo Region, *Lawrance* 446 (A, BM, E, G, GH, K, LL, MO, NY, U, UC, US); El Humbo, *Lawrance* 655 (A, E, G, K, MO, S, W), 682 (A, E, G, K, MO, S, US). Cundinamarca: Llano de San Martín, *J.J. Triana* 3396 p.p. (BM, COL, P); 6 km NW of Medina, *Grant* 10439 (NA, US, WIS). Meta: Sabanas de San Juan de Arama, *Idrobo & Schultes* 609 (COL, GH, U, US); Villavicencio, *Karsten* s.n. (NY, W); Mun. La Macarena, *Marulanda* 747 (WAG); El Mico, *Philipson* et al. 1398 (BM, COL, MO, NY, US); Sierra de la Macarena, *Philipson* et al. 2207 (BM, COL, MO, S, US), 2333 (BM, COL); ibid., Río Guapaya, *Philipson* et al. 1673 (BM, COL, MO, US); Río Güejar, *Forero* et al. 862 (COL); Duda R., *Barbosa* 5359 (FMB). Cauca: Gorongona Island, "St. George Exped." 566 (US); Mun. Guapi, Gorgona National Park, *Lozano* et al. 5651 (COL). Caquetá: SE of Tres Esquinas, *Little* 9530 (COL, US). Putumayo: Río Rumiyaco, *Soejarto* et al. 1286 (COL, GH, U, US); San Antonio del Río Guamués, *Plowman* 2089 (COL, GH, K); Río San Miguel ó Sucumbios, *Schultes* 3577 (COL, GH). Amazonas: Puerto Porvenir, *Cuatrecasas* 10613 (COL); Río Loretoyacu, *Schultes* 6594 (COL, K, US); ibid., *Schultes & Black* 8346 (COL, ECON, GH, MO, U, US); near Leticia, *Schultes* et al. 24090 (ECON).

VENEZUELA. Bolívar: Río Nichare, *Steyermark* 95660 (VEN); Mun. Cedeño, Cuenca Río Nichare, *Elcoro* 311 (MO); Río Caura, between Campamento Las Pavas and El Playón, *Morillo & Liesner* 9149 (MO); near Campamento Las Pavas, *Steyermark* et al. 117204 (F, MO, U); Río Mawela, tributary of Río Erebato, *Boom & Marin* 10565 (WAG); near Santa Maria de Erebato, *Sanoja* 2521 (MO); Río Erebato tributary, *Boom & Marin* 10374 (WAG); Guanagujaña, *Stergios & Delgado* 12606

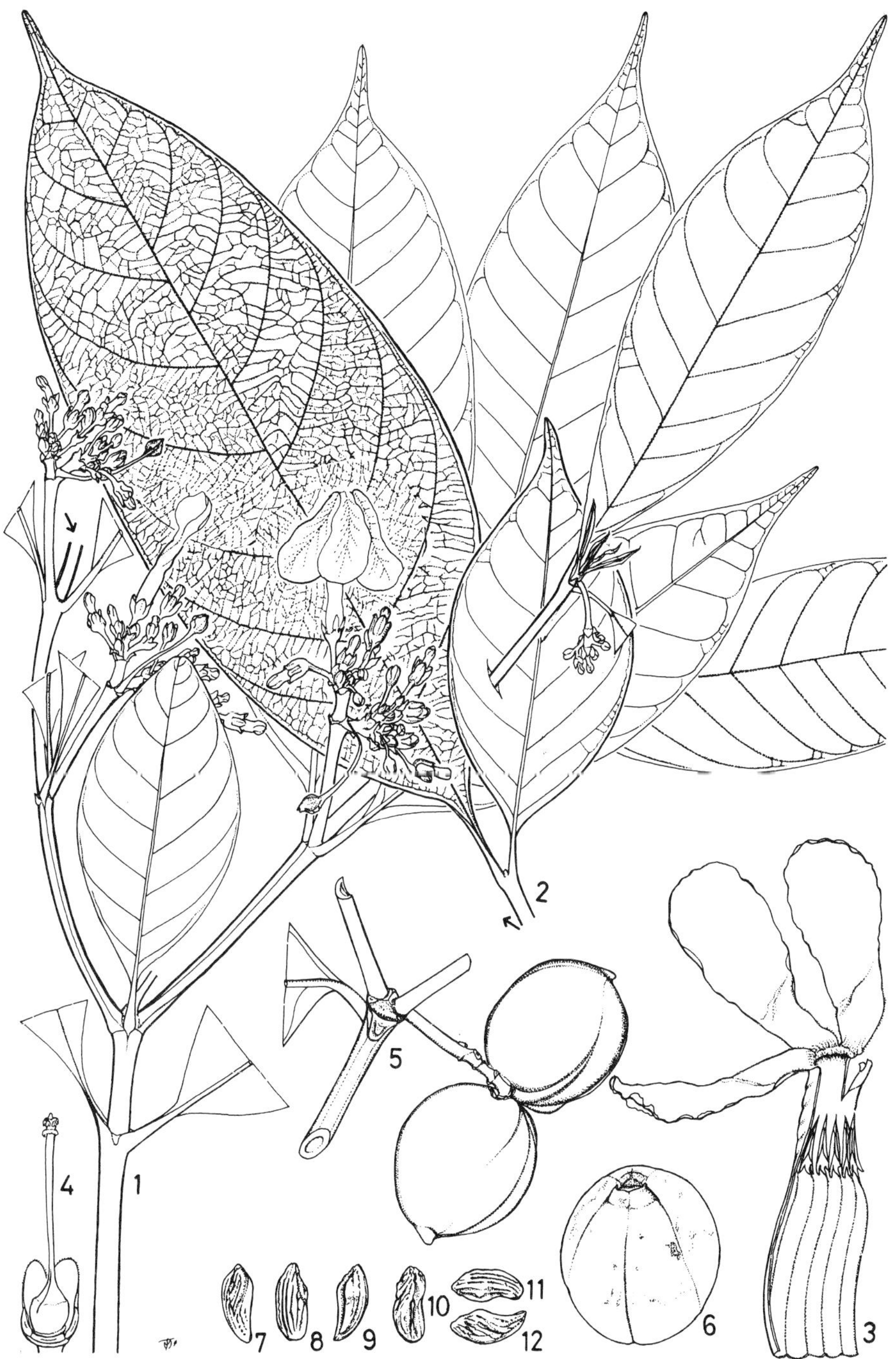

Fig. 94. *Tabernaemontana sananho.* **1,** habit (x 2/3); **2,** apex of branchlet (x 2/3); **3,** opened corolla (x 2); **4,** calyx with pistil (x 2); **5,** immature fruit (x 2/3); **6,** mature mericarp (x 2/3); **7-12,** seeds (x 1). 1-4 from Schunke 9595; 5 from Leeuwenberg 11754; 6-12 from Hijmans 36.

(MO); 2 km below Caño Guacamaya (Guaya), *Stergios* 12095 (MO); Mun. Foráneo Aripao, Upper Río Caura, Aymard & Delgado 6681 (MO); Río Paragua, *Stergios* 11152 (MO); near Minas de Manaima, *Stergios* 10239 (MO); Mun. Raul Leoní, *Aymard & Fernández* 7269 (MO); between Maihia and mouth of Río Paramichi, *Steyermark* 90548 (NY); Río Uaiparú, tributary of Río Icabarú, *Cardona* 1916 (US, VEN); NE of Serranía Pai-soi, *Steyermark* 90655 (MO, NY); El Dorado-Santa Elena de Uairén Road, *Trujillo* 3746 (MY); W of El Polo, *Liesner* 19606 (MO, WAG); 5 km S of El Pauji, *Liesner & Holst* 18839 (MO, WAG). Amazonas: Cañon Harita, Ocamo, *Lizot* 1972-16 (US, VEN); Salto Salas, *Croizat* 526 (NY); Raudal Montserrat, *Croizat* 690 (NY); Río Orinoco, above Salto Siro Vásquez, *Berry* 4779 (MO).

GUYANA. Marudi Mts, Mazoa Hill, *Stoffers* et al. 226/82 (B, CAY, F, K, MO, S, U, US); Basin of Shodikar Creek, Essequibo R. tributary, *A.C. Smith* 2842 (A, G, K, MO, NY, P, S, U, US).

FRENCH GUIANA. Toukouchipan Mt base, *Sastre* 1730 (CAY, P, WAG); N of Saül, *Oldeman* 2050 (CAY, MO, P, U); S of Saül, *Leeuwenberg* 11754 (CAY, P, WAG); Camopi R., *Oldeman & Sastre* 230 (P, U); Grand Tamouri R., *de Granville* 2137 (CAY, MO, P); Upper Oyapock R., Takululi Creek, *de Granville* 2476 (CAY, WAG); Takwana, *de Granville* B 5222 (CAY, P); Trois Sauts, *Grenand* 579 (CAY); Cayenne, herb. *Poiret* s.n. (P, phot GH).

BRAZIL. Roraima: near Auaris, *Prance* et al. 9660 (GH, K, P, S, U, US, Z); between Surucucu and Uaicá, *Prance* et al. 10532 (F, K, NY, S, U, US, W, Z), 10750 (F, K, NY, P, S, U, US, Z); Ilha de Maracá, *Milliken* et al. 655 (K, U); Serra Tepequem, *Prance* et al. 4467 (K, NY, US, Z); Posto Mucajaí, *Prance* et al. 11124 (F, K, NY, S, U, US, Z); Rio Uraricoeara, near Uaicá airstrip, *Prance* et al. 19976 (NY). Amazonas: Benjamin Constant, *Drees* 36 (INPA, MG); Esperança, Rio Javari mouth, *Ducke* 1121 (K, MO, NY, SP, US); Rio Javarí, Mun Atalaia, Estirão do Equador, *Cid* et al. 9901 (WAG); Mun. São Paulo de Olivença, near Palmares, *Krukoff* 8151 (A, BM, BR, G, K, MO, NY, P, S, U, US); Mun. Jutaí, near Bairro de São Francisco, *Cid* et al. 8348 (WAG); Rio Iaco, *Cid & Nelson* 2732 (NY); Mun. Maraá, Rio Japurá, *Amaral* et al. 330 (MO, WAG); Mun. Barcelos, near Pico Rondon, *Amaral* 1495 (WAG); Rio Demeni Basin, near Tototobí, *Prance* et al. 10222 (C, K, M, NY, S, US, Z); Mun. Humaitá, near Tres Casas, *Krukoff* 6254 (A, BM, BR, G, K, MO, NY, S, U, US). Pará: Rio Parú de Oeste, *Cavalcante* 2367 (NY, US, Z); Altamira, *Bahia* 94 (F, MO, NY); km 96 Capanema-Maranhão Road, *Prance & Pennington* 1781 (GH, NY, S, U, US, Z). Amapa: Rio Oyapoque, near Utussansain Fall, *Irwin* et al. 48103 (K, NY, S, US, Z). Maranhão: between Assutina and Carutapera, *Fróes* 11952 (NY). Acre: Mun. Mancio Lima, Serra do Divisor, *Cid* et al. 10110 (NY); km 33 Rio Branco-Porto Acre Road, *Lowrie* et al. 489 (GH, MO, NY, US, WAG); km 39 of same road, *Cid* 2884 (MG, NY, WAG); Mun. Sena Madureira, 4 km from Rio Iaco, *Cid & Nelson* 2773 (MG, NY, WAG); km 7 Sena Madureira-Rio Branco Road, *Prance* et al. 7642 (K, NY, S, US, Z); Rio Macauhan mouth, tributary of Rio Iaco, *Krukoff* 5564 (A, BM, G, K, LE, M, MO, NY, S, U, UC, US); near Serra da Moa, *Prance* et al. 12394 (F, K, M, NY, S, U, US, Z); Projeto Humaitá, *de Souza* 137 (WAG); between Igarapé Cujubim and Igarapé Jacamin, *Campbell* et al. 9315 (WAG). Rondônia: Rio Jatuarana source, *Krukoff* 1652 (A, BM, G, K, NY, P, S, U, US); BR 364, km 290 Porto Velho-Cuiabá Road, Bom Retiro, *Freitas* et al. 89 (INPA, MO); near same road, section Jarú-Ouro Preto, *Freitas* 137 (MO); km 378 of same road, Res. Ecol. CNP-INPA, *Cid* et al. 4877 (WAG); SW of Ariquemes, *Teixeira* et al. 446 (K, WAG); Rio dos Pacaás Novos, *Anderson* 12180 (K, WAG).

ECUADOR. Napo: Orellana Canton, Yasuni National Park, *Neill & Rojas* 9954 (WAG); between Cuyabeno and Ponta Avenilla, *Bravo & Gómez* 244 (QCA); between Tiputini and Lagartoeocha, *Fagerlind & Wibom* 2323 (S, UPS); Nuevo

Rocafuerte, *Jaramillo & F. Coello* 4194 (AAU, QCA); Res. Faun. Cuyabeno, *Balslev* et al. 97416 (AAU); Río Aguarico, Shushufindi, *Vickers* 229 (F); near Santa Cecilia and Río Aguarico, *MacBryde & Dwyer* 1338 (MO, QCA); Río Wai si ayá, N tributary of Río Aguarico, *Brandbyge* et al. 32662 (AAU, WAG); Auca oil-field, *Brandbyge & Asanza* 30104 (AAU, WAG); Yasuní National Park, *Cerón & Coello* 3217 (MO); ibid., *Øllgaard* et al. 57005 (QCA); Res. Biol. Jatun Sacha, *Palacios* 2754 (MO, WAG); ca 50 km NE of Coca, *Lugo* 3266 (GB, USF); Res. Ecol. Cayambe, Coca, *Jaramillo & Coello* 3063 (AAU); Payamino-Loreto Road, *Zaruma* 645 (WAG); near Volcan Sumaco, *Hurtado* et al. 199 (WAG); Río Yasuni, 80 km upriver from Nuevo Rocafuerte, *Forster* 3738 (AAU, F, MO, U, USF); between Tena and Río Napo, *Asplund* 9350 (G, K, LD, S, UPS, US). Pastaza: near Laguna Gazayacu, *Palacios & Neill* 652 (MO, WAG); Río Curaray, *Holm-Nielsen* et al. 21930 (AAU, K, WAG), 22352 (AAU, WAG); Kapawí, Amuntai, Río Pastaza, *Lewis* et al. 13904 (MO); Upper Río Bobonaza, *Fuller* 137 (MO, US); Tsurakú, *Lewis* et al. 14089 (MO); Río Chico, *Shemluck & Ness* 162 (F). *Morona-Santiago:* Taisha, *Brandbyge & Asanza* 31834 (AAU); Macuma, *Van Asdall* 82-13 (QCA); Río Mangosiza, Miazal, *Limbach* 132 (WAG); W of Macas, *M.A. Baker* 6701 (WAG); near Pumpuentza, *Brandbyge & Asanza* 32406 (AAU, WAG); km 18 Mendez-Morona Road, *Dorr & Valdespino* 6306 (QCA); between Logroño and Yaupi, *Madison* et al. 3157 (GH, US); near Bomboiza, *Jimpikit* RBAE 2024 (WAG). Zamora-Chinchipe: Nangaritza Canton, Shaime, *Neill* 9586 (WAG); near Zumbi, *Knight* 328 (WIS).

PERU. Amazonas: Río Cenepa, *Ancuash* 567 (GH, MO, Z); ibid., *Berlin* 650 (GH, MO, Z); near Kusu, *Ancuash* 72 (GH, MO, Z); Bagua, *Knapp & Alcorn* 7593 (MO, WAG). Loreto: San Felipe, *Vásquez & Jaramillo* 7271 (MO, WAG); Caballo-Cocha, *Ll. Williams* 2109 (MO); near Sucusari, *Pipoly* et al. 13170 (WAG); Maynas Prov., Indiana, *Vásquez & Jaramillo* 12923 (WAG); Quebrada Paparo, *Vásquez & Jaramillo* 11643 (WAG); Alpahuayo, *Vásquez* et al. 5975 (MO, WAG); Nuevo Nazaret, Río Morona, *Lewis* et al. 12449 (MO); Vista Alegre, Río Tigre, *Lewis* et al. 12847 (MO); Río Nanay, near Bellavista, *Rimachi* 1785 (F, MO, NY); Mazan Path, between Río Amazonas and Río Napo, *Croat* 19441 (AAU, MO, NY, Z); Río Momón, *Revilla* 434 (F, MO, NY, P, Z); Santa María de Nanay, *Vásquez & Jaramillo* 12214 (WAG); San Antonio, Río Momón, *Mathias & D. Taylor* 3779 (MO); km 44 Iquitos-Nauta Road, *Vásquez & Jaramillo* 11456 (WAG); Samito, *Plowman* 2503 (ECON); Nauta, *Vásquez & Jaramillo* 3414 (F, MO, WAG); Saquena Distr., Genaro Herrera, *Rimachi* 3641 (MO, NY); Río Tigre, Paichi Playa, *Ayala* et al. 2573 (GH, MO, WAG); Oleoducto Secondario, between Camps Batra 1 and Batra 4, *Diaz & Jaramillo* 1414 (MO, P, Z); Pampa Hermosa, *Lewis* et al. 10094 (WAG); Nueva Jerusalem, *Lewis* et al. 10893 (MO); Puranchim, *Lewis* et al. 13434 (MO, WAG); near Yurimaguas, *Mexia* 6071 (BM); Balsapuerto, *Killip & A.C. Smith* 28377 (NY, US); Pongo de Manseriche, *Tessmann* 4239 (G, NY). Ucayali: Quebrada Sapallal, base of Cerro Las Cachoeiras, *A. Gentry & Diaz* 58452 (MO, WAG); Prov. Coronel Portillo, near A. von Humboldt Arboretum, *Encarnación* 26403 (G, MO, US); Yarinacocha, *Vásquez & Jaramillo* 10451 (GH, MO, WAG); Pucallpa, *Woytkowski* 5776 (MO, US); Bosque Nacional A. von Humboldt, 14 km from Res. Headquarters, *A. Gentry* et al. 25460 (F, MO, USF); W of Iparía, *Schunke* 2656 (G, NY, US); near junction of Río Pachitea and Río Yuyapichis, *Morawetz & Wallnöfer* X31-251085 (Z); Canchahuayo, *Vásquez* et al. 6928 (MO); km 86 Pucallpa-Tingo María Road, *A. Gentry & Salazar* 29403 (F, MO); near Aguaytía, *Mathias & D. Taylor* 3542 (K, MO). San Martín: Lamas, km 66 Tarapoto-Yurimaguas Road, *Knapp* et al. 7239 (F, MO, WAG); near Tarapoto, *Schunke* 9595 (F, GH, MO, U, USF, WAG); Prov. Huallaga, near Bella Vista, *Ferreyra* 10097 (MO); Prov. Mariscal Caceres, SW of Sión, *Schunke* 3522 (F, G, NY, US); La Merced de Chaquillaquillo, *Ferreyra* 4380 (MO, US); Tocache, *Poeppig*

1923 (W; phot. A, GH, MO, US, type of *T. poeppigii*); near Río Tocache, *Schunke* 3283 (F, G, GH, NY, US); Distr. Uchiza, Nuevo Progreso, *Schunke* 3186 (G, NY, US). Huanuco: Prov. Pachitea, Distr. Honoria, 6 km from Miel de Abeja, *Schunke* 1437 (F, G, NY, US); W part of Sira Mts, *Marowetz* & *Wallnöfer* 15-7988 (Z); ibid., *Wallnöfer* 112-28988 (WAG, Z); NW of Puerto Inca, *Schunke* 2945 (G, NY, US); E of Tingo María, *Schunke* 10150 (AAU, F, MO, NY, U); km 33 Tingo María-Monzón Road, *Hijmans* 36 (WAG). Pasco: Prov. Oxapampa, Proj. Pichis Palcazu, *A. Gentry* et al. 42023 (MO); Puerto Laguna, *D. Smith* 8458 (MO); between Río Pichinas and Río Cacazu, *D. Smith* 2670 (MO, WAG). Junín: near Perene, *Killip* & *A.C. Smith* 25300 (MO, NY, S, US); Río Negro, *Woytkowski* 5852 (GH, MO, US). Ayacucho: near Kimpitiriki (= Quimpitirique), *Killip* & *A.C. Smith* 22905 (MO, NY, US). Madre de Dios: Manu National Park, near Cocha Cashu Stat., *Forster* & *Terborgh* 5062 (F, MO, US); ibid., *Nuñez* 5746 (MO, WAG); near Aguajal, *Campos* 1028 (MO); Pantiacolla, *A. Gentry* et al. 27321 (MO). San Antonio de Playa (not localized), *Ruiz* & *Pavón* s.n. (BR, F, G, G-DC, MA, US; phot. of G-DC sheet F, MO, NY, US; phot. of MA sheets WAG; phot. of US sheet WAG, type); s.n. (FI-W, G, K, P, US; phot. of G sheet GH, MO, NY, US, distributed as *T. arcuata*).

Notes. *T. sananho* is so closely allied to *T. siphilitica*, that they are often confused. They differ as is described in couplet 12 of key 6, p. 225.

95. Tabernaemontana siphilitica (L.f.) Leeuwenberg in Kisakürek et al., Chapter 5: 339 (1983) in Pelletier, Alkaloïds 1. – Type: Suriname, sin. loc., *Dahlberg* s.n. (lectotype LINN 302.3, designated by Leeuwenberg in 1983; isolectotypes S, S-LINN, UPS-THUNB.6156). Fig. 95, p. 374; map 50, p. 376

Basionym and homotypic synonym:

Echites siphilitica L.f., Suppl. 167 (1781). *Bonafousia siphilitica* (L.f.) Allorge in Mém. Mus. natn. Hist. Nat. IIB, 30: 114, pl. 51 (1985), partly, excl. synonyms *T. nervosa* Desf. ex Poir. and *T. longiflora* Rusby, non Benth. (1849).

T. speciosa Lam., Tableau Enc. 2: 300 (1792); Poiret in Lamarck, Enc. Suppl. 5: 275 (1817). *Bonafousia speciosa* (Poir.) Boiteau in Phytologia 31: 247 (1975), partly, excl. specimens cited which belong to *T. sananho*. – Type: French Guiana, Cayenne, *Martin* s.n. in herb. *Desfontaines* s.n. (holotype FI-W).

T. tetrastachya H.B.K., Nov. Gen. 3: 177 (1819). *Malouetia tetrastachya* (H.B.K.) Miers, Apoc. S. Am. 92 (1878). *Bonafousia tetrastachya* (H.B.K.) Mgf. in Pulle, Fl. Surinam 4, 1: 454 (1937). – Type: Colombia, Tolima, Río Magdalena, between Morales and Tenerife, *Bonpland* 1489 (holotype P-BO; phot. GH, US; isotype P).

T. repanda E.Mey. in Nova Acta physicomed. acad. Caes. Leop. Carolin Germ. 12: 784 (1824). – Type: Guyana, sin. loc., *Hostmann* 58 (holotype B†), 320 (neotype K; isoneotypes BM, CGE, G, OXF, TCD).

T. longifolia Benth. in Hooker, Jour. Bot. 3: 243 (1841). *Anacampta longifolia* (Benth.) Miers, op. cit. 66. – Type: Guyana, Berbice, *Robert Schomburgk* I 292 (lectotype K, designated here; isolectotypes BM, CGE, E, F, FI-W, G, G-DC, L, OXF, P, SING, TCD, UPS, US, W; phot. of G-DC sheet NY, P, US).

T. guianensis Miq. in Linnaea 18: 754 (1844). – Type: Suriname, near Paramaribo, *Kappler* 1627 (holotype U; isotypes G, MO, P, S, W; phot. of G sheet GH; phot. of W sheet MO).

T. guyanensis Muell. Arg. in Linnaea 30: 404 (1860). *B. guyanensis* (Muell. Arg.) Miers, op. cit. 51. – Type: French Guiana, sin. loc., *Poiteau* s.n. (holotype LE, not seen; isotype G).

T. hirtula Mart. ex Muell. Arg. in Martius, Fl. Bras. 6, 1: 73 (1860). *A. hirtula* (Mart. ex Muell. Arg.) Miers, op. cit. 67. *B. hirtula* (Mart. ex Muell. Arg.) Mgf. in Notizbl. Bot. Gart. Berlin 14: 166 (1938). *B. siphilitica* var. *hirtula* (Mart. ex Muell. Arg.) Allorge in op. cit. 119. – Type: Brazil: Pará, Belém, *Martius* s.n. (lectotype M, designated here; isolectotypes FI-W, K, L, W; phot. of M sheet GH, MO, US). The K sheet which had not been seen by Mueller Arg., designated lectotype by Allorge.

T. duckei Huber in Bull. Soc. Bot. Genève 1914. II, 6: 199 (1915). – Type: Brazil, Amapa, Rio Cuminá, *Ducke* 7911 (holotype B†; lectotype RB, designated here).

T. cuyabensis Malme in Arkiv Bot. Stockh. 21A, 6: 11 (1927). – Type: Brazil, Mato Grosso, Cuiabá, *Malme* II 1871 (holotype S; isotypes LD, UPS).

T. killipii Woods. in Ann. Miss. Bot. Gard. 18: 541 (1931). *B. killipii* (Woods.) Mgf. in Notizbl. Bot. Gart. Berlin 14: 167 (1938). – Type: Peru, Loreto, Iquitos, *Killip & A.C. Smith* 27414 (holotype MO; isotypes F, NY, US; phot. of US sheet WAG).

B. juruana Mgf. in op. cit. 167, 181. *T. juruana* (Mgf.) Macbride in Publ. Field Mus. Bot. Ser. 13, 5: 405 (1959). *B. siphilitica* var. *juruana* (Mgf.) Allorge in Mém. Mus. natn. Hist. Nat. IIB, 30: 119 (1985). – Type: Brazil, Amazonas, Marari, Rio Juruá, *Ule* 5178 (holotype B†; lectotype HBG, designated here; isotypes G, K, L; phot. of G sheet GH, MO, US).

B. tessmannii Mgf. in op. cit. 167, 182. *T. tessmannii* (Mgf.) Macbride in op cit. 410. – Type: Peru, Ucayali, middle Río Blanco, *Tessmann* 3022 (holotype B†; lectotype G, designated here; isotypes NY, S).

Shrub or small tree, 0.25-4(-10) m high. Trunk 1-6 cm in diameter or more; bark pale brown; wood whitish. Branches grey-brown, lenticellate; branchlets mostly with paler lenticels, terete, elliptic in section or sometimes angular when fresh, glabrous, puberulous or hirto-pubescent. *Leaves* of a pair equal or unequal (larger up to twice as long as the other and similarly shaped), petiolate; petiole glabrous, 3-20 mm long; (ocreae not widened into intrapetiolar stipules); blade coriaceous or subcoriaceous, even when fresh, elliptic or narrowly elliptic, 2-5 x as long as wide, 6-30(-40) x 1.5-17 cm, acuminate at the apex, cuneate or rounded at the base, entire or undulate, often bullate, glabrous and with scattered black dots on both sides, with 6-20 pairs of rather straight or upcurved secondary veins forming an angle of 60-80° with the costa; tertiary venation reticulate. *Inflorescence* shortly pedunculate, 3-12 x 2-9 cm, many-flowered, rather dense, 1-3 x furcate or subumbellate; branches each ending thyrsoid and densely covered with flowers. Peduncle robust, glabrous or less often puberulous or pubescent, 10-55 mm long; pedicels glabrous or less often puberulous or pubescent, 2-15 mm long. Bracts numerous, about half as long as the sepals, almost triangular, obtuse, deciduous, with flowers in axils. *Flowers* fragrant, open during the day. *Sepals* white, greenish-white or pale green, subequal (outer or inner larger), connate at the base for about 0.5 mm, erect, ovate or elliptic, 1.3-2 x as long as wide, 2.5-7 x 1.6-4 mm, rounded, entire, glabrous or less often puberulous or pubescent outside, ciliate or sometimes not, glabrous or sometimes partly puberulous inside and with 2-14 colleters in 1-3 rows in the middle at the base; colleters 0.3-1 x 0.1-0.3 mm. *Corolla* white, with light brown or yellow throat, 12-30 mm long in the mature bud and forming a comparatively rather small ovoid head 0.24-0.33 of the bud length (3.5-8 x 2.5-7 mm) with an acute or obtuse apex, glabrous or only on the part of the lobes not covered in bud, puberulous or pubescent outside, mostly ciliate at the edge of the same part of the lobes, with a 2-5 mm wide pilose-pubescent belt inside just below the insertion of the stamens, especially on the filament ridges; tube 2.4-5.6(-6) x as long as the calyx, 0.6-1.3 x as long as the lobes, 10-28 mm long, almost cylindrical, 1.5-4 mm wide at the base, mostly slightly widened above (but not around the anthers to

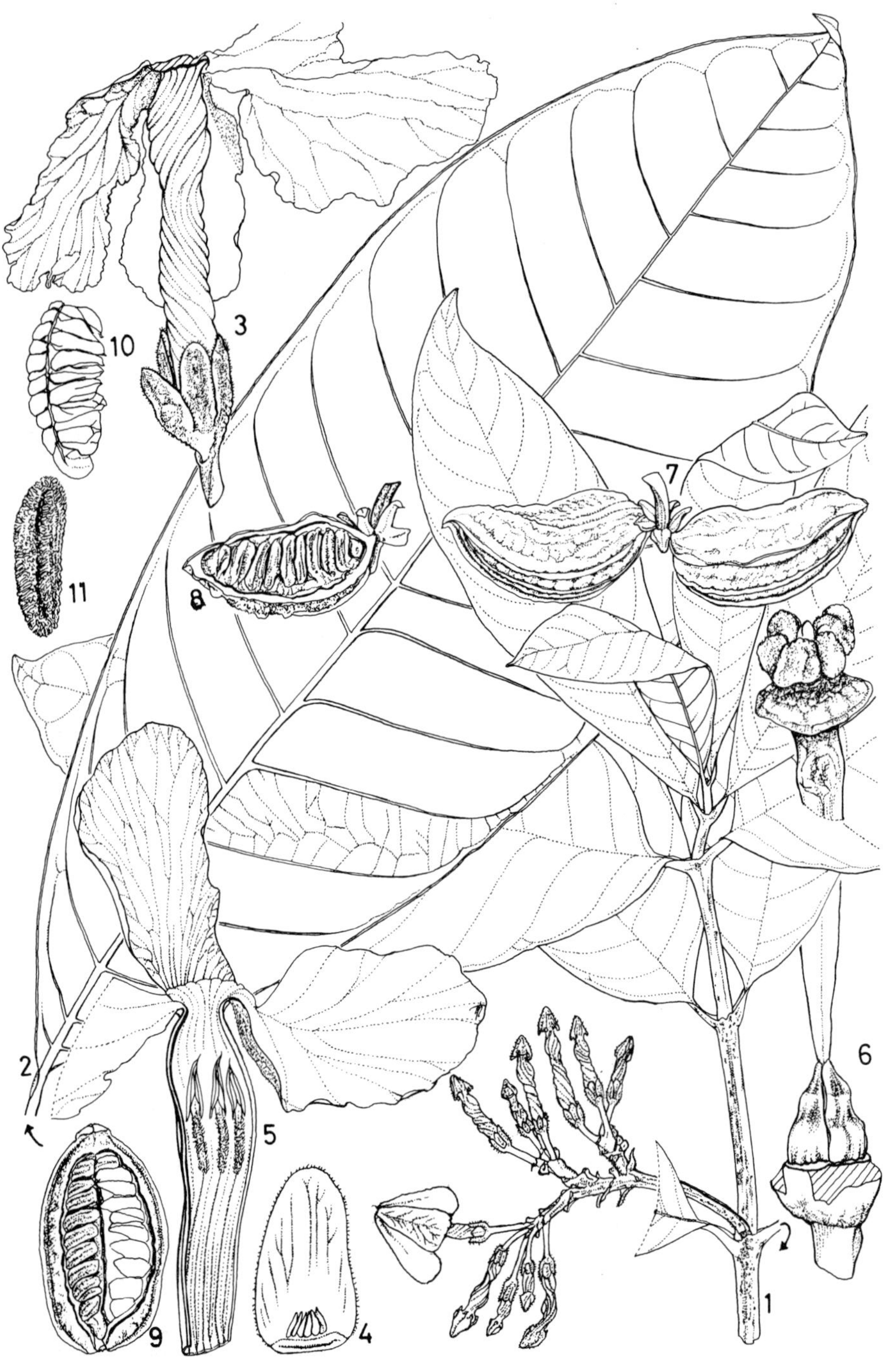

Fig. 95. *Tabernaemontana siphilitica.* **1,** habit (x 2/3); **2,** leaf (x 2/3); **3,** flower (x 2); **4,** sepal inside (x 2); **5,** opened corolla (x 2); **6,** pistil (x 14); **7,** fruit (x 2/3); **8,** section of mericarp (x 2/3); **9,** dehiscing mericarp (x 2/3); **10,** seeds with arils (x 2/3); **11,** seed (x 2). 1-2 and 7-11 from Leeuwenberg 11708; 3-6 from Prévost 238.

1.5-4 mm wide, twisted (0-)0.4-1 turn from the base to just below the insertion of the stamens to the left and from there 0.25-0.7 turn to the right, with most of the twisting near the turning point; lobes obliquely and often narrowly obovate or elliptic, 0.7-1.6 x as long as the tube, 1.5-3 x as long as wide, 10-28 x 6-14 mm, rounded, undulate, recurved. *Stamens* with apex 0.5-3.5 mm below mouth of corolla tube, inserted 0.57-0.72 of the length of the corolla tube (at 6-20 mm from the base); anthers sessile, narrowly triangular, 3-4 x as long as wide, 4-5 x 1.2-1.5 mm, apex acuminate, sterile for 0.2-0.5 mm, sagittate at the base, glabrous. *Pistil* glabrous, 8-21 mm long, with apex 0.3-0.5 way along anthers, with an abscission layer at the apex of the ovary and therefore style and pistil head shed with the corolla (the pistil head not sticking to the anthers as in *Voacanga*); ovary ovoid, 2-3 x 1-2 x 1-2 mm, acuminate, of two carpels connate at the base for 0.5-1 mm, often with a disk-like thickening at the base; style filiform, 5-16.5 mm long; pistil head composed of an entire or undulate ring 0.3-0.5 x 1.2-1.5 mm, a broadly obovoid apically 5-lobed central part 0.7-1 x 0.7-1 mm and a short bilobed apex about 0.2-0.3 x 0.2-0.3 mm. Ovules approximately 70 in each carpel. *Fruit* of 2 separate mericarps; mericarps green or yellow, obliquely ellipsoid, ovoid or almost pod-like, obtuse or shortly acuminate at the apex, 30-60 x 17-25 x 14-20 mm, slightly recurved, with a longitudinal ridge or narrow wing at each side and often a faint adaxial ridge, slightly lumpy and smooth (not dotted) when fresh and minutely tuberculate when dried, about 10-40-seeded; wall rather thin, 1.5-2 mm thick in fresh fruits; aril white, enveloping the seed. *Seed* ochraceous, obliquely ellipsoid or oblong, 7.5-10 x 4-4.5 x 3 mm, with deep longitudinal grooves in a spongy testa with long papillae; embryo 4.5-5.5 mm long; cotyledons ovate or oblong, 2 x 1.3-1.5 mm, rounded at the apex, subcordate at the base; rootlet 2.5-3.5 x 0.5-0.6 mm.

DISTRIBUTION: Northern and western South America.

ECOLOGY: On river banks in periodically inundated savanas, bush or forest. Alt. 0-1000 m. Flowering in Venezuela mainly from February to May, in Brazil, Peru and Bolivia mainly from July to November, in other countries scattered over the year. A fruiting season could not be deduced from the available data.

Geographical selection of the approximately 560 specimens examined:

COLOMBIA. Chocó: Mun. Bojayá, Quebrada del Fuerto, *Forero* et al. 9283 (MO). Bolívar: Mt Líbano, *Romero-Castañeda* 1706 (COL, MO); Mun. San Martin de Loba, *Cuadros* 3755 (MO, WAG). Norte de Santander: Ocaña, *Holton* 6 May 1854 (G, K, NY). Arauca: Río Meta, confluence with Caño Aricaporo, *Cuatrecasas & García Barriga* 4311 (COL). Casanare: km 5 Aguazul-Mani Road, *Sastre* 737 (P). Tolima: Río Magdalena, between Morales and Tenerife, *Bonpland* 1489 (P, P-BO; phot. of P-BO sheet GH, US, type of *T. tetrastachya*). Meta: Villavicencio, *J.J. Triana* 3396, p.p. (BM); Cabuyaro, Terr. San Martin, *Sprague* 10 Jan. 1899 (K, US); km 21 Villavicencio-Puerto López Road, *Sanbria & Calle* 137 (US); Río Guayabero, 10 km from Caño Lozada, *Pinto* et al. 213 (P); Agua Linda, *Haught* 2577 (COL, MO, US); Río Metica, just E of Puerto López, *Davidse & Llanos* 5481 (COL, MO, Z); Río Meta, María, *Cuatrecasas* 3771 (US). Vichada: El Tuparro National Park, *Zarucchi & Barbosa* 3476 (WAG); Upper Río Guaviare, *Cuatrecasas* 7600 (COL, F, US). Vaupés: Mitú, 2-3 km below Urania, *Zarucchi & McElroy* 1165 (COL, GH, K, USF); lower Río Kubiyú, *Zarucchi* 1252 (COL, ECON, GH, K); Soratama, *Schultes & Cabrera* 19764 (BM, GH, MO, U, US); Río Ariari and Río Guayabero confluence, *Pinto & Sastre* 911 (P); Río Inírida, *Fernández* 2264 (US). Caquetá: Río Ortequaza, San Luís, *Cuatrecasas & Soderstrom* 31743 (US); Florencia, *Juzepczuk* 6483 (LE); S of Cartagena, *Romero-Castañeda* 4070 (COL). Putumayo: Río Putumayo, La Concepción, *Cuatrecasas* 10839 (COL, F, US). Amazonas: Mun. Leticia, Amacayacu National Park, *Rudas* et al. 2157 (MO); Río Loretoyacu, about 3 km from mouth,

Plowman et al. 2371 (COL, ECON, K, P, S, US); Leticia, *Idrobo* 4692 (COL, ECON); Puerto Nariño, *Soejarto* 899 (BO, ECON, GH).

VENEZUELA. Zulia: near Río Catatumbo, *Liesner* & *González* 13214 (MO, WAG); ca km 70 Machiques-La Fria Road, Río Ariguisa, *de Bruijn* 1223 (MO, S, U, US, WAG); ca 13 km NE of intersection of Maracaibo-La Fria Highway and Río Aricuaisá, *Davidse* et al. 18423 (MO, WAG). Barinas: Ticoporo F.R., *Breteler* 3712 (S, U, UPS, US, WAG); Caño Anarú, *Trujillo* et al. 14711 (K, U). Apure: San Camilo F.R., *Steyermark* 101933 (Z); Laguna del Termino, 18 km SSW of Elorza, *Davidse* & *González* 16083 (MO, WAG); Hato Morichito, 60 km W of Achaguas, *Lock* 85/29 (K); Distr. Pedro Camejo, ca 19 km WSW of Paraqüito, *Davidse* & *González* 13954 (MO); ca 1 km SE of Yaruro, Davidse & González 15887 (MO); Distr. San Fernando, N bank of Río Orinoco, N of Isla Urbana, *Davidse* & *González* 13180 (MO, Z); near La Chorrerosa, *Davidse* & *González* 13480 (MO); between El Yagual and Cunaviche, *Fernández* 674 (GH); 4 km NE of El Betun, Río Capanaparo, *Davidse* & *González* 12962 (MO, Z); ca 8 km ESE of San Carlos del Meta, Río Meta, *Davidse* & *González* 13791 (MO, WAG); Río Orinoco, opposite Isla Pantallo, *Davidse* & *González* 14476 (MO); Isla Peladura, *Davidse* & *González* 12139A (GB, MO, Z). Amazonas: Dept. Atures, Isla Hormiga, *G.A. Romero* et al. 1954 (GH); near Isla del Raton, *Breteler* 4861 (U, WAG); San Fernando, Río Atabapo, *Spruce* 3660 (BR, K, P, TCD, W, paratype of *T. hirtula*); Caño Yureba, tributary of lower Río Ventuari, *Guanchez* 1542A (MO), 1617 (MO); Delta of Río Ventuari, *Ll. Williams* 14991 (VEN); near Ríos Casiquiari, Vasiva and Pacimoni, *Spruce* 3175 (AWH, BM, C, CGE, G, GH, GOET, K, LD, P, W, paratype of *T. hirtula* and of *Bonafousia juruana*); Río Siapa, *Stergios* et al. 9870 (MO, WAG); Río Casiquiare, *Stergios* & *Aymard* 9171 (WAG); Isla Macará, *Cardona* 1299 (VEN); near Río Ocamo, *Fuentas* 16 (VEN); Tamatama,

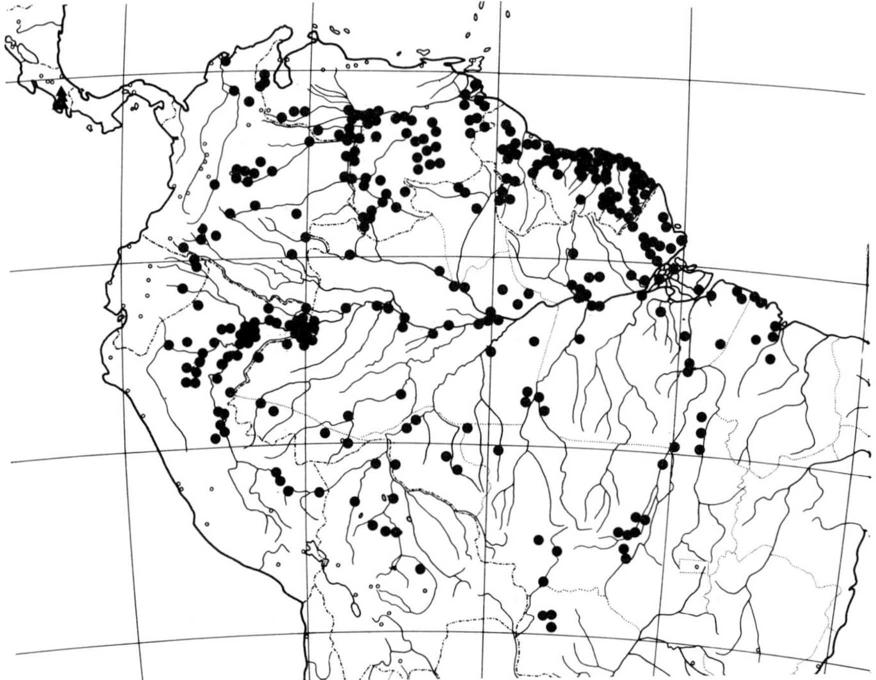

Map 50. ●. *Tabernaemontana siphilitica*, ▲. *Stemmadenia pauli.*

Ll. Williams 15271 (F, MO, NY, US); Puerto Venado, *Morillo* et al. 7344 (VEN); between Ocamo and Mavaca, *Aristeguieta* & *Lizot* 7428 (F, MO, NY, U, US). Bolívar: Caño Chaviripa, Distr. Cedeño, *Stergios* et al. 8608 (MO); Caicara-Los Pigiuaos Road, *Stergios* et al. 7250 (MO); Guayapo, *Ll. Williams* 11785 (MO, US); between Puerto Paez and Orupe, *Wurdack* & *Monachino* 39965 (G, K, NY, S, U, US); Mun. Sucre, Jabillal, *Sanoja* & *Fernández* 2661 (MO); Mun. Raul Leoní, *Delgado* 258 (MO); Río Nichare, *Steyermark* & *Gibson* 95641 (K, MO, US); Guachimaca, El Pescado, *Ll. Williams* 11527 (US, VEN); Río Caura, near Guanagujaña, *Stergios* & *Delgado* 12593 (MO); Mun. Foráneo, Caño Fatúu, *Aymard* et al. 6920 (MO); Cerro Guiaquinima, *Steyermark* 90804 (MO, NY); ibid., Río Canapo, *Boom* 9325 (WAG); Río Paragua, between La Paragua and Salto de Auraima, *Killip* 37280 (MO, US); near Macarapa, *Liesner* & *González* 5677 (MO, VEN); Raudal Guaiquinima, *Cordona* 488 (US, VEN); between Guaiquinima and Río Torono, *Killip* 37517 (NY, US); Mun. Raul Leoní, 25 km N of Macizo Ichún, *Fernández* 4401 (MO); Tumeremo, *Stergios* et al. 3801 (MO, PORT); km 50 Tumeremo-Bochinche Road, *Rutkis* & *Udris* 993 (PORT); El Dorado, *Couret* 42 (US); Cuyuni R., near Anacoco, *A. Gentry* et al. 10720 (F, LD, MO, US, WIS, Z). Delta Amacuro: Dept. Antonio Díaz, Caño Jotajana, *Steyermark* 115126 (MO); Sacupana, *Rusby* & *Squires* 76, p.p. (A, BM, E, G, M, MO, NY, UC, US, W, WU, Z; the GH and K sheet belong to *T. undulata*); Río Grande o Toro, *Breteler* 3862 (S, SP, U, UPS, US, WAG); Curiapo, *Gines* 4964 (US); below La Margarita and Puerto Miranda, *Steyermark* 87741 (NY).

GUYANA. Barima R., *Jenman* 7019 (K, U); Wenamu Creek, Barama R., For. Dept. 6987 (K, U); Cuyuni R., between Aurora and Takar-opali, *Gillespie* 2323 (P); Puruni R., Mazaruni R., *Bogan* 38 in F.D. 7722 (K, MO, NY, U); near Lower Camaria, *Sandwith* 654 (K); Mazaruni R., *Fanshawe* F.D. 4749 (K, NY); Essequibo R., Rockstone, *Gleason* 883 (NY); Berbice, *Robert Schomburgk* I 292 (BM, CGE, E, F, FI-W, G, G-DC, K, L, OXF, P, SING, TCD, UPS, US, W; phot. of G-DC sheet NY, P, US, lectotype of *T. longifolia*); Corantyne R., *im Thurn* Sept. 1879 (K); Takutu Creek to Puruni R., For. Dept. 4894 (K, MO, NY); Corantyne R., above Cow Falls, *McDowell* 2383 (P); Epira, *Jenman* 118 (K); Rupununi Distr., near Karasabai, *Knapp* & *Mallet* 2772 (MO, WAG); ca 5 km NE of Sumara, *McDowell* et al. 1980 (P); W of New R., tributary of Corantyne R., For. Dept. 7380 (NY); Kanuku Mts, A.C.Smith 3514 (A, B, G, K, MO, NY, P, S, U, US, W): S of Lethem, *Irwin* 818 (US); Puwib R., *Jansen-Jacobs* et al. 291 (CAY, K, MO, U, US, WIS); sin. loc., *Robert Schomburgk* I 41a (K, paratype of *T. longifolia*).

SURINAME. Nickerie, Wouw Canal, *Lanjouw* & *Lindeman* 3199 (NY, U); Koekri Creek, Corantyne R., *Gonggrijp* BW 2176 (U); Arawarra Creek, *Linder* 85 (GH); Nickerie R., *Tulleken* 429 (L, MO); Coppename R., *Boon* 1061 (U); Saramacca R., near Maripaston, *Pulle* 31 (U); near Paramaribo, *Kappler* 1627 (G, MO, P, S, U, W; phot. of G sheet GH; phot. of W sheet MO, type of *T. guianensis*); ibid., *Wullschlaegel* 314 (BR, GOET, NY, U, W); railroad, near km 70, *Maguire* & *Stahel* 23638a (A, K, MO, NY, U, US); Powakka Creek, *Mennega* 287 (A, C, U); Toekoemoetoe Creek, Upper Saramacca R., *Daniels* & *Jonker* 1329 (BR, U, UC, WIS); between Wia wia-bank and Grote Zwiebelzwamp, *Lanjouw* & *Lindeman* 1161 (K, NY, U); Perica, *Heyde* & *Lindeman* 293 (U); Marowijne R., *Hugh-Jones* 107 (K, U); Bakhuis Mts, *Florschütz* & *Maas* 2360 (U); near Jacob-kondre, *Maguire* 23840 (A, K, MO, NY, U, US); near Janbasigado, *Pulle* 150 (U); Sara Creek, *van Donselaar* 1099 (U); Tapanahoni R., near Manlobi Mts, *Wessels Boer* 225 (U); Kapoea Rapids, *Rombouts* 648 (AAU, P, U, US); Litani R., *Sastre* 1419 (CAY, NY, P, U); Ouarémapan Creek mouth, *Sastre* 1426 (CAY, NY, P, U); sin. loc. *Dahlberg* s.n. (LINN 302.3, S, S-LINN, UPS-THUNB. 6156, lectotype); *Hostmann* 320 (BM, CGE, G, K, OXF, TCD, neotype of *T. repanda*).

French Guiana. Godebert, *Wachenheim* 435 (P, WAG); Malmanoury Creek, *Billiet & Jadin* 1723 (BR, CAY, P); Acarouany, *Sagot* 825 (G, P); Iracoubo R., *Oldeman* B 597 p.p. (CAY); near Petit-Saut, Sinnamary R., *Prévost* 1314 (CAY, P); Kourou R., *Oldeman & Sastre* 1370 (CAY, P, U); Cayenne, *Martin* s.n. (FI-W, type of *T. speciosa*), s.n. (K, P); Maroni R., Beligui, *Sastre & Moretti* 3815 (CAY, K, MO, P, WAG); Abounamy Creek, *Petrov* 237 (CAY, P, U); Saut Stephanie, Sinnamary R. basin, *Hoff* 6025 (P); between Bélizon and Folie Creek, *Oldeman & Sastre* 1455 (CAY, MO, P, U); Approuague R., above Tortue Creek, *Oldeman* 2308 (CAY, P); Gabrielle Creek, *Leeuwenberg* 11708 (CAY, MO, P, USF, WAG); ibid., *Prévost* 238 (CAY, P, WAG); Kaw, *Billiet & Jadin* 4555 (BR); Grand Inini R., *de Granville* 7310 (CAY); Tampoc R., *Lemoine* or BAFOG SF 7856 (NY, U); Carapana Creek, *Oldeman* 3010 (CAY, P); Mandé, Oyapock R., *Oldeman* 1748 (CAY, MO, P, U); Camopi R., *Oldeman* 2719 (CAY, NY, P, U); Gabaret Creek, *de Granville* 10251 (CAY, US); Upper Marouini R., *de Granville* et al. 9593 (CAY, US, WAG); Tampoc R., Serv. For. 7856 (CAY, NY, P); Eulepoussing Creek, *Sastre* 4633 (MO, P); between Saut Petit et Saut Impossible, Camopi R., *Oldeman & Sastre* 169 (CAY, NY, P, U); Saut Moutouci, *Oldeman* B 3221 (CAY, P, WAG); sin. loc., *Poiteau* s.n. (G, type of *T. guyanensis*).

Brazil. Roraima: Ecol. Res. Maracá, *Milliken* et al. 166 (K, U); ibid., *Stannard & Arrais* 752 (K); between Pratinha and Rio Apiaú, *Prance* et al. 3986 (K, NY, S, US, Z). Amazonas: Rio Javari, N of Palmeiras airstrip, *Lleras* et al. P17089 (NY); Forte Veneza, *Lleras* et al. P16908 (AAU, NY, Z); Benjamin Constant, *Duarte* 6875 (Z); Ilha Aramaçá, almost opposite Tabatinga, *Prance* et al. 16756 (K, MO, NY, U, US, Z); near Belém, Rio Solimões, *Mori* et al. 9196 (K, NY, Z); Bôca do Acre, *Prance* et al. 2284 (GH, K, P, S, U, US, Z); Mararí, Rio Juruá, *Ule* 5178 (G, HBG, K, L; phot. of G sheet GH, MO, US, type of *B. juruana*); Tonatins, *Mori* et al. 9062 (K, NY, Z); Rio Vaupés, near Panuré, *Spruce* 2474 (K, P); Fonte Boa, *M. Silva* 2158 (MG); near Maraã, *Plowman* et al. 12351 (MG, US, WAG); Lago Ega (= Tefé), *Poeppig* 2504a (P, W); ibid., Santa Missoes, *Amaral* et al. 105 (MO, WAG); Rio Purus, Lábrea, *Prance* et al. 8069 (F, INPA, MG, NY, S, U, US, Z); Mun. Humaitá, near Livramento, *Krukoff* 6312 (NY); Mun. Coari, *L. Coêlho* 119 (INPA); Codajás, *E. Ferreira* 58-264 (INPA, MG); Rio Demeni basin, near Tototobí, *Prance* et al. 10337 (F, GH, K, NY, P, S, U, US, Z); Rio Negro, just below mouth of Rio Branco, *Prance* et al. 16262 (C, F, G, GH, K, NY, S, U, US, Z); Rio Jauaperí, *M.R. Santos* 8 (F, NY, Z); Xiborena, near Manaus, *W. Rodrigues & L. Coêlho* 6864 (INPA); Igarapé Januari, S bank of Rio Negro, opposite Manaus, *Prance* et al. 3033 (K, S, U, US, Z); Rio Manacapuru, *W. Rodrigues* 418 (MG, MO); Rio Madeira Falls, *Rusby* 2396 (BM, G, GH, K, NY, P, US, WIS; phot. of US sheet WAG; this specimen had been confused with 2376, the type of *T. cuspidata*); Mun. Careiro, Cambixe, *Mello* INPA 12729 (INPA); Rio Uatumã, Mun. Itapiranga, above Balbina Fall, *Cid* et al. 394 (MO, NY, US, WAG); Mun. Pres. Figueredo, Rio Pitinga, 2 hrs above confluence with Rio Uatumã, *Cid* et al. 6906 (K, WAG); between Morena and Balbina Falls, *Cid* et al. 144 (MO, NY, US, WAG); Mun. Maués, Rio Parauari, *Zarucchi* et al. 3011 (NY). Pará: Upper Rio Tapajós, Rio Cururú, *Anderson* et al. 10504 (F, K, NY, Z); ibid., 1-10 km upriver from Pratatí, *Anderson* et al. 10836 (F, K, NY, Z); Mun. Oriximiná, Rio Erepecurú, *Farney & Mota* 1963 (WAG); near Bocca do Parú, *Krukoff* 5927 (A, BM, BR, G, K, MO, NY, S, U, US), 5941 (A, G, K, MO, NY, S, U, US); Santarém, *Spruce* 430 (K, P); near Obidos, *Kuhlmann* 92 (RB, US); between Santarém and Obidos, *Spruce* 233 (BM, G, K, M, NY, P, TCD, W, paratype of *T hirtula*); Upper Rio Cupari, *Krukoff* 1083 (A, BM, G, K, NY, P, S, U); Taparinha, near Santarém, *Ginzberger* 330 (WU); Igarapé Maica, *Bamps* 5285 (BM, BR, WAG); Sete Varas airstrip on Rio Curua, *Strudwick* et al. 4198 (F, GH, MO, WAG); ca 23 km upstream from Lageira airstrip, Rio Maicuru, *Strudwick* et al. 3843 (F, K, MO, WAG); Mun. Almeirim, Mt Dourado, *M.J. Pires* &

N.T. Silva 1360 (MG, WAG); Lower Rio Xingu, Ilha Mucuripe, *Almeida* 363 (MG); Rio Jari, Mt Dourado, *Oliveira* 4415 (NY); Gurupá, *Killip & A.C. Smith* 30625 (US); Ilha do Marajó, Mun. Anajás, *Sobel* et al. 4904 (K, US, WAG); Rio Ararí, *Black* et al. 52-14297 (MO); Tucuruí dam area, Rio Tocantins, *A.S. Silva* et al. 125 (K, WAG); ibid., Itupiranga, *Berg & Henderson* BG 679 (MO, US, WAG); Marabá, *Fróes & Black* 24716 (MO); Rio Itacaiune, São Domingos Fall, *Fróes & Black* 24469 (MO); Belém, *Martius* s.n. (FI-W, K, L, M, W; phot. of M sheet GH, MO, US, lectotype of *T. hirtula*); Rio Capim, Lago Putiryta, herb. *Schwacke* 21 Feb. 1862 (GOET); Road BR 22, Capanema to Maranhão, km 80, *Prance & N.T. Silva* 58779 (K, S, US, Z); same road, Rio Piria, N of km 90, *Prance & Pennington* 1731 (GH, K, P, S, U, US, Z); near Miraselvas, *Davidse* et al. 18121 (MO, NY, US, WAG). Goias: Belém-Brasilia Highway, 5-10 km N of Nova Colinas, *Prance & N.T. Silva* 58489 (GH, K, NY, S, U, US, Z); 2 km S of Guará, *Irwin* et al. 21430 (GH, MO, US, Z); Rio Araguaia, between Rio Caiapós and Santana do Araguaia, *N.T. Silva* 4798 (NY). Amapá: Rio Oiapoque, *Fróes* 26684 (MO); ibid., Mt Tipac base, *Irwin* 48763 (GH, K, NY, S, U, US, Z); Rio Mutaquere, *Irwin* et al. 48060 (NY); Colonia do Torrao, *J.M. Pires & Cavalcante* 52660 (NY, Z); Rio Cuminá, lower Rio Trombetas, *Ducke* 7911 (RB, type of *T. duckei*); Rio Calçoene, *Ducke* 2509 (RB; phot. of lost B sheet GH, MO, NY, US, paratype of *T. duckei*); Igarapé Carnot Pequeno, *Austin* et al. 7358 (MO, NY, US, WAG); Rio Araguari, near Pedra Finca Camp, *J.M. Pires* et al. 51654 (NY, Z); Serra do Navio, *Cowan* 38492 (NY, US); Rio Araguari, between Travessao and Santa Maria Falls, *J.M. Pires* et al. 50382 (NY); Rio Falsino, *Pruski* et al. 3365 (K, WAG); Fazenda Capoeira do Rei, *Egler* 657 (MG, NY); Rio Cupixi, *Rosa* 1017 (MO, NY); Igarapé do Lago, Mun. Macapá, *Fróes & Black* 27532 (MO). Maranhão: 3 km NW of Lago do Junco, *Daly* et al. 537 (F, K, US, WAG); Mineirinho, Rio Pindaré, *Jangoux & Bahia* 812 (Z); Santa Inês, *Rosa & Vilar* 2813 (F, NY); Rio Alto Turiaçu, Nova Esperança, *Jangoux & Bahia* 206 (NY, Z); Rio Maracaçumé Region, *Fróes* in *Krukoff* 1744 (A, BM, F, G, K, MO, P, S, U, US, WIS, paratype of *B. juruana*). Acre: Cruzeiro do Sul, near Serra da Moa, *Prance* et al. 12404 (F, K, M, NY, S, U, US, Z); near Tarauacá, *Prance* et al. 7354 (B, K, NY, S, US, W, Z); 4 km E of Madureira, *Prance* et al. 7608 (NY, Z); Rio Branco, *Lowrie* et al. 197 (NY). Rondônia: Rio Pacaás Novos, *Prance* et al. 6845 (G, K, M, NY, S, US, Z); island in Rio Madeira, near mouth of Rio Jaciparaná, *Prance* et al. 5336 (GH, NY, S, U, US, Z); Porto Velho, *Cordeiro & J.F. da Silva* 200 (MO); Porto Velho-Cuiaba Highway, between Nova Vida and Rondônia, *Maguire* et al. 56759 (F, NY, Z); Rio Ji-Paraná, Rolim de Moura Road, *Vieira* et al. 585 (WAG). Mato Grosso: near Tabajara, *Krukoff* 1449 (A, BM, G, K, MO, P, S, U, UC, US, WIS); Rio Aripuanã, above Andurina Falls, *Berg* et al. P18672 (F, GH, NY, S, Z); Pantanal, *Rossi & I .Cordeiro* Aug. 1980 (HBG); Rio Teles Pires, *dos Passos* 6 (MG); Santa Cruz (= Barra do Bugres), *S. Moore* 304 (BM); Mun. Poconé, 16 km NW of Porto Jofre, *W. Thomas* et al. 4622 (F, K, MO, US, WAG); Cuiabá, *Malme* II 1871 (LD, S, UPS, type of *T. cuyabensis*); near Barra do Garças, *Irwin & Soderstrom* 6849 (B, GH, K, U, US, Z); between Barra do Garças and Xavantina, *Onishi & Fonseca* 335/1104 (K); km 47 Xavantina-Aragarcas Road, *Harley* et al. 10969 (K, P, U, UC); Serra do Roncador, *Prance & N.T. Silva* 59320 (GH, K, S, U, US, Z); Araguaia National Park, Ilha Bananal, *Ratter* et al. 4403 (K). Mato Grosso do Sul: Rio Paraguai, ca 80 km from Corumbá, *Elias* et al. 1663 (US, WAG); Mun. Corumbá, *Pereira* et al. 240 (SP, Z); Fazenda Salina, Pantanal do Rio Negro, *Dubs & Kramer* 1057 (WAG, Z).

ECUADOR. Napo: near Puerto Montúfar, *Holm-Nielsen* et al. 21224 (AAU, WAG); near Río Aguarico, *Holm-Nielsen* et al. 21636 (AAU, K, WAG). Pastaza: Río Papayacu at Río Curaray, *Holm-Nielsen* et al. 22615 (AAU, WAG).

PERU. Amazonas: Pongo de Manseriche, *Tessmann* 4405 (G, S). Loreto: Washintsa, Río Huasaga, *Lewis* et al. 11308 (MO); Maucallacta, *Klug* 3942 (BM, F, GH, K, MO, NY, S, UC, US, WIS); near Yurimaguas, *Mexia* 6071 (BM, G, GH, K, L, MO, NY, S, US, Z); Quebrada Cuinico, *Croat* 17793 (NY, USF, WAG); NW of Chazuta, *Schunke* 6349 (F, MO, US); Río Ampyacu, *Croat* 20696 (MO); Prov. Maynas, District Fernando Lores, Tamshiyacu, *Rueda* & *J. Ruiz* 811 (WAG); District Fernando Lores, Tamshiyacu, *Rueda* & *J. Ruiz* 811 (WAG); Río Amazonas, Isla Rondiña, opposite Leticia, *Plowman* 6388 (GH, K); Río Yavari, opposite Brazilian village Paumari, *A. Gentry* & *Revilla* 20800 (F, MO); between Estiron Pucaurquillo and Pebas, *Ayala* 2787 (MO, NY, WAG); Brillo Nuevo, *Plowman* et al. 6864 (GH); San Alejandro, *Vásquez* & *Jaramillo* 19 (F, LL, MO); Fatima, *Vásquez* et al. 425 (K, MO); Quebrada Tahuayo, above Tamishiyaco, *Croat* 19721 (AAU, K, MO, U, UC, US, WIS, Z); Río Mazán, *Schunke* 60 (A, NY, UC, US); Iquitos, *Killip* & *A.C. Smith* 27414 (F, MO, NY, US; phot. of US sheet WAG, type of *T. killipii*); Río Itaya, *Killip* & *A.C. Smith* 29388 (MO, NY, US); Buena Vista, Río Tahuayo, *Vásquez* & *Jaramillo* 1215 (MO, P); Nauta, *A. Gentry* et al. 29946 (LL, MO); Prov. Requena, Becerro, *Spichiger* & *Encarnación* 1072 (G, MO, WAG); Santa Clara, Río Nanay, *Góngora* & *Oliveira* 2612 (MO, US); Río Marañon, near mouth of Río Tigre, *Killip* & *A.C. Smith* 27533 (NY, US); Cocha Paucate, *Vásquez* & *Jaramillo* 891 (MO, P); Lago Yarinacocha, *Maas* et al. 4597 (U, WAG); Cocha Supai, *Spichiger* & *Encarnación* 1147 (G, WAG); Prov. Marsella, Río Tigre, *Lewis* et al. 12718 (MO); Quebrada Nucuray, tributary of Río Marañon, *Diaz* & *J. Ruiz* 888 (MO, P); Lower Río Huallaga, *Ll. Williams* 4656 (MO); La Victoria, *Ll. Williams* 2539 (F, paratype of *B. juruana*). Ucayali: middle Río Blanco, *Tessmann* 3022 (G, NY, S, type of *T. tessmannii*); Río Aguaytia, *McDaniel* et al. 2538 (US); Río Ucayali, San José de San Martín, one hour below Pucallpa, *McDaniel* 2529 (LL, US); Prov. Coronel Portillo, Bosque Nac. Iparía, *Schunke* 2801 (G, US). San Martín: near Juan Guerra, *Ferreyra* 17351 (MO, US). Huánuco: Prov. Pachitea, Quebrada Miel de Abeja, *Schunke* 2051 (G, US); ca 26 km S of Puerto Inca, *Wallnöfer* 17-301088 (Z). Madre de Dios: Manu National Park, Cocha Cashu Stat., *Forster* 5893 (F); ibid., *Emmons* 88 (MO); Río Manu, playa 16 above the Boca, *Forster* 3156 (K, MO, US); Tambopata Prov., 1.15 km NNE of Puerto Maldonado, *Nuñez* et al. 10556 (WAG).

BOLIVIA. Pando: Prov. Vaca Diez, ca 3 km SW of Riberalta, *Solomon* 6350 (MO, U, WAG). Beni: Río Chaparé-Mamoré, *Werdermann* 2247 (MO, S, paratype of *B. juruana*); Prov. Iturralde, Luisita, *Beck* & *Haase* 9954 (HBG); Prov. Ballivian, Espiritu, *Beck* 5253 (WAG, Z); Prov. Cercado, ca 11 km NW of Trinidad, *Solomon* et al. 8148 (MO, WAG); Est. Biol. Beni, 50 km E of Río Maniqui, *Solomon* 14654 (MO, WAG). Santa Cruz: Chaparal, *Herzog* 474 (Z). Cochabamba: Todos Santos, *Steinbach* 382 (MO).

CULT. Brazil, Bot. Garden Pará, from Rio Purus, *C.F. Baker* 25 June 1908 (B, LD, NSW, P, UC); Rio de Janeiro, *Glaziou* 9936 (C, K, P).

Note. *T. siphilitica* is closely allied to *T. sananho* (q.v.).

96. Tabernaemontana solanifolia A. DC., Prod. 8: 365 (1844). – Lectotype: Brazil, Bahia, Serra Jacobina, *Blanchet* 2724 (lectotype G-DC, designated here; isolectotypes BM, BP, BR, FI-W, G, K, LE, P, W; phot. of lost B sheet GH, MO, NY, US).

Fig. 96, p. 382; map 51, p. 384

Homotypic synonym:

Peschiera solanifolia (A. DC.) Miers, Apoc. S. Am. 46 (1878); Markgraf in Notizbl. Bot. Gart. Berlin 14: 173 (1938).

Heterotypic synonyms:

T. fallax Muell. Arg. in Martius, Fl. Bras. 6. 1: 84 (1860). *P. fallax* (Muell. Arg.) Miers, op. cit. 40. P. *solanifolia* var. *fallax* (Muell. Arg.) Allorge in Mém. Mus. natn. Hist. Nat. IIB. 30: 138 (1985). – Type: Brazil, Minas Gerais, sin. loc., *A. Saint-Hilaire* 944 (holotype P; isotype F; phot. of holotype F, GH, US).

T. accedens Muell. Arg. in Kjoeb. Vidensk. Meddel. 1869: 107 (1869). – Type: Brazil, Lagoa Santa, *Warming* s.n. (holotype C, not seen; phot. F, GH, US).

T. warmingii Muell. Arg., l.c. – Type: Brazil, Lagoa Santa, *Warming* s.n. (holotype C, not seen; phot. F, GH, US).

T. nervosa Glaziou in Bull. Soc. Bot. France 57. Mém. 3: 453 (1910), non Desf. ex Poir. (1817). – Type: Brazil, Minas Gerais, sin. loc., *Glaziou* 15212 (holotype P; isotypes C, G, K; phot. of C sheet F, GH, NY).

T. affinis var. *campestris* Rizz. in Simposio sobre o Cerrado, ed. Univ. São Paulo 165 (1963). *Peschiera campestris* (Rizz.) Rizz., Arq. Jard. Bot. Rio de Janeiro 25: 181 (1981). – Type: Brazil, *Rizzini* s.n. (holotype not traced).

Shrub or small tree 0.50-8 m high. Branches pale or dark brown, lenticellate; branchlets terete, glabrous or pubescent. *Leaves* sessile or with petiole up to 3 mm long; petiole glabrous or pubescent; (ocreae not widened into intrapetiolar stipules); blade papery when dried, variable in size, elliptic, narrowly elliptic or narrowly obovate, 1.3-3 x as long as wide, 3-24 x 1.5-10.5 cm, rounded to acuminate at the apex, rounded or cordate at the base, less often cuneate and decurrent into the petiole, entire, glabrous or sparsely pubescent above, glabrous or pubescent beneath, without black dots beneath, with 7-21 pairs of rather straight or upcurved secondary veins forming an angle of 60-80° with the costa; tertiary venation (rather) conspicuous, reticulate beneath. *Inflorescence* pedunculate, 3-12 x 3-12 cm, many-flowered, rather dense. Peduncle 5-75 mm long, glabrous or pubescent; pedicels 2-14 mm long, glabrous or pubescent. Bracts sepal-like, 0.5-1 x as long. Flowers slightly fragrant, open during the day. *Sepals* pale green (?), connate at the base for 0.3 mm, ovate or strap-shaped, 2-5.5 x as long as wide, 3-6 x 1-2 mm, obtuse or acute, glabrous on both sides and not ciliate, or pubescent all over outside and at apex inside and ciliate, with 2-9 colleters in 1 row inside in the middle at the base; colleters variable in size, 0.3-0.5 x 0.1-0.3 mm. *Corolla* white or cream(?), with a yellow throat, 9-15 mm long in the mature bud and forming a comparatively large ovoid head 0.4-0.5 of the bud length and 2-4 x as wide at the narrow part of the tube (4-6 x 2.5-4 mm) often with a blunt or sometimes acute apex, glabrous outside, pilose inside in stripes 0-1.5 mm long just below the insertion of the stamens and in a belt from there to the extreme base of the lobes; tube 1.2-2.4 x as long as the calyx, 0.44-1 x as long as the lobes, 5-11 mm long, flask-shaped, 2.5-3.5 mm wide above the base, narrowed above the insertion of the stamens and higher up even more to 1-2 mm wide, widened again at the throat to 1.5-3 mm wide, not twisted or sometimes twisted 0.25 turn from above the insertion of the stamens; lobes obliquely obovate to dolabriform, 1-2.2 x as long as the tube, 1.3-2 x as long as wide, 9-14 x 5-10 mm, rounded, if dolabriform at both apices, not auriculate nor undulate, spreading. *Stamens* with apex 0-3 mm below mouth of corolla tube, inserted 0.35-0.5 of the length of the corolla tube, (at 2.5-4.2 mm from the base); anthers sessile, narrowly triangular, 3-4 x as long as wide, 3.5-4.5 x 1-1.5 mm, apex acuminate, sterile for 0.4-0.7 mm, sagittate at the base, with straight tails. *Pistil* glabrous, with apex almost halfway along anthers, 3.5-5.5 mm long; ovary ovoid, 1.5-2 x 1-1.5 x 1-1.5 mm; style short, 1-2 mm long; pistil head 1-1.2 mm high, composed

Fig. 96. *Tabernaemontana solanifolia.* **1-2,** habit (x 2/3); **3,** leaf (x 2/3); **4,** opened corolla (x 4); **5,** sepal inside (x 8); **6,** pistil (x 8); **7,** fruit (x 2/3). 1 and 4-6 from Pereira 345; 2-3 from Blanchet 2724; 7 from Coradin et al. 6467.

of a basal ring 0.1-0.2 x 0.8-1 mm, consisting of 10 acute spreading or upcurved lobes, a stipitate 5-lobed depressed globe 0.3-0.4 x 0.6-0.8 mm and a stigmoid apical part 0.2-0.2 mm. Ovules approximately 50-100 in each carpel. *Fruit* or 2 separate mericarps; mericarps purplish-green or brown, obliquely ellipsoid, 30-60 x 15-35 x 15-30 mm, recurved, rounded or shortly acuminate at the apex, not ridged, densely covered with short blunt prickles, approximately 15-40-seeded; wall 1-3 mm thick in dried fruits; aril only at hilar side, not enveloping the seed. *Seed* dark or pale brown, obliquely ellipsoid, 9-12 x 4-7 x 3-4 mm, with longitudinal grooves, dull, papillose; embryo 7-8 mm long; cotyledons broadly ovate or oblong, 1.2-1.4 x as long as wide, 3.5 x 2.5-3 mm, rounded at the apex, rounded or subcordate at the base; rootlet 3.5-4.5 x 0.8-1 mm.

DISTRIBUTION: Brazil (Distrito Federal, Bahia, Minas Gerais, Goias).

ECOLOGY: Secondary forests or deciduous forests. Alt. 600-1100 m (or less?). Flowering mainly September-November and fruiting in March.

Geographical selection of the approximately 50 specimens examined:

Brazil. Bahia: Serra Jacobina, *Blanchet* 2724 (BM, BP, BR, FI-W, G, G-DC, LE, P, W; phot. of lost B sheet GH, MO, NY, US, lectotype), 3638 (BM, C, FI-W, G, P, W, paratype); Serra do Tombador, *Furlan* et al. CFCR 7475 (K); Mun. Morro do Chapéu, Ferro Doido Fall, *Martinelli* 5315 (RB, WAG); Mun. Utinga, Fazenda Lajinha, *Sarmento & Bautista* 835 (GUA, RB); 2 km from Lençois, Coradin et al. 6467 (K, WAG); Rio de Contas, *Luetzelburg* 16024 (M); km 5-15 Vitória da Conquista-Barra da Choça Carrasco, *T.S. dos Santos* 2538 (P, Z); 3-6 km from Barra do Choça, *Mori* et al. 11322 (K, NY, P); sin. loc., *Blanchet* 247 (BM, G, P, W). Goias: Rio São Bartolomeu, *Pereira* 7423 (RB); Rio Gama, *Glaziou* 21737 (BR, C, G, K, MPU, P, S). Minas Gerais: Itaobim, *Pereira* 10142 = Duarte 9232 (Z); Arrasuahy (= Araçuaí), *Glaziou* 15212 (C, G, K, P; phot. of C sheet F, GH, NY, type of *T. nervosa*); Mucurí, *Pereira* 10145 = *Duarte* 9235 (Z); Capelinha, *Brazao* 244 (RB); 8 km N of Teófilo Otôni, *Magelhães* 15667 (MO, WAG); 10 km N of Campanario, *Hatschbach* 44295 (HBG, MU, Z); Mun. Patrocínio, *Hatschbach & Nicolack* 54135 (WAG); Mun. Indianópolis, Lagoa dos Bexiguentes, *Magelhães* 289 (MO); Lagoa Santa, *Warming* s.n. (C, not seen; phot. F, GH, US, type of *T. accedens*), s.n. (C, not seen; phot. F, GH, US, type of *T. warmingii*); Mun. Manieria, Parque Estudial do Rio Doce, *Heringer* 18615 (K, SP, US); Rio Casca, *Magelhães* 15669 (MO); Sertao, *Reinhardt* Oct. 1855 (C); sin. loc., *Saint-Hilaire* C1 944 (F, P; phot. of P sheet F, GH, US, type of *T. fallax*). D.F.: Brasília, *Duarte & R.S.Santos* 95 (HBG, K, M, NY, Z); ibid., *Irwin* et al. 10722 (F, G, NY, US, W, Z); ibid., *B.A.S. Pereira* 345 (WAG).

Notes. The lectotype of *T. solanifolia*, *Blanchet* 2724, has leaves 6-21 x 2-8 cm, the paratype, *Blanchet* 3638, 7-19 x 3.5-9 cm, the type of *T. fallax, A. Saint-Hilaire* 944, 3.5-5 x 1.2-1.7 cm, the type of *T. nervosa, Glaziou* 15212, 7-11 x 3.7-6.5 mm, *Magelhães* 15669, 15-17 x 6-8 cm, *Mori* et al. 11322, 7.5-16 x 3.5-7.5 cm. The leaves of the two *Blanchet* collections are pubescent beneath, those of *A. Saint-Hilaire* 944, *Glaziou* 15212 and *Mori* et al. 11322 are glabrous, while those of *Magelhaes* 15669 are partly pubescent beneath. Therefore *T. fallax*, maintained as variety within this species by Allorge (1985), is not kept up as a separate taxon here.

97. Tabernaemontana undulata Vahl, Eclog. Amer. 2: 20 (1798) and Icon. 1: t. 6 (1798); G.F.W. Meyer, Prim. Fl. Esseq. Goettingen 135 (1818). – Type: Trinidad, sin. loc., Ryan anno 1778 (holotype C-VAHL; phot. A, F, GH, MO, P, US; isotypes BM, P-LA). Fig.97, p. 386; phot. 10, 11; map 51, p. 384

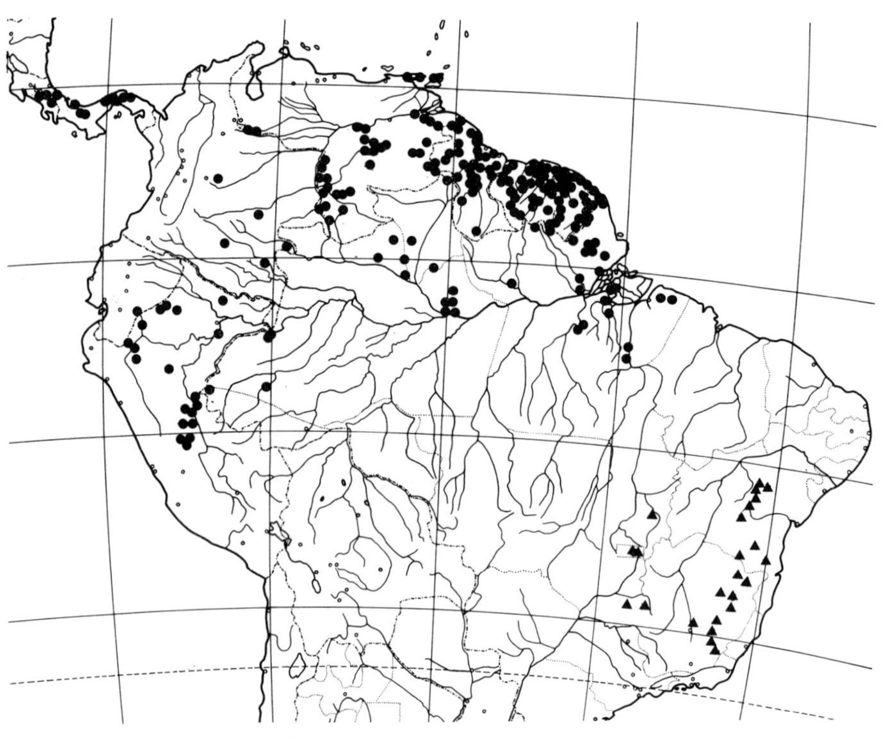

Map 51. ▲. *Tabernaemontana solanifolia,* ●. *T. undulata.*

Homotypic synonym:

Bonafousia undulata (Vahl) A. DC., Prod. 8: 359 (1844).

Heterotypic synonyms:

Echites brasiliensis Thunb., Diss. Echit. 5 (1819). – Type: Brazil, sin. loc., Herb. *Thunberg* (holotype UPS-THUNB 6135 and 6136).

T. meyeri G.Don, Gen. Syst. 4: 89 (1837). *Anartia meyeri* (G.Don) Miers, Apoc. S. Am. 80 (1878), partly, as for type. Types: Guyana, Essequibo R., *G.F.W. Meyer* s.n. (holotype apparently not preserved; not in GOET); ibid., near Rockstone, *Maas & Westra* 3936 (neotype U, designated here; isoneotypes K, MO, NY, WAG).

T. perrottetii A. DC., op. cit. 362; *B. perrottetii* (A. DC.) Miers, op. cit. 51. – Type: French Guiana, sin. loc., *Perrottet* 275 (holotype G-DC; isotype G).

Peschiera surinamensis Miq. in Linnaea 18: 742 (1844). – Type: Suriname, near Victoria, *Kappler* 1398 (holotype U; isotypes FI-W, G, P, S, W).

B. undulata var. *ovalifolia* Miers, op. cit. 49. – Type: Guyana, Essequibo R., Berbice, *Robert Schomburgk* I 42bis (holotype BM; isotypes G, K).

B. obliqua Miers, op. cit. 49. *T. obliqua* (Miers) Leeuwenberg in Meded. Landbouwh. Wageningen 83, 7: 60 (1984), **syn. nov.** – Type: Venezuela, Amazonas, Río Casiquiare, Vasivia and Pacimoni, *Spruce* 3199 (holotype BM; isotypes BR, CGE, G, K, NY, P, TCD, W).

T. albescens Rusby, Descr. 300 new sp. S. Am. Pl. 83 (1920). *Anacampta albescens* (Rusby) Mgf. in Pulle, Fl. Surinam 4, 1: 452 (1937). – Type: Venezuela, Delta Amacuro, Santa Catalina, *Rusby & Squires* 300 (holotype NY; phot. P; isotypes A,

BM, E, F, G, GH, K, M, MO, US, W, WU, Z; phot. of US sheet WAG).
Stemmadenia nervosa Standl. & L.Wms. in Ceiba 3: 126 (1952). – Type: Costa Rica, Puntarenas, Esquinas For., *Allen* 5834 (holotype US; phot. F, GH, NY, WAG; isotypes BM, F, GH, UC).

Shrub or small tree 1-10 m high. Trunk 1-10 cm in diameter; bark pale or dark brown or grey, rather smooth, shallowly longitudinally and transversely fissured, lenticellate, 1.5-2 mm thick; wood pale brownish. Branches pale brown, lenticellate; branchlets terete when fresh, glabrous. *Leaves* of a pair equal or subequal, petiolate; petiole glabrous, 3-15 mm long; (ocreae not or obscurely widened into intrapetiolar stipules); blade coriaceous or subcoriaceous also when fresh, elliptic or narrowly elliptic, 2.5-4.5 x as long as wide, 4-25 x 1-10 cm, acuminate to caudate at the apex, cuneate or sometimes rounded at the base or decurrent into the petiole, undulate or entire, glabrous and with scattered black dots on both sides, dull, seemingly velvety and much paler green or glaucous beneath, with 9-25 pairs of rather straight or upcurved secondary veins forming an angle of 60-85° with the costa; tertiary venation reticulate, rather conspicuous. *Inflorescence* solitary, erect, shortly pedunculate, 3-5 x 2-3 cm, 1-17-flowered, dense. Peduncle robust, glabrous, 2-8 mm long; pedicels glabrous, 3-6 mm long. Bracts very small, 0.2-0.3 x as long as the sepals, triangular, obtuse, deciduous. *Flowers* fragrant, open during the whole day. *Sepals* white and often with pink or greenish, less often red or green, equal or unequal (larger up to 1.5 x as long as smaller), connate at the base for 0.2-0.5 of their length (1-3.5 mm), erect (closely clasping the corolla base when fresh), ovate or elliptic, 1.5-2.8 x as long as wide, 2.5-7(-8) x 1.5-5 mm, rounded, sometimes subcordate at the base of the free part, entire, glabrous or sometimes puberulous outside, ciliolate or not, glabrous inside and with 3-12 colleters in 1 row in the middle of the base or slightly above; colleters 0.5-1.2 x 0.1-0.3 mm. *Corolla* tube white, mostly with pink or purple short and long longitudinal lines and often also spots, or less often red; lobes white or creamy, or with these short lines at the base, mostly with pink, purple or yellow, sometimes red; throat dark yellow or coloured as lobes; 20-25 mm long in the mature bud and forming a comparatively small transversely ellipsoid or sometimes broadly ovoid head 0.15-0.26 of the bud length (3-6 x 4-6 mm) with a rounded or sometimes obtuse apex, glabrous or more often puberulous on the part of the lobes not covered in bud, and often also on the apex of the tube outside, often ciliate on the edge of the same part of the lobes, with a 1-8 mm wide interrupted pilose-pubescent belt inside from 0-1 mm below the insertion of the stamens up to the level of 2.5 mm below to 3 mm above the apices of the anthers or less often that of the insertion itself, and often (if hairy outside) puberulous on the base of the lobes; tube 2.4-5.2 x as long as the calyx, 1-3 x as long as the lobes, 17-30 mm long, almost cylindrical, 1.5-5 mm wide, often narrowed below the insertion of the stamens in the throat to 1.5-4 mm wide, not twisted; lobes obliquely and narrowly elliptic or obovate, 0.36-1 x as long as the tube, 2-4.3 x as long as wide, 6-25 x 2.5-8 mm, rounded, undulate, with mostly recurved margins and therefore seemingly narrower, auriculate at the left side of the base, upcurved (arching upwards and touching at tip). *Stamens* with apex 1.5-11.5 mm below mouth of corolla tube, inserted 0.4-0.7 of the length of the corolla tube (at 10-22 mm from the base); anthers sessile, narrowly triangular, 3-5 x as long as wide, 4-6 x 1-1.5 mm, apex acuminate, sterile for 0.2-0.5 mm, sagittate at the base, glabrous. *Pistil* glabrous, 12-25 mm long, with apex 0-3 mm below apices of anthers; ovary ovoid, 2-3 x 1-2 x 1-2 mm, acuminate, of 2 carpels united at the base by a disk-like thickening 1-1.5 mm high; style filiform, 7-21 mm long, persistent; pistil head composed of an entire or undulate ring 0.2-0.7 x 1-1.2 mm, a broadly obovoid apically 5-lobed central part 0.5-0.8 x 0.6-1 mm and a stigmoid apex 0.2-0.3 x 0.2-0.3 mm.

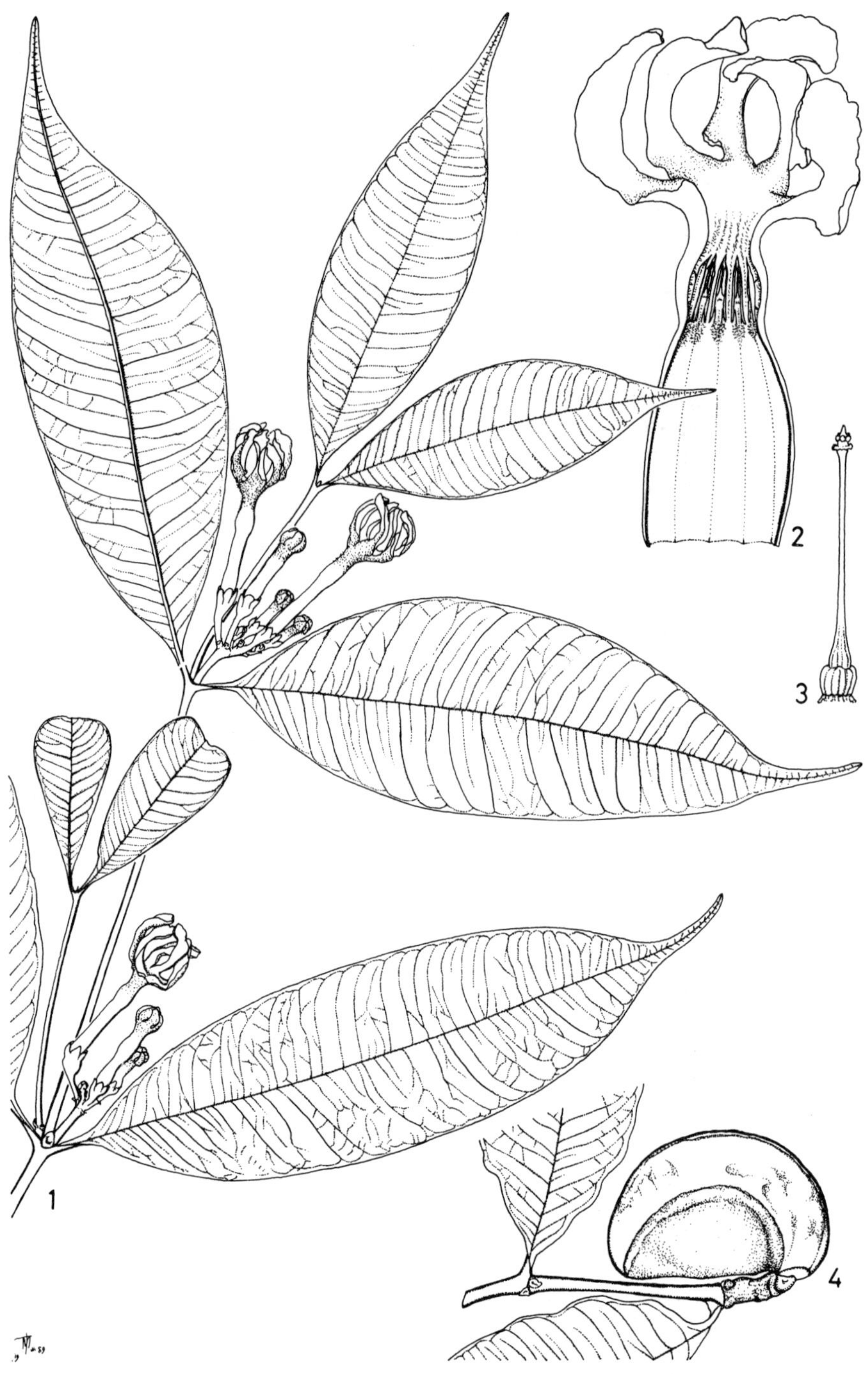

Fig. 97.*Tabernaemontana undulata.* **1,** habit (x 2/3); **2,** opened corolla (x 2); **3,** pistil (x 2); **4,** fruiting branchlet (x 2/3). 1 from Jacquin 1892; 2-3 from Feuillet et al. 9929; 4 from Leeuwenberg 11816.

Ovules approximately 70-200 in each carpel. *Fruit* of 2 separate mericarps; brown, purplish or purple, obliquely ellipsoid or sometimes pod-like, 20-60 x 10-30 x 10-30 mm, recurved or straight, with 2 faint lateral ridges, smooth when fresh, minutely tuberculate or papillose and sometimes slightly torulose when dried, rounded or, especially when pod-like, acuminate at the apex, about 20-30-seeded; wall 3-5 mm thick; aril orange, enveloping the seed. *Seed* pale brown, obliquely ellipsoid, 10 x 4-5 x 3 mm, with longitudinal grooves, papillose; embryo 7 mm long; cotyledons ovate, 3 x 2 mm, rounded at both ends; rootlet 4 x 0.7 mm.

DISTRIBUTION: Tropical America from Costa Rica to Peru and northern Brazil.
ECOLOGY: Forest understorey, usually not on river banks. No clearcut flowering or fruiting season could be deduced from the collecting notes, but there are peaks in flowering: in March in Panama, in April-May in Venezuela and Guyana, in October in Suriname and in November in French Guiana and Brazil. Alt. 0-1000 m. Flowers visited by bees (*Mori & Kalunki* 2125, *Sytsma* 1005).
Uses: Fruits edible (teste *Greand* 1349).

Geographical selection of the approximately 710 specimens examined:
COSTA RICA. Puntarenas: Quepos, Cerro Nara, *Zamora* 1463 (F, MO); Rincón de Osa, *Grayum* et al. 4028 (WAG); ibid., *Liesner* 1795 (MO, NY, U); Corcovado National Park, *Kernam & Phillips* 899 (MO, WAG); Esquinas For., *Allen* 5834 (BM, F, GH, UC, US; phot. of US sheet F, GH, NY, WAG, type of *Stemmadenia nervosa*); Road to Puerto Jiménez, *Gomez* 19490 (MO, WAG). Alajuela: Río Naranjo, *Tonduz* 7962 (BR). Limón: 7 km SW of Bribri, *Gomez* et al. 20369 (MO, WAG, WIS).
PANAMA. Bocas del Toro: Chiriqui Lagoon, Cayo Agua, *McPherson* 11470 (MO). Coclé: base of Cerro Tife, *Sytsma* et al. 2535 (MO, WAG); Coclecito Road, *de Nevers* et al. 6719 (MO, WAG); near Caño Sucio, *Antonio* 3639 (MO). Colón: 3 km SE of Puerto Pilon, *Nee & Hale* 9685 (AAU, GH, MO, USF, WAG, WIS); Santa Rita Ridge Road, *Mori & Kallunki* 2125 (AAU, GH, MO); ibid., 13 km E of Transisthmian Highway, *McPherson* 7476 (L, MO, WAG); between Cerro Brewster and Alto Pacora, *de Nevers* et al. 5572 (MO). Panamá: km 19 El Llano-Carti Road, *Busey* 892 (C, MO, NY, USF, WAG); same road, 16 km from maingate near Gamboa, *Croat* 33717 (MO, NY); same road, km 17, Sytsma 1005 (MO, WAG); same road, ca 10 km N of Interamerican Highway, *A. Gentry* et al. 8883 (LD, MO, US, WIS). San Blas: 19 km from same highway, on El Llano-Carti Road, *de Nevers* et al. 3910 (HBG, MO, WAG); Nusagandi, *van der Werff* 7025 (MO, WAG).
COLOMBIA. Cundinamarca: Mun. Medina, Boquerón de Santa Inés, *Pinto & Bernal* 1608 (P). Vaupés: Río Inírida, *Fernández* 1983 (COL); Río Piraparaná, *Schultes & Cabrera* 17307 (US); Río Guayabero, *Cuatrecasas* 7527 (COL); near Bacaricuara, *Allen* 3096 (COL, MO). Amazonas: Río Cahuinari, road to La Chorrera, *Sastre* 2328 (COL, P).
VENEZUELA. Apure: Cutufí For., *Davidse & González* 21877 (MO); San Camilo F.R., *Steyermark* et al. 101567 (Z). Tachirá: near Represa Dorado, *Liesner & González* 10335 (MO, WAG). Sucre: Peninsula de Paria, Cerro Espejo, *Steyermark & Rabe* 96047 (K, MO, NY, U, US); Aricagua For., *Broadway* 559 (GH, NY, US); Cerro Patao, *Milliken* et al. 27 (MO). Amazonas: Dept. Atures, Santa Rosa de Ucata, *G.A. Romero* et al. 1887 (GH); Cerro Sipapo, *Maguire & Politi* 28770 (NY, US); Victorino, *Ll. Williams* 14842 (G, US, VEN); Yavita, *Ll. Williams* 13978 (F, MO, US, VEN); Río Cataniapo, *Steyermark* et al. 122260 (MO, WAG); 9 km from Raudal Remo, *Foldats & Velazco* 9553 (MO); near Macuruco, *Berry* 744 (Z); ibid., *Morillo* 7556 (VEN); Cucurital de Caname, *Davidse* et al. 17022 (MO); Cucurital de Yagua, *Davidse* et al. 17431 (MO); Río Puruname, *Tillet & O. Huber* 825-32 (MO); Cerro Huachamacari, *Liesner* 25591 (MO); Río Casiquiare, Vasiva and Pacimoni, *Spruce* 3199 (BM, BR,

CGE, G, K, NY, P, TCD, W, type of *B. obliqua*); Capihuara, *Ll. Williams* 15647 (MO, US). Bolívar: W of Maripa, *Morillo & Szewant* 9005 (MO); Guayapo, *Ll. Williams* 11812 (K, MO, S, US); Mun. Cedeño, 30 km W of Río Caura, *Fernández* 5199 (MO); Río Nichare, *Horner* et al. 174 (MO); Dept. Sucre, Caño Cani, *Morillo & Liesner* 8987 (MO); SE of Pie de Salto, Río Caura, *Y. Fernández & Yanez* 421 (MO); Río Erebato, *Stergios & Delgado* 13006 (MO); ibid., *Boom & Marin* 10462 (WAG); Río Oris, Turumban, *Stergios* 10743 (MO); Distr. Piar, Amaruay Tepui, *Liesner & Holst* 20305 (MO, U, WAG); Sierra de Lema, *Steyermark* 89526 (MO, NY, US); near El Palmar, *Wessels Boer* 2096 (U, WAG); km 38 El Dorado-Santa Elena de Mairen Road, Trujillo 3667 (MY); SE of El Dorado, *Fernández* 1068 (MY); Imataca F.R., *Stergios* et al. 5333 (MO, PORT); Gran Sabana, ca 10 km SW of Karaurin Tepui, *Liesner* 23618 (MO); Mt Roraima, *Robert Schomburgk* II 734 (BM), 754 (= *Richard Schomburgk* 1375B) (BM, CGE, FI-W, G, K, NY, P, W); near border with Delta Amacuro, Río Grande o Toro, *Breteler* 3803 (WAG). Delta Amacuro: 13 km ESE of Sierra Imata, *Davidse & González* 16584 (VEN); Santa Catalina, *Rusby & Squires* 300 (A, BM, E, F, G, GH, K, M, MO, NY, US, W, WU, Z; phot. of NY sheet P; phot. of US sheet WAG, type of *T. albescens*); Sacupana, *Rusby* 76a (GH, K; other sheets of *Rusby* 76 seen belong to *T. siphilitica*); 34 km ENE of El Palmar, *Steyermark* 93052 (K, MO, NY, P, U, US).

TRINIDAD. St. George: Maracas Valley, *Britton & Hazen* 1628 (GH, NY, US); Arima Valley, *A.C. Smith* 10059 (A, MO, NY, S, UC, US); ibid., *Riley* 205 (BM, C, K, NY). St. Andrews: Matura F.R., *Philcox* et al. 7330 (K, P); near Valencia, *Eggers* 1159 (AMD, HBG, M, P, UC, W, WU, Z). Sin. loc., *Ryan* anno 1778 (BM, C-VAHL, P-LA; phot. of C-VAHL sheet A, F, GH, MO, US, type).

GUYANA. Waini R., *de la Cruz* 3609 (GH, MO, NY, UC, US); Wanama R., NW Distr., *de la Cruz* 3898 (GH, K, MO, NY, UC, US); Barima-Waini Region, *McDowell* et al. 4461 (US, WAG); 3 km S of Barima R., *Pipoly & Lall* 8141 (P); Pomeroon R., *Jenman* 7540 (K); Acqueero Landing, *de la Cruz* 1107 (GH, NY, US); Upper Mazaruni R., *de la Cruz* 2049 (GH, MO, NY, US); Aurora, *Gillespie & Tiwari* 2147 (P); Kamakusa, *de la Cruz* 4184 (GH, MO, NY, US); Essequibo R., Berbice, *Robert Schomburgk* I 42 (BM, CGE, E, FI-W, G, G-DC, K, P, SING, TCD, US, W, narrow and/or wide leaves), 42bis (BM, G, K, wide leaves, type of *B. undulata* var. *ovalifolia*); Kamuni Creek, Groete Creek, *Maguire & Fanshawe* 22846 (A, K, MO, NY, US); near Bartica, *de la Cruz* 1937 (BH, F, GH, MO, NY, UC, US); Darah Creek, tributary of Arranami R., *Beckett & Kortright* July 1906 (K); Essequibo R., near Rockstone, *Maas & Westra* 3936 (K, MO, NY, U, WAG, neotype of *T. meyeri*); Christianburg, Demerara R., *Bartlett* Jan. 1907 (K); Waranama, Berbice R., *Harrison* 10 June 1958 (K); Karowtipu Mt, *Boom & Gopaul* 7694 (US, WAG); Tukait, Potaro R., *Graham* 177 (K); Tumatumari, *Gleason* 387 (GH, K, NY); Upper Demerara Region, Mabura Hill, *W. Hahn* 5816 (P, US); ibid., *Maas* et al. 5912 (F, K, MO, S, U, WAG, Z) & 5929 (K, MO, U, WAG); Kurupukari Creek, Essequibo R., *Grewal* 1 (U); E of Digitima Creek, Canje R., *Pipoly* et al. 11434 (P); Corantyne R., above Cow Falls, *McDowell & Gopaul* 2346 (P); Potaro-Siparuni Region, E of Maikwak, *W. Hahn* et al. 4196 (P); near end of Mabura Road, *McDowell* 3258 (US, WAG); near Lethem, *Acevedo* 3428 (US, WAG); Mt Makarapan, *Maas* et al. 7423 (U, WAG); Kanuku Mts, *Jansen-Jacobs* et al. 1273 (U, US, WAG); Gunn's, *Jansen-Jacobs* 1920 (U, WAG).

SURINAME. Corantyne R., Winanna Creek, *Maas & Tawjoeran* 3180 (U, USF); Nickerie Distr., near Kabalebo Dam project, *Lindeman* et al. 110/80 (K, S, U, US, WAG), 155/80 (MO, U, US, WAG); Upper Nickerie R., *Tulleken* 382 (L); Saramacca R., *Stahel* 332 (A, B, K, MO, NY, S, U, UC, WAG, WIS); near Paramaribo, sectie O, BW (= F.D.) 10a (BR, L, NY, P, U, US); near Republiek, *Mennega* 326 (A, BR, C, U); Zanderij I, *Maguire & Stahel* 25057 (A, K, MO, NY, U, US); near Blakawatra, 60 km

SE of Paramaribo, *den Outer* 869 (L, U, WAG); Jodensavanne-Mapane Creek area, *Kramer & Hekking* 2403 (U); near Victoria, *Kappler* 1398 (FI-W, G, P, S, U, W, type of *Peschiera surinamensis*); 2 km W of Albina, *Hekking* 1104 (A, U); Anjoemara Creek, *Jonker* 436 (U); Bakhuis Mts, *Florschütz & Maas* 2521 (U), 3126 (U); Coppename R., Raleigh Falls, *Stahel & Gonggrijp* BW 6351 (BR, P, U); Emmaketen, *Daniëls & Jonker* 863 (A, NY, U, WIS); Saramacca R. Headwaters, *Maguire* 24125 (A, BR, G, K, MO, NY, P, U, UC, US, W); Suriname R., near Gansee, *Lanjouw* 1294 (MO, S, U); near Gran Dam, *van Donselaar* 3339 (U); Nassau Mts., *Lanjouw & Lindeman* 2566 (U); Tapanahoni R., *Versteeg* 634 (U); Lely Mts, *Lindeman* et al. 471/75 (C, F, K, MO, NY, U, Z); Wilhelmina Gebergte, Frederik Top, *Maguire* et al. 54413 (C, F, K, MO, NY, U, US); Zuid R., *Maguire* et al. 54011 (C, F, K, MO, NY, U, US); Upper Suriname R., near Awaladam, *Tresling* 387 (U); Upper Gran Rio, *Hulk* 245 (U); near Pepejoe, *Schmidt* 304 (AAU, MO, U); near Oelemari R. airstrip, *Wessels Boer* 939 (U); Mt Icholi-Epoyan, *Sastre* 1458 (CAY, NY, P, U); sin. loc. *Hostmann* 564 (G, MO, P, S, U, W), 1200 (BM, E).

French Guiana. Maroni R., *Mélinon* 384 (G, HBG, K, P, WAG); near Mousse Creek, between St. Laurent and Paul Isnard, *Billiet & Jadin* 1707 (BR, P); Margot Creek (= Maipouri Creek), Serv. For. 7047 (CAY, NY, P, U); Iracoubo R., *Oldeman* B 613 (CAY, NY, P, U); SW of Sinnamary, Piste St. Elie, km 16, *Leeuwenberg* 14044 (CAY, WAG); ibid., *Prévost* 212 (CAY, P, WAG) & 256 (CAY, P, WAG); Grégoire Creek, *Deward* 33 (CAY, K, P, WAG); Roura, *Sagot* 1309 (BM, G, K, NY, P, S, W, WAG); Mt Française, *Sastre & Moretti* 4105 (CAY, MO, P); Paul Isnard Region, *Feuillet* 324 (BR, CAY, P); Baboune Creek, ca 15 km from its confluence with Mana R., *de Granville* 4705 (CAY); Approuague R., Arataye R., Parare Fall, *Sastre* 5588 (CAY, MO, P); Galibi Creek, *Oldeman* 1124 (CAY, P, U); Ipoussin Creek, *Oldeman* 2253 (CAY, P); Cayenne, *Poiteau* annis 1819-1821 (G, GH, K, W); Mt Tortue, *Feuillet* et al. 9929 (CAY, US, WAG), 10024 (CAY, US, WAG); Mt Kaw, *Leeuwenberg* 11816 (CAY, P, WAG); Inini R., Stat. Inst. Pasteur, *Sastre* 1379 (CAY, P, U); Upper Maroni R., Litani R., Toulalapata Fall, *Sastre & Moretti* 3948 (CAY, MO, P, WAG); Saül, *Oldeman* 4044 (CAY, P); Upper Oyapock R., opposite Miriaflor, *Oldeman* 1291 (CAY, MO, P, U); Approuague R., near Grand Canori Fall, *Oldeman* 2749 (CAY, K, P, WAG); Yaoué R., *Oldeman* T 870 (CAY, P, WAG); Notaye Creek, *Oldeman* T 354 (CAY, P, WAG); near Mitaraka, *Sastre* 1570 (CAY, P, U); Koutou Rock, *de Granville* et al. 9487 (US); Oyapock R., Oscar, *Oldeman* B 3434 (CAY, P, WAG); Tumuc Humac Mts, near Koulé-Koulé Creek, *de Granville* 982 (CAY, P, WAG); Trois Sauts, *Grenand* 1349 (CAY, P); ibid., *Jacquemin* 1892 (BR, CAY, P, WAG); sin. loc., *Martin* s.n. in coll. *Rudge* s.n. (BM); *Perrottet* 275 (G, G-DC, type of *T. perrottetii*).

Brazil. Amapá: Oiapóque R., *Fróes* 25726 (MO, U, UC); Oiapóque Airfield, *Cowan* 38679 (NY, S); Rio Iaue, *Irwin & Westra* 47746 (K, NY, US, Z); Mun. Macapá, Riozinho, *Mori & Sousa* 17617 (NY); Rio Araguari, *J.M. Pires* et al. 51277 (K, NY, S, US, Z); Serra do Navio, *Pereira & Egler* 3421 (SP, U, Z); ibid., *Vogel* 137 (MJG); ibid., *J.M. Pires* et al. 50264 (NY, US, Z); Gaúcho Area, *Austin* et al. 7153 (NY, WAG); Vila Água Branca, *Rabelo & Cardodo* 2737 (WAG). Pará: Ourém, *J.M. Pires & N.T. Silva* 4508 (MO, US); from Belém, BR 22, km 64, *Prance & N.T. Silva* 58879 (K, NY, S, US, Z); Rio Tocantins, Jacundá, *M.G. Silva & Bahia* 3601 (NY); km 55 Maraba-Altamira Road, *Bamps* 5137 (BR, U); Beira do Rio Mapuá, *Black* et al. 50-9839 (MO); near Antonio Lemos, *Krukoff* 5886a (A, NY); Rio Tapacú, Portel, *Fróes* 32794 (MO); SW of Ilha de Breu, *Prance* et al. 1534 (G, GH, K, NY, S, U, US, Z); Mun. Almeirim, *M.R. Santos* 681 (NY); ibid., Mt Dourado, *M.J.P. Pires* et al. 527 (WAG); Altamira, *Bahia* 105 (MG, NY); ibid., *Vasconcelos* et al. 291 (MG); Tumuc Humac, 3 km NW of Mitakara, *de Granville* B 4526 (CAY, P, WAG); ibid.,

Kouïapane, *Sastre* 1494 (CAY, NY, P, U); Mun. Oriximiná, W of Lago Erepecu, *Cid* et al. 1669 (GH, NY, US, WAG). Amazonas: Manaus, *Luiz* 2874 = *W. Rodrigues* INPA 2874 (MO, NY); km 43 Manaus-Caracaraí Road, *Plowman* et al. 12683 (F, K, WAG); km 56 of same road, *W. Rodrigues* et al. 3648 (NY, P); km 70 Manaus-Itacoatiara Road, *W. Rodrigues & Lima* 3427 (NY); Distr. Agropecuário, Res. 1501, Smithsonian/INPA Biol. Dynam. For. Fragm. Proj., *Mori* et al. 21459 (NY); Mun. Pres. Figueredo, 0.8 km from Atroarí, *Cid* et al. 8207 (K, US, WAG); Barcelos, *Fróes* 28477 (MO); Rio Curicuriari and Igarapé Cariua, *Alencar* 625 (NY, WAG); Rio Xie, *Fróes* 12412/156 (NY, S); Upper Rio Negro, Ilha das Flores, *W. Rodrigues* 909 (MO, NY); near Paumari, *A. Gentry & Revilla* 20544 (MO); near mouth of Rio Embira (= Envira), tributary of Rio Tarauaca, *Krukoff* 4795 (A, BM, G, K, LE, M, MO, NY, S, U, UC, US). Roraima: Rio Branco, Paraná do Mararà, *M.R. Santos* 181 (F, NY, WAG); km 346 Manaus-Caracaraí Road, Igarapé Jundiá, *Prance* et al. 24327 (GH, K, NY, S, U, US, WAG).

ECUADOR. Morona-Santiago: Santiago, *Little* et al. 796 (US).

PERU. Amazonas: 1 km from Caterpiza, *Huashikat* 324 (MO); Puerto Nazaret, *Ellenberg* 3512 (WAG); Prov. Bagua, Monterrico, *Knapp & Alcorn* 7604 (F, MO, WAG); Valley of Río Marañon, above Mayasi Falls, *Wurdack* 1838 (K, MO, NY, UC, US). Loreto: Maynas Prov., Sargento Lores, *Vásquez & Jaramillo* 13234 (WAG); Nueva Jerusalem, *Lewis* et al. 10885 (WAG); Distr. Pastaza, Andoas, *Ayala* 2230 (GH, MO, WAG); Distr. Tigre, Paichi Playa, *Ayala* et al. 2602 (MO, WAG). Ucayali: near Pucallpa, *Skutch* 20 Oct. 1940 (MO); Aguaytia, *Woytkowski* 34431 (G, MO, S, UC, US); Bosque Nacional von Humboldt, km 86 Pucallpa-Tingo María Road, *A. Gentry* et al. 18685 (F, MO, US); ibid., *Maas* et al. 4550 (U, WAG); SE of Pucallpa, *Morawetz & Wallnöfer* 110-281085 (Z). San Martín: Pongo de Cainarachi, *Klug* 2740 (A, BM, G, GH, K, MO, NY, S, US). Huánuco: Prov. Pachitea, Bosque Nacional de Iparia, *Schunke* 1211 (F, G, GH, K, MO, NY, P, US); Pucallpa Region, Sira Mts, *Wallnöfer* 14-20588 (WAG, Z). Pasco: Prov. Oxapampa, Palcazu Valley, *Pariona & Pedro* 1018 (MO, WAG); ibid., *Salick* et al. 7600 (WAG); Puerto Bermudez, *Killip & A.C. Smith* 26416 (MO, NY, US).

Notes. G.F.W. Meyer (1818) named a specimen collected in Guyana near the Essequibo River as *T. undulata* and gave a description of it, as he supposed that it was different from the type of the species. Without having seen it G. Don (1837) coined the name *Tabernaemontana meyeri* for it. Miers (1878) combined *T. meyeri* with his genus *Anartia*. Allorge (1985) followed Miers and added *Bonafousia attenuata* Miers (= *T. attenuata* (Miers) Urb. = *A. attenuata* (Miers) Mgf.) as a synonym. The characters given in Meyer's description all can be found in specimens of *T. undulata* and most of them also in specimens of *T. attenuata*. Only one character, the indumentum inside the corolla, is never found in *T. attenuata* but may occur in *T. undulata*. In *T. attenuata* the indumentum is present only below the stamens and in *T. undulata* a pilose-pubescent belt is present around the anthers while the lobes are frequently puberulous at the base. Therefore Meyer's "Faux pubescens" most probably refers to the indumentum on the lobes, as the other indumentum usually ends below the throat. Moreover, *T. attenuata* is not known from the area of the Essequibo R., where the collection studied by Meyer had been collected.

As the specimen seen by Meyer apparently has been lost (it is not in the Goettingen Herbarium, where it was supposed to be preserved), the present author has selected a neotype for *T. meyeri*.

98. Tabernaemontana vanheurckii Muell. Arg. in van Heurck, Obs. Bot. 168 (1870). – Type: Peru, San Martín, near Tarapoto, *Spruce* 4209 (holotype AWH; isotypes BM, BP, BR, C, CGE, E, F, G, GH, GOET, K, LD, LE, MPU, NY, P, TCD, W; phot. of G sheet F, GH, MO, NY). Fig. 98, p. 392; map 52, p. 394

Homotypic synonyms:

Peschiera blanda Miers, Apoc. S. Am. 44 (1878). *Stenosolen vanheurckii* (Muell. Arg.) Mgf. in Notizb. Bot. Gart. Berlin 14: 177 (1938). *P. vanheurckii* (Muell. Arg.) Allorge in Mém. Mus. natn. Hist. Nat. IIB, 30: 154, pl. 68 (1985). The BM sheet is the holotype of *P. blanda*.

Heterotypic synonyms:

P. lingulata Miers, op. cit. 42 . – Type: Peru, San Martín, Tarapoto, *Matthews* 1542 (holotype BM; isotypes CGE, E, G, K).
T. macrosiphon Herzog in Fedde, Repert. 7: 66 (1909). *S. macrosiphon* (Herzog) Mgf. in Notizbl. Bot. Gart. Berlin 14: 177 (1938). – Type: Bolivia, Santa Cruz, Prov. Velasco, near Yotaú, Misiones de Guarayos, *Herzog* 374 (holotype Z).

Shrub or small tree 1-12 m high. Trunk up to at least 20 cm in diameter. Branches pale brown, lenticellate; branchlets terete, glabrous or occasionally puberulous. *Leaves* of a pair equal or unequal and larger up to 4 x as long as other and similarly shaped, petiolate; petiole glabrous or occasionally puberulous, 4-17 mm long; (ocreae not widened into intrapetiolar stipules); blade thinly papery when dried, elliptic, narrowly elliptic or sometimes obovate, 2-3.7 x as long as wide, 3-21 x 1.5-9 cm, acuminate or less often caudate at the apex, cuneate at the base, entire, glabrous on both sides or occasionally pubescent beneath, with scattered black dots beneath or not, with 5-13 pairs of upcurved secondary veins forming an angle of 50-60° with the costa; tertiary venation reticulate, often rather inconspicuous. *Inflorescence* pedunculate, 3-5 x 3-9 cm, (3-)8-25-flowered, lax. Peduncle short, rather slender, 3-6 mm long, glabrous or occasionally puberulous; pedicels 3-10 mm long, slender, glabrous or occasionally puberulous. Bracts sepal-like, 0.3-0.5 x as long as them. *Flowers* fragrant, open during the day. *Sepals* pale green, connate at the base for 0.3 mm, closely clasping the corolla base, usually also when dried, ovate or broadly ovate, 1.2-1.7 x as long as wide, 1.5-3 x 1.2-2.5 mm, rounded, glabrous or occasionally puberulous outside, ciliate, glabrous inside and with 3-8 colleters in 1 row in the middle at the base; colleters 0.3-0.7 x 0.1-0.2 mm. *Corolla* white, sometimes(?) with a yellow throat and a pale green tube, 11-25 mm long in the mature bud and forming a comparatively narrow, narrowly ovoid head 0.25-0.35 of the bud length (3.5-7 x 2.5-3.5 mm) with a blunt apex, glabrous outside, pilose inside in a belt from the insertion of the stamens to the base of the lobes and often also in stripes alternating with the stamens up to 0.5-1.5 mm below the anther tails (filament ridges glabrous); tube 5-7 x as long as the calyx, 1.1-2.4 x as long as the lobes, 10-19.5 mm long, cylindrical, 1.5-2.5 mm wide above the base around the anthers, slightly narrowed above to 1-2 mm wide, also slightly narrowed below or not, twisted 0-0.25 turn at the throat; lobes obliquely dolabriform, more or less falcate, 0.4-0.9 x as long as the tube, 1.5-2.4 x as long as wide, 7-13 x 3-7.8 mm, obtuse or rounded at the apex and at the other side, neither auriculate nor undulate, spreading. *Stamens* with apex 3.8-11 mm below mouth of corolla tube, inserted 0.13-0.3 of the length of the corolla tube, (at 1.5-5 mm from the base); anthers sessile, with tails 0.5 mm below insertion, narrowly triangular or narrowly oblong, 5-7 x as long as wide, 4-5 x 0.7-1 mm, apex acuminate, sterile for 0.4-0.7 mm, sagittate at the base, glabrous. *Pistil* glabrous, 3.5-7 mm long; ovary ovoid, 1.5-2 x 0.8-1.3 x 0.8-1 mm; style short, 0.5-3.2 mm long; pistil head 1.3-

Fig. 98. *Tabernaemontana vanheurckii.* **1-2,** habit (x 2/3); **3,** opened corolla (x 2); **4,** pistil (x 6); **5,** sepal inside (x 8); **6,** fruit (x 2/3). 1 and 3-5 from Belshaw 3261; 2 from Knapp & Alcorn 7424; 6 from Schunke 6641.

1.8 mm high, composed of a whorl of 10 acute lobes forming a ring 0.2 x 0.8-1 mm, a stipitate 5-lobed depressed globe 0.2-0.3 x 0.5-0.8 mm and a stigmoid apical part 0.2-0.8 x 0.1-0.2 mm. Ovules approximately 50 in each carpel. *Fruit* of 2 separate mericarps; mericarps green or brown, often turning yellow-orange, orange or red inside, obliquely ellipsoid or pod-like, densely covered with acute or acuminate excrescences, 20-50 x 10-25 x 10-20 mm, recurved or not, rounded or acute at the apex, not ridged, approximately 5-25-seeded; wall about 3 mm thick in dried fruits; aril orange or red, halfway enveloping the seed. *Seed* black, obliquely ellipsoid or nearly so, 8-10 x 4-6 x 3-4 mm, with longitudinal grooves, dull, papillose; embryo 6.5-7.5 mm long; cotyledons ovate, 1.1-1.4 x as long as wide, 3.5-4.5 x 2.5-3.5 mm, rounded or obtuse at the apex, cordate at the base, rootlet 3.5-4 x 0.8-0.9 mm.

DISTRIBUTION: Western South America.

ECOLOGY: Wet forest understorey, often on periodically inundated river banks. Alt 0-1000 m. Flowering and fruiting mainly from July to December, at least in Peru and Bolivia.

Geographical selection of the approximately 100 specimens examined:

COLOMBIA. Valle del Cauca: Bajo Calima, *Monsalve* 167 (WAG), 1727 (WAG).

BRAZIL. Acre: Mun. Seno Madureira, Bonsucesso Road, Rio Caeté, *Cid & Nelson* 2654 (GH, MO, NY, US, WAG); km 45 Rio Branco-Brasileia Road, *Lowrie* et al. 325 (NY, US, WAG); near Xapuri, *Daly* et al. 7182 (WAG).

PERU. Loreto: Portal, *Killip & A.C. Smith* 29262 (F, MA, MO, NY, US); Yurimaguas, *Ferreyra* 5044 (MO, US). Ucayali: Canchahuayo, *Vásquez* et al. 6948 (WAG); San Carlos, between Pucallpa and Contamana, *Tessmann* 3129 (NY); Neshuya, *Schunke* 895 (F, NY, US); 10 km E of Pucallpa, *Lamar* June-July 1981 (US); Prov. Coronel Portillo, Leoncio Prado, *Vásquez* 5022 (MO, WAG); Cocha Muerte, Pucallpa, *Schunke* 976 (F, G, NY, US); Est. Exp. A. von Humboldt, *Begazo* 173 (MO, U, WAG); km 11 Nueva Requena Road, Distr. Campo Verde, *Arana* 183 (K); Bosque Nacional von Humboldt, km 86 Pucallpa-Tingo María Road, *A. Gentry* et al. 18748 (F, MO). San Martín: Lamas, *Knapp & Alcorn* 7424 (MO, WAG); Tarapoto, *Matthews* 1542 (BM, CGE, E, G, K, type of *Peschiera lingulata*); ibid., *Spruce* 4209 (AWH, BM, BP, BR, C, CGE, E, F, G, GH, GOET, K, LD, LE, MPU, NY, P, TCD, W; phot. of G sheet F, MO, NY, US, type); 1-4 km NE of Tarapoto, *Belshaw* 3261 (BH, GH, K, LL, MO, NY, U, UC, US, WIS); near Puente Colombia, *Ferreyra* 17524 (MO, US); Pucacaca, *Zegarra* 10 (MO, USF, Z); Saposoa, *Woytkowski* 5455 (MO, P, US); Prov. Rioja, Sabana, *Ferreyra* 18466 (MO); between Bella Vista and Banos, *Ferreyra* 4761 (MO, US); Juanjuì, *Klug* 3805 (BM, F, GH, K, MO, NY, S, US, WIS), 4215 (BM, F, GH, K, MO, NY, S, U, UC, US); Prov. Mariscal Caceres, Mashuyacu, *Schunke* 4220 (F, G, GH, K, MO, NY, US); Distr. Tocache Nuevo, 5 km NE of Puerto Pizana, *Schunke* 6641 (F, GH, MO, Z); NE of Tocache, *Schunke* 8404 (MO, USF, Z). Huánuco: Puerto de Tournavista, *Schunke* 2214 (F, US). Pasco: Oxapampa, near Paujil, *Forster* 8891 (MO, P). Cuzco: Río Mapitunuari, ca 5 km from Río Apurimac, *Dudley* 11554 (NA); Prov. Paucartambo, Villa Carmen, *Vargas* 14902 (US). Madre de Dios: Manu National Park, Río Sotileja, *Forster* et al. 11686 (WAG); Cocha Cashu Biol. Stat., *Núñez* 7997 (MO); Río Manu. playa 16 above the Boca, *Forster & Augspurger* 3092 (F, MO, US); 15 km NE of Puerto Maldonado, *Núñez* et al. 11149 (WAG); Tambopata Res., *Reynel* et al. 5320 (MO).

BOLIVIA. Pando: W bank Río Madeira, opposite Abuña, *Prance* et al. 5699a (GH, K, P, S); junction of Ríos Beni and Madre de Dios, *Rusby* 929 (NY), 930 (GH, NY, US, WIS); Prov. Manuripi, Isla Gargántua, *Solomon* 10864 (MO, WAG). Beni: Prov. Vaca Diez, Tumi Chucua, *Solomon* 6521 (MO); Río Beni, sin. loc., *Rusby* 2378 (BM, E, F, G, GH, K, MO, NY, P, US, WIS); Loma Suárez, ca 15 km N of Trinidad,

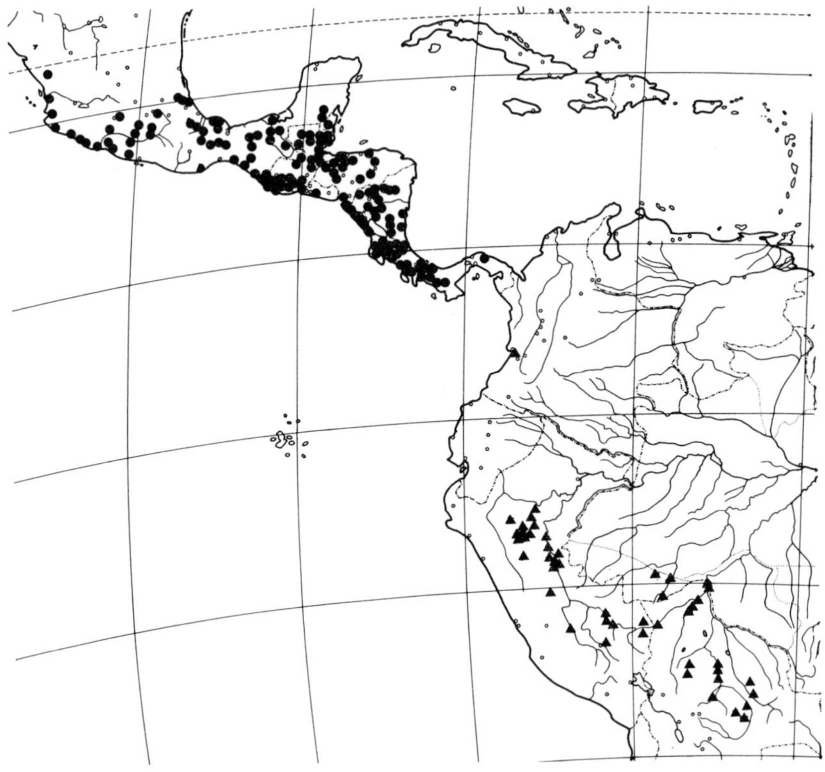

Map 52.▲. *Tabernaemontana vanheurckii,* ●. *Stemmadenia donnell-smithii.*

Solomon et al. 8124 (MO, WAG); Río Beni, above confluence with Río Quiquibey, *Daly* et al. 6584 (WAG); Río Inicua, *Rusby* 753 (K, NY, US, paratype of *T. unguiculata*); Exp. Stat. Naranjito, 25 km SSE of Trinidad, *Solomon* et al. 8115 (MO, U, WAG). La Paz: Río Kaka, tributary of Upper Río Beni, *Evans* 6 (BM); Santa Cruz: Prov. Velasco, Misiones de Guarayos, near Yotaú, *Herzog* 374 (Z, type of *T. macrosiphon*); Prov. Nuflo de Chávez, San Ramón, *Quevedo & Centurión* 516 (MO); Prov. Santiesteban, 12 km NE of crossroads at Montero, *Nee* 39463 (WAG); Gral. Savedra, *Krapovickas & Schinini* 31533 (Z); Prov. Ichilo, Vila Yapacani, *Nee* 39595 (WAG); Santa Cruz, *Brooke* 5843 (BM, F, NY). Cochabamba: Prov. Todos Santos, Chapare, *Steinbach* 382 (GH, S, U, US, WIS).

Note. *Tabernaemontana vanheurckii* is easily confused with *T. heterophylla*, as both species resemble each other strikingly in their leaves, corollas and fruits. They differ mainly in the sepals. Those of *T. vanheurckii* are 1.2-1.7 x as long as wide, rounded and ciliate, while those of *T. heterophylla* are 1.5-7 x as long as wide, acute or obtuse and not or only occasionally ciliate.

The paratype of *T. unguiculata, Rusby* 753 is placed here, while the holotype is named *T. heterophylla*. Therefore, the name *T. unguiculata* should be a synonym of *T. heterophylla* and not of *T. vanheurckii*, as was supposed by Allorge (1985).

99. Tabernaemontana wullschlaegelii Griseb., Fl. Brit. W. Ind. Isl. 409 (1861); Stearn in Journ. Arn. Arb. 52: 617 (1971). – Type: Jamaica, Manchester, near Fairfield, *Wullschlaegel* 918 (holotype GOET; isotypes BR, M, W; phot. of M sheet in BM). Fig. 99, p. 396, map 53, p. 397

Homotypic synonym:

Anartia wullschlaegelii (Griseb.) Miers, Apoc. S. Am. 81 (1878), as *wulfschaegelii.*

Heterotypic synonyms:

T. lactea Urb., Symb. Antill. 6: 34 (1909). – Type: Jamaica, Trelawny, near Troy, *Harris* 8815 (holotype BM; isotypes F, NY, US; phot. of US sheet in WAG).

T. glaucescens Urb., op. cit. 35. *T. wullschlaegelii* var. *glaucescens* (Urb.) Allorge, Mém. Mus. natn. Hist. Nat. IIB. 30: 27 (1985), **syn. nov.** – Type: Jamaica, Hanover, Fray Woods, *Harris* 10342 (lectotype BM; isolectotypes C, F, NY, P, US, designated by Allorge in 1985).

T. calcicola Urb. in Fedde, Repert. 13: 472 (1915). – Type: Jamaica, Clarendon, Peckham Woods, *Harris* 11045 (holotype BM; isotype NY).

T. rendlei Stearn, op. cit. 618. *T. wullschlaegelii* var. *rendlei* (Stearn) Allorge, l.c., **syn. nov.** – Type: Jamaica, St. Thomas, SE end of John Crow Mts., *Stearn* 509 p.p. (holotype BM; isotypes IJ, K). The A sheet of this number belongs to *T. laurifolia* and should be a duplicate of *Stearn* 316 (BM).

Shrub or small tree 2-10 m high. Branches pale grey-brown, not lenticellate, branchlets terete, glabrous. *Leaves* petiolate; petiole 5-25 mm long, glabrous; (ocreae widened into intrapetiolar stipules); blade coriaceous when dried, elliptic or narrowly elliptic, 1.5-4 x as long as wide, 4-21 x 1.5-8 cm, acuminate to obtuse or almost rounded at the apex, cuneate at the base, entire, with revolute margin, glabrous on both sides, often regularly dotted beneath and also with scattered black dots on both sides, with 6-14 pairs of rather straight secondary veins forming an angle of 60-90° with the costa; tertiary venation inconspicuous. *Inflorescence* pedunculate, 1.5-6 x 1.5-6 cm, 5-50-flowered, lax, 2-3 x branched. Penduncle rather slender, glabrous, 2-30 mm long; pedicels 4-9 mm long, slender, glabrous. Bracts scale-like, 0.2-0.5 x as long as the sepals. *Flowers* fragrant, open during the day. *Sepals* green, persistent, even under the fruit, connate at the base for about 0.3 mm, erect, ovate or broadly ovate, 1-1.5 x as long as wide, 1.5-2.5 x 1-2 mm, rounded, glabrous outside, not ciliate, glabrous inside and with 3-5 colleters in 1 row in the middle at the base; colleters 0.2-0.4 x 0.1-0.15 mm. *Corolla* white, cream, yellow or limb white and tube greenish, 7-11 mm long in the mature bud and forming an ovoid or broadly ovoid head 0.25-0.36 of the bud length (2-4 x 2-3 mm) with an acute or blunt apex, glabrous on both sides; tube 3-4.7 x as long as the calyx, 0.7-1.4 x as long as the lobes, 5-9 mm long, almost cylindrical, 1.2-2.5 mm wide above the base, narrowed below the anthers or not to 1-1.7 mm wide, widened again around them to 1.5-2.2 mm wide, slightly narrowed above to 1.2-2 mm wide, not twisted; lobes obliquely elliptic, 0.7-1.4 x as long as the tube, 1.7-3 x as long as wide, 5-11 x 2-4 mm, obtuse or rounded, with an acute or obtuse lateral lobe about halfway along the length and at the same time auriculate at the base at the left side, not undulate, with recurved margin, spreading. *Stamens* with apex 0-1 mm below mouth of corolla tube, inserted 0.5-0.64 of the length of the corolla tube, (at 2.5-5.5 mm from the base); anthers sessile, with tails 0.5 mm below insertion, narrowly triangular, 2.5-4 x as long as wide, 2.5-3.5 x 0.6-1 mm, apex acuminate, sterile for 0.3 mm, sagittate at the base with straight tails, glabrous. *Pistil* glabrous, 3.5-6.5 mm long, with apex about halfway along anthers; ovary ovoid, 1-2 x 1-1.2 x 0.8-1 mm, with an obscure disk-like thickening 0.3-0.7 mm high; style

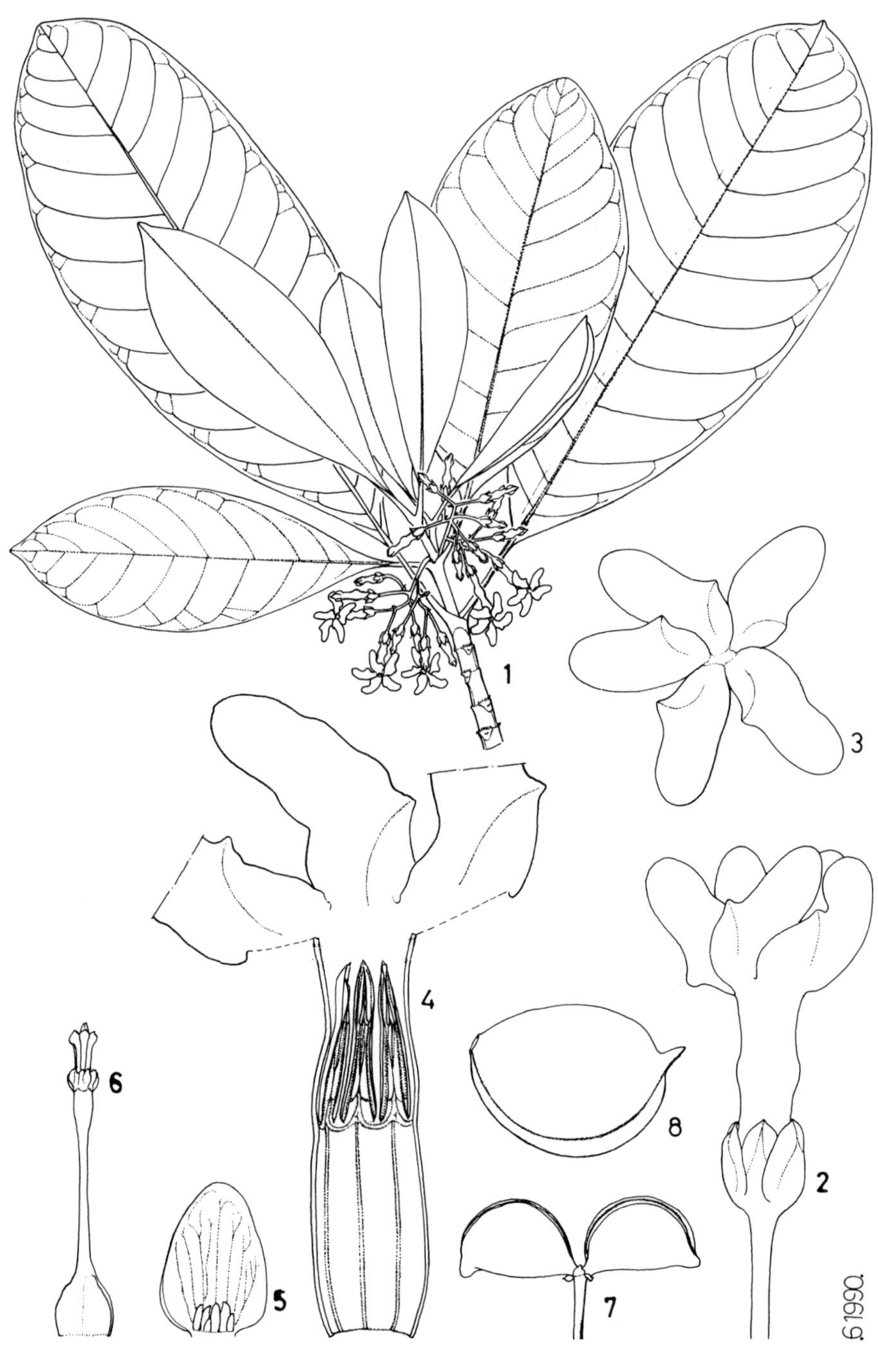

Fig. 99. *Tabernaemontana wullschlaegelii.* **1,** habit (x 2/3); **2,** flower (x 4); **3,** flower above (x 4); **4,** opened corolla (x 8); **5,** sepal inside (x 10); **6,** pistil (x 8); **7,** fruit x 2/3); **8,** mericarp (x 2/3). 1-6 from Adams 12491; 7 from Proctor 26494; 8 from Harris 8412.

rather thin, 1.3-3.3 mm long; pistil head 1.2-1.5 mm high, composed of a basal ring 0.3-0.7 x 0.7-1.2 mm, consisting of 10 rounded lobes, a 5-lobed stipitate depressed globe 0.3 x 0.5-0.8 mm and a stigmoid apex 0.2 x 0.2 mm. Ovules approximately 70 in each carpel. *Fruit* of 2 separate mericarps; mericarps green, obliquely ellipsoid, 1.2-1.7 x as long as wide, 25-40 x 20-25 x 20-25 mm and rounded or apiculate, or (only in *Britton* 2424 and *Harris* 10342) pod-like, 45 x 10 x 10 mm and long-acuminate, with lateral ridges, mostly recurved, smooth, approximately 10-20-seeded; wall about 1-2 mm thick in dried fruits; aril surrounding the seed. *Seed* dark brown, obliquely ellipsoid, 8-9 x 3-4 x 3 mm, with longitudinal grooves, dull, papillose.

DISTRIBUTION: Endemic to Jamaica.

ECOLOGY: Forest on limestone. Alt. 350-900 m. Flowering mainly March-April and fruiting probably from October to December.

Geographical selection of the approximately 40 specimens examined:

JAMAICA. Hanover: Fray Woods, *Harris* 10342 (BM, C, F, NY, P, US; phot. of US sheet WAG, lectotype of *T. glaucescens*); near Kempshot, *Britton* 2424 (NY). Trelawny: near Troy, *Harris* 8815 (BM, F, NY, US; phot. of US sheet WAG, type of *T. lactea*); between Burnt Hill and Spring Garden, *Proctor* & Bro. *Alain* 24906 (BM, IJ); ca 1.5 km N of Warsop, *Proctor* 24823 (IJ). Manchester: near Fairfield, *Wullschlaegel* 918 (BM, BR, GOET, M, W, type); near Christiana, *Harris* 8267 (A, BM); near Hibernia, *Purdie* Dec. 1843 (K); Banana Ground, *Adams* 10123 (BM, M); Olive R., *Harris* 8412 (BM, F, NY); Contrivance, *Adams* 12491 (BM). Clarendon: Spaldings, *Dignum* 4 July 1949 (IJ); between Ritchies and Balcarres, *Proctor* 33786 (BM); between Balcarres and Sunbury, *Proctor* 33627 (IJ), 35641 (MO, U); Peckham Woods, *Harris* 11045 (NY, type of *T. calcicola*); ibid., *Proctor* 35674 (MO, U). St. Ann: Mt Diablo, *Proctor* 26494 (BM); near Moneague, *Britton & Hollick* 2809 (K, NY); Roaring River, *Harris* 10358 (BM, F, K, NY, US, paratype of *T. glaucescens*); Mason R., *Proctor* 16200 (BM). St. Catherine: 1.6 km N of Mt Rosser, *Howard & Proctor* 15079 (A, BM, IJ). Portland: NW slopes of Joe Hill, *Proctor* 10112 (BM, IJ, US); E slope of John Crow Mts, 1.6-2.5 km SW of Ecclesdown, *Proctor* 8593 (A, IJ), 9982 (IJ, paratype of *T. rendlei*); 5 km SW of Priestmans R., *Proctor* 7643 (IJ, paratype of *T. rendlei*); ca 16 km SW of Priestmans R., *Proctor* 5270 (IJ). St. Thomas: SE end of John Crow Mts, *Stearn* 509 p.p. (BM, IJ, K, type of *T. rendlei*); ibid., *Harris & Britton* 10728 (BM, K, NY, US, paratype of *T. rendlei*).

Notes. *Tabernaemontana wullschlaegelii* is remarkably uniform in all characters except for the fruit. Two specimens, *Britton* 2424 and *Harris* 10342 are known with

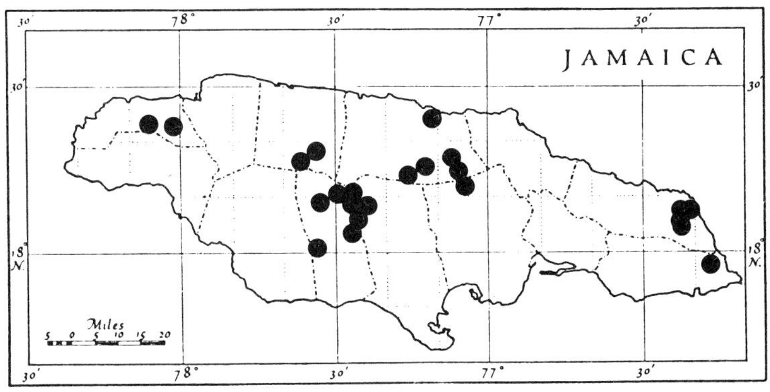

Map 53. *Tabernaemontana wullschlaegelii.*

very narrow and long-acuminate instead of ellipsoid and rounded or apiculate follicles. The corolla tube varies in length from 5 to 9 mm; 5-7 mm has been observed in *Britton* 4135, 7 m in e.g. *Harris* 8815, type of *T. lactea*, *Stearn* 509, type of *T. rendlei*, *Harris* 10728 and *Proctor* 9982, paratypes of *T. rendlei* and *Adams* 12491 cited as *T. wullschlaegelii* by Stearn. The corolla tube of *Howard & Proctor* 15079, cited as *T., wullschlaegelii* is even 9 mm long like that of *Proctor* 36741, collected after Stearn's publication. The lobes are longer or shorter than the tube, 8 mm in *Harris* 10728 and *Proctor* 9982, 9 mm in *Howard & Proctor* 15079 and up to 11 mm in Proctor 36741. Therefore couplet 5 in Stearn's key does not differentiate the specimens cited in his paper. As for his couplet 6, the leaves do not dry clearly differently. After careful comparative studies, for which almost all flowers of the specimens examined were measured, the present author is absolutely convinced that *T. wullschlaegelii*, *T. glaucescens* and *T. rendlei* are one species. Therefore, he has reduced the two latter names to synonyms with sincere apologies to Dr. W.T. Stearn.

Stemmadenia Benth., Bot. Voy. Sulph. 124, t. 44 (1844); Woodson in Ann. Miss. Bot. Gard. 15 :352 (1928). – Type species: *S. glabra* Benth. (= *S. obovata* (Hook. & Arn.) K. Schum.).

Homotypic synonyms:

Stemmadenia subgenus *Eustemmadenia* Woods. in op. cit. 353. *S.* section *Eustemmadenia* (Woods.) Pichon in Mém. Mus. natn. Hist. Nat. II, 27 : 230 (1949). S. subgenus E. section *Obovatae* Woods., l.c.

Heterotypic synonyms:

Odontostigma A.Rich., Fl. Cub. Fanerog. 3: 86 (1853), non Zoll. & Morr. (1845). S. subgenus *Eustemmadenia* section *Galeottiae* Woods. in op. cit. 359. – Type species: *O. galeottianum* A.Rich. (= *S. galeottiana* (A.Rich.) Miers = *S. litoralis* (H.B.K.) Allorge).

S. subgenus *Ochrodaphne* Woods. in op. cit. 363. *S.* section *Ochrodaphne* (Woods.) Pichon, l.c. – Lectotype species: *S. grandiflora* (Jacq.) Miers, designated here.

Shrubs or trees with much white latex in all parts. Branches usually pale brown, lenticellate; branchlets terete. *Leaves* of a pair equal or unequal, the smaller one mostly similarly shaped to the larger one, petiolate; blade usually membranaceous when fresh, elliptic or obovate or narrowly so, mostly acuminate to caudate at the apex, entire; tertiary venation mostly reticulate and conspicuous. *Inflorescence* mostly 1-4-flowered. *Flowers* large and showy, usually fragrant, open during the day. *Calyx* often subtended by bracteoles; sepals free or connate at the extreme base, subequal or unequal and inner larger than outer, with colleters inside at the base only. *Corolla* yellow, creamy or white, glabrous outside, not ciliate, with 5 longitudinal wings inside just above the anthers or less often continuing below them; tube infundibuliform to almost cylindrical, twisted around the anthers only; lobes obliquely obovate, ovate or less often dolabriform. *Stamens* deeply included, inserted mostly in the lower half of the corolla tube just below the wings or on them; anthers sessile, narrowly triangular, acuminate at the apex, sagittate at the base, with tails sometimes curved towards each other, glabrous. *Pistil* with apex approximately halfway along anthers; ovary ovoid or conical, mostly gradually narrowed into the style, of two carpels, connate at the base with a shallowly 5-lobes disk-like thickening; style cleft at the base, persistent; pistil head with a basal ring or veil, a more or less stipitate 5-lobed mostly depressed globe and a bilobed stigmoid apex. *Fruit* of two separate mericarps; mericarps yellow, orange, green or brown, often lenticellate, obliquely ellipsoid, subglobose or sometimes pod-like, mostly acuminate to caudate, dehiscent; aril orange or red, completely or almost completely enveloping the seed. *Seed* obliquely ellipsoid, mostly dark brown or black, with deep hilar groove and with longitudinal grooves at the other side, papillose, with ruminate endosperm; embryo straight or nearly so, spathulate; cotyledons usually ovate and rootlet slightly longer than cotyledons.

DISTRIBUTION: 10 species in Central and northern South America.

A more or less natural sequence of the species is as follows:
S. brasiliensis Leeuwenberg
S. stenoptera Leeuwenberg
S. alfari (Donn. Sm.) Woods.
S. macrophylla Greenm.
S. grandiflora (Jacq.) Miers
S. pauli Leeuwenberg
S. donnell-smithii (Rose ex Donn. Sm.) Woods.
S. litoralis (H.B.K.) Allorge
S. obovata (Hook. & Arn.) K. Schum.
S. tomentosa Greenm.

KEY TO THE SPECIES

The complex species *S. macrophylla* can be confused with *S. grandiflora* when its sepals are large, although they are narrower. It can also be confused with *S. alfari* when the sepals are small, especially as for its collections known from Panama.

1. Corolla tube 11-13 mm long. Mexico (Colima) .... **9. S. stenoptera**
   Corolla tube at least 19 mm long 2
2. Sepals 2-4 mm long; corolla tube 22-32 mm long, mature flower bud with a sub-globose rounded head. Costa Rica and Panama .... **1. S. alfari**
   Sepals longer than 5 mm, if less then corolla tube 35-45 mm long; mature flower bud with an ovoid or ellipsoid acute or obtuse head (unknown from *S. brasiliensis*).. 3
3. Corolla tube 19-22 mm long and leaves papery when dried. Brazil .... **2. S. brasiliensis**
   Corolla tube longer than 22 mm, if less then leaves membranaceous when dried (see *S. macrophylla*).. 4
4. Sepals (2-)3-6 mm long; corolla tube 33-45 mm long, infundibuliform, 5.5-14 x as long as the calyx .... **10. S. tomentosa**
   Sepals longer than 6 mm, if less then corolla tube less than 32 mm long (see *S. macrophylla*) 5
5. Calyx not or occasionally subtended by bracteoles, outer sepals often subcordate at the base and at least 8 mm long and up to 1.5 x as long as wide, persistent, even under the fruit; inner sepals 11-23 x 5-20 mm; corolla tube 1.5-3 x as long as the longest sepals and 25-35 mm long. Panama and northern South America .... **4. S. grandiflora**
   Calyx usually subtended by bracteoles, outer sepals not subcordate or otherwise more than twice as long as wide 6

6. Sepals free, shed with the corolla, inner often torn at the apex at anthesis; calyx subtended by 2-3 persistent bracteoles, longest of which approximately half as long as sepals and about twice as long as shortest; corolla tube almost cylindrical, up to 34 mm long; leaves often subcordate at one of the unequal sides of the base; fruit mostly large, mericarps 5-10 cm long and thick-walled. Mexico to Panama..**3. S. donnell-smithii**
   Sepals connate at the base, not shed with the corolla; calyx subtended by 0-2(-3) bracteoles which are up to half as long as the sepals when the corolla tube is infundibuliform and 35-67 mm long; leaves equal-sided and not subcordate at the base; mericarps mostly less than 5 cm long..7
7. Sepals retuse to obtuse, 1.2-3 x as long as wide; calyx subtended by 1-2(-3) bracteoles 8
   Sepals obtuse or sometimes rounded, (2-)2.5-10 x as long as wide; calyx subtended by 0-1 bracteole; bracteole usually less than 0.3 x as long as the sepals which are in that case at least 3 x as long as wide and obtuse. Mexico to Colombia...**6. S. macrophylla**
8. Corolla tube 26-27 mm long. Costa Rica (southern Puntarenas)...........**8. S. pauli**
   Corolla tube 35-66 mm long 9
9. Corolla yellow, tube infundibuliform, not narrowed at the throat; anthers with tails 15-19 mm above the corolla base; plants usually at least partly hairy, especially on leaves beneath and on bracteoles; mericarps green, smooth. Mexico to Bolivia...**7. S. obovata**
   Corolla white, often(?) with an orange throat, tube narrowly infundibuliform, slightly narrowed at the throat or not; anthers with tails 7-11 mm above the corolla base; plants entirely glabrous or sometimes with some minute white hairs on the leaves beneath along the costa; mericarps orange or yellow, often lenticellate . Mexico to Colombia ....**5. S. litoralis**

1. Stemmadenia alfari (Donn. Sm.) Woods. in Ann. Miss. Bot. Gard. 15 : 360 (1928), partly, excl. *Tonduz* 7645 of which the sheet seen belongs to *Tabernaemontana heterophylla*. – Type: Costa Rica, Puntarenas, Limoncito and Vuelta, *Pittier* 11094 (lectotype US, designated here, cited as type by Woods.; isolectotypes BR, F, G, M). Fig. 100, p. 402; map 54, p. 403

Basionym:

Tabernaemontana alfari Donn. Sm. in Bot. Gaz. 24: 396 (1897).

Shrub or tree, 1.50-9 m high. Trunk 2-15 cm in diameter. Branches pale brown, lenticellate; branchlets terete, glabrous. *Leaves* of a pair equal or unequal (larger up to 3 x as long as other and similarly shaped or smaller comparatively broader and from 1.5 x as long as wide), petiolate; petiole glabrous, 2-12 mm long; (ocreae widened into intrapetiolar stipules or not); blade membranaceous, also when fresh, elliptic or narrowly elliptic, (1.5- in smaller leaves of a pair) 2-3 x as long as wide, 2-14 x 1-7.2 cm, acuminate to caudate at the apex, cuneate or rounded at the base, glabrous on both sides, with scattered black dots beneath, with 5-12 pairs of upcurved secondary veins forming an angle of 60-80° with the costa; tertiary venation more or less reticulate, not very conspicuous, not impressed above. *Inflorescence* shortly pedunculate, 3-5 cm long; 1-4-flowered. Peduncle glabrous, 3-9 mm long; pedicels glabrous, 2-5 mm long. Bracts sepal-like and approximately half as long as them. *Flowers: Calyx* sometimes subtended by one sepal-like bracteole, smaller than the sepals; sepals pale green, connate at the extreme base, orbicular or broadly ovate, 0.8-1.3 x as long as wide, 2-4 x 2-4 mm, rounded, glabrous on both sides, not ciliate, equal or subequal, inner 1-1.3 x as long as outer, not or obscurely veined, with 3-8 colleters in 1-2 rows in the middle at the base; colleters small, 0.3-0.5 x 0.15-0.3 mm. *Corolla* yellow, with a depressed-globose head in the mature bud, about 0.1-0.2 x as long as the bud, rounded at the apex, glabrous outside, not ciliate, with 5 longitudinal wings 7-8 x 1 mm from the insertion of the stamens just above them, pubescent inside on the filament ridges just below the anthers for 1-2 mm and on the wings up to 0-2 mm below their apices; tube 8-13 x as long as the longest sepals, 1.4-2.1 x as long as the lobes, 22-32 mm long, narrowly infundibuliform, 3-6 mm wide above the base, narrowed to 2-4 mm wide above and/or below the anthers, from there gradually widened towards the throat and 6-11 mm wide, not narrowed at the throat, twisted 0.7-1 turn around the anthers; lobes obliquely obovate or ovate, 0.45-0.8 x as long as the tube, 1.4-2 x as long as wide, 13-18 x 8-11 mm, not undulate, spreading. *Stamens* deeply included, with tails of anthers about 1 mm below insertion, inserted 0.3-0.4 of the length of the corolla, (at 7-12 mm from the base); anthers sessile, just below the wings, narrowly triangular, 3-4 x as long as wide, 5-6 x 1.5 mm, apex sterile for 0.3-0.8 mm, (tails sometimes curved towards each other), glabrous. *Pistil* glabrous, 9-14 mm long; ovary ovoid, 2.5-4 x 2 x 1.5 mm, gradually narrowed into the style, of two carpels connate at the base with a 1.2 mm high slightly 5-lobes disk-like thickening; style 5-9 mm long; pistil head 1.5-2 x 1.4-2 mm, of a basal recurved veil 0.6-0.8 x 1.4-2 mm, a stipitate 5-lobed depressed globe 0.5-0.6 x 0.7-0.8 mm and a stigmoid apex 0.2 x 0.2 mm. Ovules approximately 25 in each carpel. *Fruit* of 2 separate mericarps; mericarps yellow or orange, obliquely ellipsoid or pod-like, 25-40 x 7-20 x 7-20 mm, recurved, apiculate to caudate, without lateral ridges, partly and irregularly lenticellate, otherwise smooth; wall 0.5-1 mm thick in dried fruits; aril orange. *Seed* dark brown, obliquely ellipsoid, 11-13 x 5-6 x 4-5 mm, with longitudinal grooves, papillose; embryo 8-9 mm long; cotyledons broadly ovate, 1-1.2 x as long as wide, 3.5-4 x 3-4 mm, rounded at the apex, cordate at the base; rootlet 5 x 0.8-0.9 mm.

DISTRIBUTION: Costa Rica, Panama.

ECOLOGY: Moist forest. Alt. 300-1300 m. Flowering mainly from January to March and fruiting mainly from August to November.

Geographical selection of the approximately 80 specimens examined:

COSTA RICA. Guanacaste: Volcán Orosi, *Wibur & Stone* 10704 (F, MO, US); Rincón de la Vieija National Park, *Garwood* et al. 800 (BM, MO); 5 km E of Tilarán, *L.O. Williams* 24567 (F, LL, NY). Puntarenas: Limoncito and Vuelta, *Pittier* 11094 (BR, F, G, M, US, lectotype); 6 km S of San Vito de Java, *Raven* 22019 (F, MO). Alajuela: Santa Maria National Park, *Liesner* 5178 (MO); San Ramón F.R., *Herrera & Chacón* 493 (WAG); Buenavista, San Carlos, *Jimenez* 421 (F, MO, NY); Monteverde Cloud F.R., *Haber* et al. 5436 (MO). San José near Alfombra, *Burger & R. Baker* 10128 (F). Limón: Río Suerte, Llanuras de Santa Clara, *J.D. Smith* 6648 (US; phot. WAG, paratype); near Pacuare, *L.O. Williams* 16206 (BM, F, MO, US); Río Parismina, 8 km W of Dos Bocas, *Lent* 2448 (F); 1 km W of Cerro Mucilla, *Herrera & Solís* 2541 (WAG).

PANAMA. Veraguas: Cerro Tute, *Mori* et al. 7543 (MO). Coclé: N rim of El Valle de Antón, *Allen* 1783 (F, GH, MO, NY, US); Cerro Pilón, *Porter* et al. 4628 (AAU, GH, MO). Panamá: Cerro Campana area, *Dwyer* 7848 (GH, MO, US).

2. Stemmadenia brasiliensis Leeuwenberg, **sp. nov.**

Fig. 101, p. 404; map 54, p. 403

Frutex foliis petiolatis anguste ellipticis longe acuminatis utroque latere glabra. Inflorescentia biflora brevissime pedunculata bracteis pedicella obtectantibus. Sepala foliacea ovata rotundata glabra. Corolla alba tubo cylindraceo intus alis hirsutis limbo

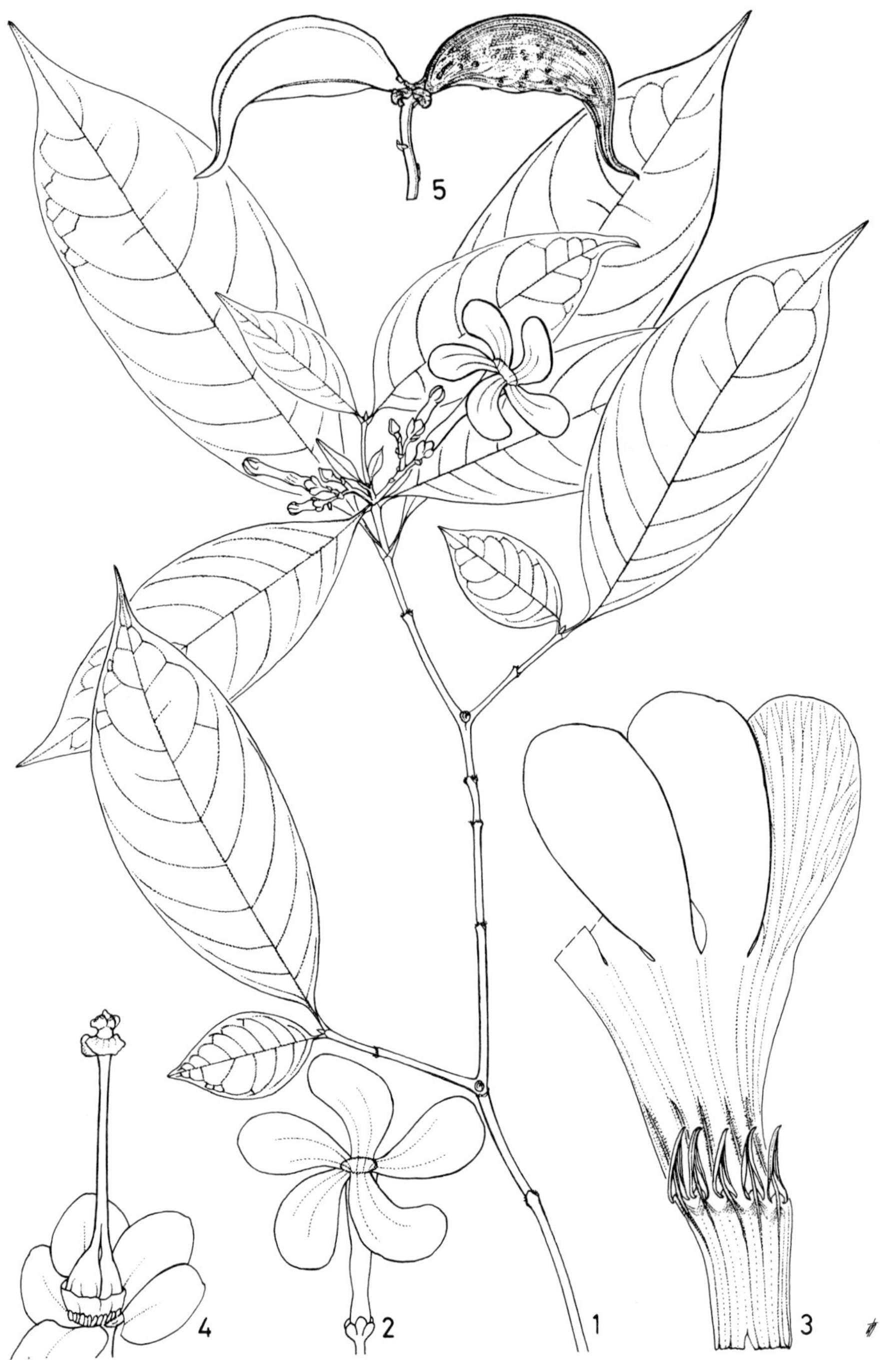

Fig. 100. *Stemmadenia alfari.* **1,** habit (x 2/3); **2,** flower (x 1); **3,** opened corolla (x 2); **4,** calyx with pistil (x 3); **5,** fruit (x 1). 1 from Liesner 5178; 2-4 from Herrera 493; 5 from Haber et al. 5436.

Map 54. ●. *Stemmadenia alfari*, ▲. *S. brasiliensis*, ■. *S. stenoptera.*

patente. Stamina sub alis insertis antheribus anguste triangularibus. Pistillum glabrum ovario ovoideo basi annulo cirumdato. Fructus ignotus.

Typus: Brazil, Pará, Municipio Oriximiná, 7-km N of Cachoeira Porteira, 20 August 1986, *C.A. Cid Ferreira* et al. 7935 (holotypus INPA; isotypi NY, WAG).

Shrub 4 m high. Branches pale brown, with fissured bark; branchlets with paler lenticels, especially at the base, terete, glabrous. *Leaves* petiolate; petiole glabrous, 2-5 mm long; (ocreae not widened into intrapetiolar stipules, blade papery when dried, narrowly elliptic, 3.3-3.7 x as long as wide, 5 11 x 1.5 3.5 cm, long-acuminate at the apex, decurrent into the petiole, glabrous on both sides, with or without scattered black dots beneath, with 9-14 pairs of secondary veins; tertiary venation rather conspicuous, reticulate. *Inflorescence* sessile or nearly so, 2-flowered. Peduncle glabrous, 0-1 mm long; pedicels 2-3 mm long, glabrous, covered with many almost sepal-like bracts, 0.2-0.4 x as long as the sepals. *Flowers* open during the day. *Sepals* green(?), foliaceous, veined, free, erect, not clasping the corolla base, ovate, unequal, 1.7-1.9 x as long as wide, largest 7,5 x 4 mm, smallest 5 x 3 mm, rounded, glabrous outside, not ciliate, glabrous inside and with 4-5 colleters in 1 row at the base; colleters 0.3 x 0.1 mm. *Corolla* white, bud with an ellipsoid blunt head about as wide as the widest part of the tube, glabrous outside, hirto-pilose for 1 mm on filament ridges between anther tails, with 5 hirsute wings about 1 mm wide from just above the insertion of the stamens to 1 mm above the apices of the anthers, otherwise glabrous; tube 2.7-3 x as long as the calyx, 1.6-1.8 x as long as the lobes, 19-22 mm long, almost cylindrical, 3 mm wide above the base, narrowed below the insertion of the stamens to 2-2.2 mm wide, slightly widened around the anthers to 2.2-2.5 mm wide, narrowed again above to 2 mm wide, not twisted; lobes obliquely elliptic, 0.5-0.6 x as long as the tube, 2.4-4 x as long as wide, 10-12 x 3-5 mm, rounded, not undulate, spreading. *Stamens* with apex 4-5 mm below mouth of corolla tube, inserted 0.55-0.6 of the length of the corolla tube, (at 11.5-23.5 mm from the base); anthers sessile, with tails 0.5 mm below insertion, narrowly triangular, 4 x 1 mm, apex sterile for 0.4 mm, with tails curved towards each other, glabrous. *Pistil* glabrous, 13-15 mm long; ovary ovoid, 2 x 1.5 x 1.5 mm, composed of 2 separate carpels connate at the base with a disk-like ring, 1 mm high; style filiform, 8-10 mm long; pistil head composed of an undulate basal ring 0.2 x 0.8 mm, a stipitate 5-lobed depressed globe 0.3 x 0.8 mm and a stigmoid apical part 0.2 x 0.2 mm. *Fruit* unknown.

Fig. 101. *Stemmadenia brasiliensis.* **1,** habit (x 2/3); **2,** flower (x 2); **3,** opened corolla (x 4); **4,** sepal inside (x 4); **5,** pistil (x 4). 1-5 from Cid et al. 7935.

DISTRIBUTION: Only known from the type specimen.
ECOLOGY: Forest understorey. Alt. low.

3. Stemmadenia donnell-smithii (Rose ex Donn. Sm.) Woods. in Ann. Miss. Bot. Gard. 15: 369 (1928). – Type: Guatemala, Retalhuleu, San Felipe, *J.D. Smith* 2763 (lectotype US; phot. WAG, designated here, cited as type by Woodson; isolectotypes A, F, G, GH, K, M, MO, NY). Fig. 102, p. 406; map 52, p. 394

Basionym:

Tabernaemontana donnell-smithii Rose ex Donn. Sm. in Bot. Gaz. 18: 206 (1893).

Heterotypic synonyms:

T. donnell-smithii var. *costaricensis* Rose ex Donn. Sm. in op. cit. 24: 397 (1897). – Type: Costa Rica, San José, Río Toro Amarillo, Llanuras de Santa Clara, *J.D. Smith* 6646 (holotype US; phot. WAG).

Shrub or tree 1.5-25 m high. Trunk 3-80 cm in diameter; bark pale grey or light brown, shallowly fissured or smooth, with protruding lenticels; inner bark white. Branches pale brown, lenticellate; branchlets terete, glabrous. *Leaves* of a pair equal or unequal (and then larger up to 3 x as long as other and similarly shaped), mostly shortly petiolate; petiole glabrous, 2-6(-10) mm long; blade membranaceous when fresh, membranaceous or thinly papery when dried, narrowly elliptic or obovate, 2-4 x as long as wide, 2.5-22 x 1.5-8 cm, acuminate or caudate at the apex, gradually narrowed towards the base, base acute, rounded or subcordate and often unequal-sided, glabrous on both sides or white-pilose in or near the axils of the secondary veins beneath; without black dots, with 7-16 pairs of upcurved secondary veins forming an angle of 45-80° with the costa; tertiary venation reticulate, not impressed above. *Inflorescence* pedunculate, 1-4-flowered, 15-40 mm long, excluding the flowers. Peduncle glabrous, 7-25 mm long; pedicels glabrous, 3-10 mm long. Bracts like outer bracteoles. *Flowers* fragrant. *Calyx* subtended by 2-3 more or less persistent bracteoles (persistent when the calyx is shed), sepal-like, obtuse or rounded, glabrous on both sides, outer very small, broadly to very broadly ovate, 0.6-1 x as long as wide, 3-5 x 3-6 mm, other broadly ovate, about as long as wide, 5-12 x 5-10 mm, all of them up to half as long as longest sepals. Sepals pale green, shed with the corolla (as in most *Voacanga* spp.), free, subequal, inner 1.1-1.5 x as long as other, obovate, elliptic or ovate, 1-1.7 x as long as wide, (6-)15-25 x (6-)10-18 mm, rounded at the apex, inner often torn at anthesis and therefore seemingly retuse, entire, glabrous on both sides, not ciliate, clearly veined, with approximately 30 colleters in dense rows inside at the base, forming a continuum in the entire calyx; colleters 0.5-1 x 0.1-0.2 mm. *Corolla* bright or pale yellow, sometimes creamy, with an ovoid obtuse or acute head in the mature bud, glabrous outside, not ciliate, with 5 longitudinal wings inside, 14-18 x 1-2 mm from 6-10 mm above the base, which is 8-9 mm below the anthers tails, pubescent inside from 3-8 mm above the base in 10 stripes, in a belt or on the wings only to the insertion of the stamens and above on the wings, often only on their edges to 0-3 mm below their apices, 2-4 mm of the wings above the anthers; tube 1.2-1.7(-4) x as long as the longest sepals, 1.3-2.4 x as long as the lobes, (15-)25-34 mm long, almost cylindrical, 6-11 mm wide at the base, narrowed to 3-5 mm wide below the insertion of the stamens, widened again above, 5-12 mm wide at the throat and not contracted, twisted 0.25-0.5 turn around the anthers; lobes obliquely obovate or nearly so, 0.6-0.7 x as long as the tube, 1.1-2.5 x as long as wide, 11-22 x 8-13 mm (sometimes wider?), rounded, not undulate, spreading. *Stamens* deeply included, inserted 0.4-0.6 of the length of the corolla tube, (at 15-20 mm from the base); anthers sessile,

Fig. 102. *Stemmadenia donnell-smithii.* **1,** habit (x 2/3); **2,** flower (x 1); **3,** opened corolla (x 2); **4,** calyx with pistil (x 1); **5,** sepal (x 1); **6,** fruit (x 2/3); **7,** mericarp (x 2/3); **8,** mericarp in section (x 2/3). 1 from J.D.Smith 2763; 2-5 from Sallee 41; 6 from Hansen et al. 1698; 7-8 from Croat 47532.

with tails about 1 mm below insertion, narrowly triangular, 3-4 x as long as wide, 5-6 x 1.5-2 mm, apex sterile for 0.3-0.7 mm. *Pistil* glabrous, (9-)12-22 mm long; ovary broadly ovoid or conical 3 x 3 x 2-3 mm, gradually narrowed into the style, of two carpels connate at the base with a slightly 5-lobed disk-like thickening 1-1.5 mm high; style (4-)7-17 mm long; pistil head 1.5-2.2 x 1.5 mm, composed of a basal cup 0.5-0.7 x 1.5 mm, a ring 0.3 x 1.5 mm, a stipitate 5-lobed depressed globe 0.3-0.4 x 0.6-0.7 mm and a stigmoid apex about 0.2 x 0.2 mm. Ovules approximately 200 in each carpel. *Fruit* of 2 separate mericarps; mericarps green or brown, sometimes yellow, orange inside, subglobose or obliquely ellipsoid, 50-100 x 35-60 x 35-60 mm, acuminate, apiculate or rounded at the recurved apex, with a faint lateral ridge or not, with small lenticels; wall up to 30 mm thick in fresh fruits, 3-10 mm in dried; aril orange. *Seed* medium or dark brown, obliquely ellipsoid, 9-11 x 3-5 x 3-4 mm, with longitudinal grooves, papillose; embryo 7.5-9 mm long; cotyledons ovate, 1.2-1.3 x as long as wide, 3.5-4 x 3 mm, obtuse or rounded at the apex, cordate at the base; rootlet 1-1.25 x as long as the cotyledons, 4-5 x 0.7-0.8 mm.

DISTRIBUTION: Mexico to Panama.

ECOLOGY: Wet forest or thicket. Alt. 0-1800 m. Flowering and fruiting throughout the year with a peak for flowering in Mexico from April to September and south of it from February to June, and for fruiting from January to March.

Geographical selection of the approximately 500 specimens examined:

MEXICO. Sinaloa: foothills of the Sierra Madre, near Coloma, *Rose* 1711 (US); 25 km NE of Chilillos, *Martínez* et al. 4065 (MEXU). Nayarit: near Singaita, E of San Blas, *Philbrick* 784a (BH); 8 km NE of Las Varas, *Webster & Lynch* 17121 (GH, MO). Jalisco: 18 km N of El Tuito, *Lott* et al. 1029 (MEXU); Est. Biol. Chamela, *Lott* 497 (MEXU). Colima: 25 km SE of Manzanillo, *Lott & Magellanes* 854 (MEXU, USF). Michoacan: Ostula, *Emrick* 106 (F); 2 km SW of Aquila, *Soto Nuñez* 2624 (MEXU); 2 km E of Cruz de Campos, *Martínez* et al. 4454 (MEXU, MO); ca 6 km from Arteaga, *H.E. Moore & Bunting* 8757 (LL); La Correa, *Langlassé* 427 (G, GH, K, P, US); 5 km NW of La Huacana, *Martínez* 458 (MEXU, MO). Guerrero: Guadelupe, Distr. Mts de Oca, *Hinton* et al. 10186 (F, K, MO, US); Vallecitos, *Hinton* 10245 (K, US, USF); 11 km NW of Petatlán, *Martínez* et al. 5127 (MEXU); Coyuca, Hinton 5872 (K); Calavera, *Hinton* et al. 10043 (K, LL, NY, P, UC, US, USF). Mexico: Guayabal, *Hinton* 4569 (BM, G, GH, K, MO, S, US); Tlatlaya, *Matuda* et al. 31111 (MO). Morelos: near Cuernavaca, *Föderström & Hultén* 150 (S). Puebla: Agua Fria, Chapingo, *Pennington & Sarukhán* 9454 (FHO, K, MEXU, NY); Patla, *Troll* 331 (M); Tuzamapan de Galeana, *Espadas & Zito* 27 (MEXU). Oaxaca: near Temascal, *Cortes & Torres Colin* 422 (MEXU); Chiltepec, *G. Martínez* 236 (MEXU, MO, SP); Ubero, *Ll. Williams* 9499 (F); near Pluma Hidalgo, *Leyva* anno 1942 (US). Veracruz: Cerro Quebrado, between Misantla and Martínez de la Torre, *Gómez Pompa* 936 (A); Coetzala, *Vaszquez* 390 (F); near Santiago Tuxtla, *Nee & K. Taylor* 26472 (F, USF); Zacuapan, *Purpus* 4413 (UC); Est. Biol. Los Tuxtlas, *A. Gentry* et al. 32548 (MO, P, WAG); Laguna Escondida, *Lot* 327 (F, GH, MEXU); S of Tuxtepec, *Croat & Hannon* 65486 (WAG); Fortuño, Río Coatzacoalcos, *Ll. Williams* 8453 (F); Mun. Jesús Carranza, 3 km from border of Oaxaca, *Perino* 3223 (MEXU, NY); 2 km from Campamento Hnos Cedillo, Hidalgotitlan, *Valdivia* 1339 (MO). Chiapas: NE of Ocote, *Miranda* 6250 (MEXU); 22 km N of Tonalá, *Hansen* et al. 1620 (US, WIS); Mun. Pichucalco, Campamento San Joaquin, *Gilly & Hernandez* 207 (GH, MEXU); Piedra Rajada, *Enriquez* 6 Dec. 1950 (MEXU); km 33 of highway from Palenque to ruins, *Sallee* 41 (MEXU, NA); near zona Arquelógica de Palenque, *Cabrera* et al. 1921 (MEXU); km 70 Palenque-Ocosingo Road, *Breedlove & Almeda* 48437 (LL, MO); km 6 Coronado-El Retiro Road, *Calzada* et al. 3843 (MEXU); ruins of

Bonampak, *Hoover* 265 (GH, MO, US); Escuintla, *Matuda* 415 (MEXU, MO, US); 5 km N of Escuintla, *Croat* 47532 (MO); 2 km W of Cacahoatan, *Martínez* et al. 19879 (MO); Unión, Miramar, Mun. Tapachula, *Ventura* & *López* 1944 (P). Tabasco: 18 km NE of Pichucalco, near El Azufre, *Hansen* et al. 1698 (LL, US, WIS); Ejido Soberano, Balancan, *Novelo* et al. 198 (MO). Quintana Roo: Mun. Chetumal, 18.6 km N of Tomás Garrido, *Sanders* et al. 10015 (LL); 1 km W of Tres Garantias-Tomás Garrido Road, on road to Dos Aguadas, *Cabrera* 6573 (LL). Sin. loc., *Sessé* et al. 1443 (F, distributed as *S. grandiflora*).

BELIZE. Indian Church, *Arnason* & *Lambert* 17061 (MO). El Cayo: ca 10 km SW of San Ignacio, *Balick* et al. 1931 (WAG); El Cayo, *Kinloch* 12831 (NY, UC, US); Valentin, *Lundell* 6243 (C, F, GH, LL, MO, NY, S, US, WIS). Stann Creek (= Dangriga): Middlesex, *Schipp* 411 (A, BM, F, G, GH, K, MO, NY, S, UC, Z); 35 km Stann Creek Valley, *Schipp* 958 (A, BM, F, G, GH, K, MO, NY, S, UC, Z); Big Creek, *Schipp* 39 (BM, F, G, GH, K, MO, NY, UC, US, Z). Toledo: Salamanca Camp, *Whitefoord* 1885 (BM); Solomon Camp, *Davidse* & *Brant* 32187 (WAG).

GUATEMALA. Petén: near Carmelita, *Egler* 42-319 (F); Tikal National Park, *Lundell* 16150 (F, GH, LL, US); 35 km E of Santa Elena, *Molina* 15478 (F); Mt Polol, *Lundell* 3444 (GH, MO, S, US); Cerro Ceibal, *Steyermark* 46078 (F, MO); Cadenas, *Lundell* & *Contreras* 20338 (LL). Alta Verapaz: Cerro Chinajá, *Steyermark* 45567 (F); near Cobán, *Standley* 90896 (F, MO); Telemán, *Tenorio* et al. 14451 (WAG). Izabal: Cadenas (Puerto Mendez), *Contreras* 9214 (LL); Mico Mts, *Stevens* et al. 25610 (MO). San Marcos: 1.5-3 km N of Ocós, *Steyermark* 37871 (F). Quezaltenango: Coatepec, *L. Rodriguez* 1794 (P); near Santa María de Jesús, *Standley* 68197 (F). Retalhuleu: 9 km N of Champerico, *Harmon* 2300 (AAU, MO, NY, U); San Felipe, *J.D. Smith* 2763 (A, F, G, GH, K, M, MO, NY, US; phot. of US sheet WAG, lectotype). Suchitepéquez: near Tiquisate, *Steyermark* 47644 (F, MO). Escuintla: near Santa Lucía Cotzumalguapa, *Hürlimann* Aug. 1957 (Z); Finca El Baul, *Tonduz* & *Rojas* 36 (BR, US); Naranjo, *J.D. Smith* 2764 (G, US, paratype); San José, *J.D. Smith* 2465 (US, paratype); Escuintla, *J.D. Smith* 2404 (GH, M, US, paratype); Barranca de Eminencia, *J.D. Smith* 2762 (G, GH, K, M, MO, NY, P, US, paratype); Finca El Salto, *Almeda* 747 (F, MO); San Juan Mixtan, *J.D. Smith* 2405 (G, K, NY, US, paratype). Santa Rosa: E of Cuilapa, *Standley* 78146 (F); E of Los Cerritos, *Standley* 79609 (F). Solalá: Santa Barbara, *Shannon* 152 (F, US, paratype).

EL SALVADOR. Hacienda Santo Tomás, *Carlson* 1137 (F).

HONDURAS. Santa Barbara: 20 km NE of Nueva Arcadia, *Harmon* & *Dwyer* 3796 (MO); Lake Yojoa, *Clewell* 3113 (MO, NY). Cortés: Tulián, *Molina* 5219 (F, US); San Pedro Sula, *Bangham* 305 (A); 20 km S of San Antonio de Cortés, *Nelson* et al. 8113 (MO). Atlántida: near Tela, *Standley* 53519 (A, F, US); near La Ceiba, *Yuncker* et al. 8205 (BM, F, G, GH, K, MO, S). Yoro: near Progreso, *Standley* 55025 (F). Camayagua: Lake Yojoa, *L.O. Williams* & *Molina* 14630 (F); Jardines, *Barkley* & *M.L. Smith* 40878 (IJ); between La Misión and Taulabé, *Molina* 13013 (F, LL, NY). Colón: 11 km E of Trujillo, *Saunders* 306 (BM, F, MO, NY, Z). Olancho: between San Estéban and Bonito Oriental, *Croat* & *Hannon* 64516 (WAG); 8 km NE of Catacamas, *A. Ortega* 261 (MO); Río Esperanza, near Elvirá, *Wilson* 444 (NY). Choluteca: 10 km N of Jicaro Galán, *Nelson* et al. 3203 (MO).

NICARAGUA. Zelaya: Caño Majagua, *Stevens* 6915 (BM, MO); Amparo, *F. Ortiz* 1278 (MO); Caño Sansangwás, *F. Ortiz* 1744 (MO); Cukra, *Long* 184 (F); W of Nueva Guainea, *J.S. Miller* & *Sandino* 1249 (MO, WAG); 1 km N of El Zapote, *Nee* 27915 (MO, WAG); Caño Montecristo, *Moreno* 14966 (MO, WAG). Jinotega: La Bujona, *Ll. Williams* 17419 (WIS); between Wiwili and El Carmen, *Araquistain* & *Moreno* 1574 (MO). Madriz: Cerro Volán de Somoto, *Stevens* & *Grijalva* 16431 (MO). Chinandega: Ameya, *Maxon* et al. 7112 (US); near Chichigalpa, *Standley*

11491 (F). Managua: Las Nubas, *Maxon* et al. 7502 (US); near San Francisco Libre, *Laguna* 246F (MO); ca 5.4 km NE of El Crucero, *Stevens* 3493 (BM, MO, U). Matagalpa: NW of Cerro Musún, *Araquistain & Moreno* 2417 (MO, WAG); 20 km E of Matagalpa, *Neill* 2380 (MO); 20 km NE of El Tuma, *Grijalva* 1384 (MO, WAG); 11 km SW of Río Blanco, *Moreno* 23967 (MO). Granada: Volcán Mombacho, *Moreno & Sandino* 6486 (MO, WAG). Masaya: Laguna de Apoyo, *Moreno* 21305 (MO). Carazo: 2 km N of Jinotepe, *Moreno* 10748 (MO, WAG). Rivas: Hacienda Fatíma, *Sandino* 4152 (MO); ibid., SW of Sapoá, *Sandino* 3582 (MO, WAG). Boaco: SE of Mombachito, *Sandino* 2866 (MO, WAG). Chontales: 4 km N of Cuapa, *Nee* 28461 (MO, WAG). Río San Juan: La Palma, *Stevens* et al. 23406 (MO, WAG).

COSTA RICA. Guanacaste: Rincón de la Vieija National Park, *Garwood* et al. 804 (BM, MO); Comelco, near Bagaces, *Bawa* 176 (MO); 1 km E of Río Tenorio, *Grayum & Herrera* 4837 (WAG); Nicoya For., *Tonduz* CR 13904 (G, GH, K, NY, P, US); ca 2 km SE of Est. Exp. Enrique Jimenez Nuñez, *Diarmid* 73-41 (USF). Puntarenas: Monteverde, *Haber* 295 (MO, P); S side of Quebrada Bonita, Carara Res., *Grayum* 4711 (WAG); Río Cebo, near Buenos Aires, *Pittier* 667 (BR); Los Barrancas, *A. Gentry* 764 (F, MO, NY); Canton de Osa, between Palmar Norte and Cañablancal, *Allen* 5227 (F, GH, MO, UC, US); Corcovado National Park, *Kernam* 134 (CR), 366 (MO); Golfito, *Lems* 5188 (F, NY, US). Alajuela: Naranjo, *Ørsted* 15521 (C); 4 km SE of Fortuna, *Liesner* et al. 15185 (WAG, WIS); near Atenas, *Schubert & Madriz* 1005 (GH, US); Río Sarapiquí, *Jiménez* 4145 (F, MO, US). San José: Río Toro Amarillo, *J.D. Smith* 6646 (US; phot. WAG, type of *T. donnell-smithii* var. *costaricensis*); Basin of El General, *Skutch* 4882 (A, F, MO, NY, US); Villa Colon, *Sastre* 4834 (P). Heredia: Finca La Selva, *McDowell* 420 (GH, LL, MO, US). Cartago: 10 km ENE of El Palmar Norte, *Burger & Matta* 4653 (F); Turrialba, *J. León* 244 (US), 595 (US). Limón: Tortuguero National Park, *Robles & Flores* 1644 (CR); Río Catarata, *Burger* et al. 10396 (AAU, F, MO).

PANAMA. Bocas del Toro: Sanchez Potrero, *Dunlap* 586 (US); Changuinola-Altamirante Railroad, *Croat & Porter* 16393 (MO, USF); Water Valley, *von Wedel* 1379 (GH, MO, US). Chiriquí: Burica Peninsula, *Croat* 22006 (F, MO, NY, US, USF, WAG); near Gualaca, *Allen* 5022 (BR, C, F, FI, G, GH, K, MO, NY, P, S, U, UC, US); near Río Tabasara, *Woodson* et al. 411 (MO). Veraguas: W of Sona, *Allen* 1048 (MO). San Blas: Cerro Habú, *Sytsma* et al. 2670 p.p. (MO).

CULT. Mexico, Veracruz, Est. Biol. Las Tuxtlas, *Gómez Pompa* 4471 (GH, WIS). Guatemala, Guatemala, Bot. Garden, *Steyermark* 46458 (F). Honduras, Santa Barbara, Lago de Yojoa, W shore, near El Rincón, *Blackmore & Chorley* 3693 (BM, MO). Togo, Lomé, *Warnecke* 481a (BM, P). Cameroun, Victoria (= Limbe), Bot. Garden, *Winkler* 104 (Z).

4. Stemmadenia grandiflora (Jacq.) Miers, Apoc. S. Am. 75 (1878). – Types: Holotype apparently not preserved. Neotype: Colombia, Magdalena, near Santa Marta, *H.H. Smith* 1639 (neotype NY, designated here; isoneotypes A, BM, BR, E, F, G, GH, K, L, MO, S, U, US, WIS). Fig. 103, p. 411; map 55, p. 415

Basionym:

Tabernaemontana grandiflora Jacq., Enum. Pl. Carib. 14 (1760).

Heterotypic synonyms:

T. riparia H.B.K., Nov. Gen. 3: 228 (1819). *Malouetia riparia* (H.B.K.) A. DC., Prod. 8: 380 (1844). – Type: Colombia, Cundinamarca, Río Magdalena, near El Peñon, *Bonpland* 1528 (holotype P-BO; phot. F, GH, US; isotype P).

S. pauciflora Woods. in Ann. Miss. Bot. Gard. 15: 366, pl. 49, fig. 1 (1928). – Type: Colombia, Tolima, between Espinal and Cuamo, *Pennell & Rusby* 186 (holotype NY).

S. pennellii Woods. in op. cit. 367. – Type: Colombia, Bolívar, Turbaco, *Pennell* 4755 (holotype GH; isotypes K, MO, NY, US; phot. of US sheet WAG).

Shrub or small tree 1-8 m high. Trunk 2-7 cm in diameter; bark grey or pale grey-brown, rough; wood light brown. Branches pale or dark brown, lenticellate; branchlets glabrous. *Leaves* of a pair equal or unequal (smaller down to 0.17 of the length of the other and similarly shaped or about as wide as larger and ovate), petiolate; petiole glabrous, 1-12 mm long; (ocreae not or slightly widened into intrapetiolar stipules); blade herbaceous when fresh, membranaceous when dried, elliptic, obovate or narrowly so or, in smaller leaves of a pair, often ovate, 1,5-3.5 x as long as wide, 2-18 x 1-9 cm, acuminate or caudate at the apex, cuneate, less often rounded or in smaller leaves of a pair often subcordate at the base, glabrous and often with scattered black dots on both sides or sometimes pubescent beneath, with 5-15 pairs of rather straight or upcurved secondary veins forming an angle of 60-80° with the costa; tertiary venation reticulate. *Inflorescence* shortly pedunculate, 4-7 x 3-5 cm, 1-4-flowered. Peduncle slender, glabrous, 3-14 mm long; pedicels glabrous, 5-10 mm long. Bracts very small, 0.1-0.2 x as long as the sepals and more or less similar. *Flowers* slightly fragrant or not. *Calyx* not or occasionally subtended by bracteoles; sepals pale green or cream, persistent under the fruit even when ripe, free, erect, unequal, longer sepals 11-23 x 5-20 mm, shorter from 8 x 5 mm, ovate, broadly ovate or elliptic, 1-6 x as long as wide, longer often narrower than shorter, imbricate in bud, variable at the apex, from rounded to acuminate, entire, outer often subcordate at the base, clearly veined, glabrous outside, not ciliate, glabrous inside and with 1-2 dense rows of 6-20 colleters at the base in middle only or forming a continuum in the whole calyx; colleters 0.3-0.7 x 0.1-0.2 mm. *Corolla* yellow, orange or sometimes white, with an ovoid obtuse or acute head in the mature bud, glabrous outside, not ciliate at lobes, with 5 longitudinal wings inside 10-16 mm long and about 1 mm wide, from 5-6 mm below the insertion of the stamens to 3-6 mm above the apices of the anthers and with a villose or pilose interrupted belt, especially on the wings, starting at the base of the wings or slightly above and ending 2 mm below to 1.5 mm above the apices of the anthers; tube 1.5-3 x as long as the longer sepals, 0.8-1.7 x as long as the lobes, 23-35 mm long, almost cylindrical, widest just above the base and 7-8 mm wide, narrowed below the insertion of the stamens to 3-5 mm wide, slightly widened around the anthers and at the throat or not and 3-7 mm wide, twisted 0.25-0.5 turn; lobes obliquely elliptic or obovate, 0.6-1 x as long as the tube, 1.5-2.5 x as long as wide, 18-30 x 10-20 mm, rounded, not undulate, spreading. *Stamens* with apex 4-13.5 mm below mouth of corolla tube, inserted 0.43-0.65 of the length of the corolla tube, (at 13-19 mm from the base); anthers sessile on the wings, narrowly triangular, 3.5-4.5 x as long as wide, 5-6 x 1.2-1.5 mm, apex sterile for 0.2-0.5 mm. *Pistil* glabrous, 15-22 mm long; ovary broadly ovoid, 2-3 x 2-2.5 x 2-2.5 mm, more or less gradually narrowed into the style, of two carpels connate at the base by the 1-1.5 mm high slightly lobed disk-like thickening; style filiform, 9-18 mm long; pistil head composed of an undulate ring 0.3-0.8 x 1.1-1.5 mm, a broadly obovoid apically 5-lobed central part 0.8 x 0.8 mm and a stigmoid apex about 0.2 x 0.2 mm. Ovules approximately 100 in each carpel. *Fruit* of 2 separate mericarps; mericarps green or sometimes yellow when mature, obliquely ellipsoid or ovoid, 20-55 x 15-35 mm, acuminate or caudate at the recurved apex, not ridged, smooth or sometimes partly verrucose when dried; wall 3-5 mm thick; aril orange or red. *Seed* black or brown, obliquely ellipsoid, 6-10 x 2-5 x 2-4 mm; embryo 4-7 mm long; cotyledons ovate, 1.2-1.6 x as long as wide,

Fig. 103. *Stemmadenia grandiflora.* **1,** habit (x 2/3); **2,** flower (x 2/3); **3,** opened corolla (x 1.6); **4,** pistil (x 4); **5,** sepal inside (x 2); **6,** fruit (x 2/3). 1 from Engsted 11; 2-5 from Steyermark 99913; 6 from de Bruijn 1261.

1.5-3 x 1.2-1.5 mm, rounded at the apex, truncate at the base; rootlet 1-2 x as long as the cotyledons, 2.5-4 x 0.5-0.7 mm.

DISTRIBUTION: Panama and northern South America.

ECOLOGY: Forest understorey or edge, or thicket. Alt. 0-1250 m. Flowering throughout the year, but most flowering specimens collected between March and July. A fruiting season could not be deduced form the available collections.

Geographical selection of the approximately 570 specimens examined:

PANAMA. Chiriquí: Progreso, *Cooper & Slater* 233 (F); near David, *Pittier* 2824 (US); near Remedios, *Pittier* 3388 (BM, NY, US); 3 km SW of Guabala, *Tyson* et al. 4255 (MO). Veraguas: Isla de Coiba, *Dwyer* 2380 (MO, NY); near Cañanas, *Sullivan* 265 (MO, WAG); 8 km NW of Santa Fe, *Liesner* 977 (MO, US); 8 km E of Santiago, *Tyson* et al. 4276 (MO); 5 km N of Olá, *Folsom* 3097 (MO, WAG). Los Santos: between Los Santos and Guarare, *Woodson* et al. 1198 (A, MO, NY); between Tonosí and Guanico, *Stern* et al. 1897 (MO, US); Pedasi, *Dwywer* 2478 (MO). Herrera: 20 km S of Ocú, *Lewis* et al. 1650 (GH, MO, UC). Coclé: N of El Cope: *D'Arcy* 11297 (MO, USF, WAG); between Paso del Arado and Olá, *Pittier* 5011 (US); El Valle de Antón, *Alston* 8798 (BM, F, P); Potrero, *Puga* 24 (F, MO). Colón: near San Miguel de la Borda, *Croat* 9898 (MO); 4.5 km SW of Pina, *Nee* 11685 (MO); mouth of Río Piedras, *Lewis* et al. 3192 (AAU, GH, MO); between Fató and Playa de Damas, *Pittier* 3940 (NY, US). Canal Zone: Barro Colorado Island, *Ebinger* 275 (F, MO, US); Madden Dam area, *Stern* et al. 41 (G, GH, MO, US); Ft. Kobbe, *Duke* 4226 (BM, GH, MO, UC, US). Panamá: 3 km N of La Chorrera, *Lewis* et al. 5188 (AAU, GH, MO); Cerro Campana, *D'Arcy* 9601 (AAU, MO, NY); near Corozal, *Allen* 913 (G, GH, MO, NY, P, S, U); Chagres, *Fendler* 234 (CGE, FI-W, GH, K, MO, P, TCD, US); E of Pacora, *Woodson* et al. 736 (A, MO, NY); between Red Hill and East Bay, *I.M. Johnston* 152 (BM, FI, GH, MO, US); road to Jesus Maria, just W of El Llano, *Tyson* 6865 (MO, NA); Isla Taboga, *Woodson* et al. 1514 (A, MO, NY); Perlas Islands, S tip of Isla del Ray, *Knapp & Mallet* 2942 (MO, WAG); 5.5 km E of Cañasas, *Knapp* 1169 (MO, WAG). San Blas: Isla Soskatupu, *Duke* 8941 (MO, US); Perme, *Cooper* 643 (F, K); 3-8 km SW of Puerto Obaldia, *Mori* 6803 (MO, NY, U). Darién: near Garachiné, *Hammel* 7312 (MO, Z); Río Balsa, between Río Areti and Manane, *Duke* 8763 (MO, US); Río Pirre, *Kennedy* 2882 (MO, NY, US, USF, WAG); near Paya, *Stern* et al. 189 (GH, MO, US); between Manene and mouth of Río Cuasi, *Kirkbridge & Bristan* 1432 (MO, NY); Río Cocalito, *Whitefoord & Eddy* 38 (BM, MO); Serrania del Sapo, *W. Hahn* 238 (MO); near Santa Fe, *Duke* 12890 (GH, MO).

COLOMBIA. Chocó: Mun. Acandí, Unguía, *Forero* et al. 1909 (COL, GH, MO, NY); between Río Tigre and Unguía, *A. Gentry & Aguirre* 15306 (AAU, COL, MO, NY); 20 km W of Ríosucio, *Bernal* 39 (COL); Río Truando, *Duke* 11170 (GH, MO, NY); Bahía Solano, *Fernández* 304 (US); between El Valle and Bahía Solano, *A. Gentry & Fallen* 17498 (COL, MO); Corcovado Region, Río San Juan, *Killip* 35362 (BM, COL, MO, US); Quibdó, *Archer* 2032 (MO, US); near Tutunedo, *Forero & Jaramillo* 2599 (COL, MO); Río Pató, *Forero* et al. 5457 (COL, MO, P); Río Iro, *Juncosa* 2485 (MO, WAG); Condoto, *J.L. Fernández* et al. 8474 (MA); Río Baudó, 15 km from mouth, *Fuchs* et al. 22389 (COL, K, MO, US); San Juan de Palmar-Nóvita Road, near Río Irgara, *Forero* et al. 2329 (COL, MO); near Noanamá, *Forero* et al. 4636 (COL, MO, P); 16-25 km W of San José del Palmar, *Luteyn* et al. 10466 (HBG). Antioquia: 5 km S of Arboletes, *Zarucchi* et al. 4952 (WAG); Punta Piedra, *Haught* 4829 (COL, MO, US); near Río Ampurrumiadó, *Gutiérez & F.A. Barkley* 17C169 (F); Chigorodo, *Plowman* 3174 (F, GH, K, P, S, US); between Yarú and Tarazá, *Espina* 643 (COL); near Planta Providencia, *Alverson* et al. 213 (COL, MO, NY, WIS); Mun. Mutatá, Vila Arteaga, *Schultes & Cabrera* 18646 (BM, GH, MO, U, US); near Segovia,

Rentería et al. 1661 (COL); km 27 Dabeiba-Mutatá Road, *Roldán* et al. 492 (MO, WAG); 6 km E of Santa Fe de Antioquia, *F.A. Barkley* 36C475 (GH, LL); Puerto Berrio, *Pennell* 3761 (NY); near Medellin, Toro 1107 (NY); Mun. San Luís, Cañon of Río Claro, *Cogollo* 1298 (COL, MO); road to Salgar, 4 km from Bolombolo-Bolívar Road, *Zarucchi* et al. 6106 (WAG). Cordoba: Palotal, *Romero-Castañeda* 1651 (COL); Mun. Tierralta, Río Esmeralda, *Bernal* et al. 1187 (COL); Quimarí, *von Sneidern* 5754 (A, AAU, C, COL, F, LL, S, WAG). Sucre: between Colosó and Res. de Primatas, *A. Gentry* et al. 34778 (COL, MO). Bolívar: Mun. Cartagena, Hacienda Canalete, *Zarucchi & Cuadros* 4088 (K, MO, USF, WAG); Turbaco, *Pennell* 4755 (GH, K, MO, NY, US; phot. of US sheet WAG, type of *S. pennelli*); Deconsolado, *Sonntag* 39 (M, P, WU); San Martín de Loba, *Curran* 396 (S); 150 km N of Barranca Bermeja, *de Bruijn* 1062 (WAG). Atlantico: Puerto Colombia, Bro. *Elias* 1143 (F, FI, MO, US); Manati, *Dugand* 748 (F). Magdalena: Fundación, *Romero-Castañeda* 43 (COL); Santa Marta, 10 km E of Bonda, *H.H. Smith* 1639 (A, BM, BR, E, F, G, GH, K, L, MO, NY, S, U, US, WIS, neotype); Tayrona National Park, *Kirkbridge* 2508 (NY, US, USF, Z). César: near Codazzi, *Haught* 2314 (COL, MO, US). Guajira: Bosque de la Cueva, *A. Gentry & Cuadros* 55411 (MO); Manaure, *Cuatrecasas & Romero-Castañeda* 25010 (US); Serranía La Macuira, *Cuadros* 2000 (COL); La Guaira, *Funck* 401 (P). Norte de Santander: between Gramalote and Cúcuta, *Cuatrecasas & García-Barriga* 10145 (COL, F, US). Santander: near Puerto Berrio, *Haught* 1721 (COL, MO, US); ca 2 km W of Jordan, *Langenheim* 3102 (COL, UC, US); near Tablazo, *Haught* 4030 (COL, MO, US); Yarima, *Rentería* et al. 2044 (MO); 17 km NE of Socorro, *A. Gentry & L. Forero* 15337 (COL, MO, NY). Boyacá: Puerto Boyacá, *Avellaneda* 13 (COL); near Puerto Sogamoso, *Rentería* et al. 2165 (COL). Casanare: E of El Yopal, *Blydenstein & Saravia* 1248 (COL, G); along Río Pauta, near Trinidad, *Haught* 2696 (COL, MO, NY, US); between Agua Azul and Maní, *Jaramillo* et al. 3978 (COL); between Maní and Santiago Camp at Belgrado, *Heath* 87 (K, WAG). Cundinamarca: between Valle de Guaduas and Alto de Ficalito, *García-Barriga* 12364 (COL, US); Mun. Nariño, Pitalito, *J.L. Fernández* et al. 8075 (MA) & 8109 (MA); Mun. Tocaima, Hacienda El Cucharo, *Jaramillo* 227 (COL, US); Río Magdalena, near El Peñon, *Bonpland* 1528 (P, P-BO; phot. of P-BO sheet F, GH, US, type of *T. riparia*); Llano de San Martín, *J. Triana* 1895 (BM); Mun. Fusagasugá, Hacienda La Puerta, *Ramírez* in Dow Chem. Int. S.A. 54 (COL); Vegas, *Goudot* 1 (P). Caldas: 13 km N of La Dorada, *A. Gentry* et al. 18196 (COL, MO, US); Santa Cecilia, *von Sneidern* 5146 (US). Tolima: Mts Quindio, *André* 1851 bis (K); Armero, *Cuatrecasas* 10530 (COL, F, MO, US); near La Cruz Pié, *Idrobo* 11055 (U); near Honda, *Hartweg* 1275 (CGE, K, LD); Beltran, *Woronow & Juzepczuk* 4889 (LE); Chicoral, *Haught* 6346 (COL, P, UC, US); between Espinal and Cuamo, *Pennell & Rusby* 186 (NY, type of *S. pauciflora*). Meta: Restrepo, *García-Barriga* 5101 (AAU, COL, US); Sierra de la Macarena, Plaza Bonita, *Philipson* et al. 1487 (BM, COL); Río Duda, *Polanco & Barbosa* 96A (FMB); Río Guayabero, *Echeverry* 2060 (COL). Huila: between Villavieja and Fortalecillas, *Juzepczuk* 5462 (LE); Mun. La Plata, Hacienda La Limona, *Lozano* et al. 4828 (COL, WIS); NE of Neiva, Croat 55416 (Z); Juntas, W of Algeciras, *Little* 7811 (COL, US). Valle: Río Calima, La Trojita, *Cuatrecasas* 16234 (F, MO); San Isidro, *Devia & Prado* 2710 (US).

VENEZUELA. Zulia: SW of Machiques, *Steyermark* 99913 (M, Z); 20 km N of Machiques, *de Bruijn* 1261 (M, MO, NY, S, U, US, WAG, Z); 1 km W of Los Angeles de Tucuco, *Davidse* et al. 18501 (MO); between Cerro La Culebra and Cerro Caracara, *Steyermark* et al. 122597 (MG, MO); San Martín, Río del Palmar, *Pittier* 10516 (GH, US); near Perijá, *Tejera* 14 (US); Maracaibo, *Plée* s.n (P); 20 km NE of Mene Grande, *de Bruijn* 1014 (MO, WAG); 20 km NW of El Vigia, *de Bruijn* 1331 (K, MO, NY, US, WAG, Z). Falcon: Cerro Mampostal, *González* 1065 (MO, WIS);

Río Ricoa, S of Las Dos Bocas, *Steyermark & González* 113640 (MO); ca 1.5 km SW of El Candado, *van der Werff & Wingfield* 3471 (MO, U); near Tocuyo de la Costa, *Steyermark* 94484 (MO, NY); Distr. Silva, between Lizardo and Mallorquinas, *Steyermark & Manará* 110708 (MO, NY). Lara: near Barquisimeto, Saer 271 (US); Valle de El Altar, *Steyermark* 109915 (Z). Trujillo: near Valera, *Farenholtz* 911 (B). Mérida, *Bernardi* 2089 (NY, S); Estanques-Puenta la Victoria Road, *Bernardi* 519 (FI, K, NY). Portuguesa: Agua Blanca Exp. Stat., *Killip* 37099 (BH, MO, US). Barinas: Mun. Ciudad Bolivia, Distr. Pedraza, *L. Ruiz Terán* 1768 (MO); Ticoporo F.R., *Breteler* 3897 (WAG); 5 km NW of El Pescado, 15 km N of Barinas, *Liesner & Guariglia* 11498 (MO, WAG); near Barinitas, *de Bruijn* 944 (M, MO, NY, S, U, US, WAG); Santa Barbara de Barinas, *Valverde & Peña* 1176 (MO). Apure: San Camilo F.R., *Steyermark* et al. 101508 (Z). Yaracuy: Guama, *Tamayo* 277 (US); Distr. San Felipe, 5 km S of Bella Vista, *Agostini* et al. 1797 (GH, NY, U, Z). Carabobo: sin. loc., *Funck & Schlim* 532 (BM, G, LD, P); Puerto Caballo, *Engstedt* 11 (S); N of Mariara, *Trujillo* 3251 (U). Aragua: Puerto La Cruz, *Tyson* 92 (US). D.F.: near Maiquetia, *Pittier* 13385 (A, F, G, MO, NY, US); Caracas, *Linden* 280 (BM, CGE, G, K, P); near Colonia Tovar, between Caracas and La Guayara, *Fendler* 1027 (GH, K, NY). Miranda: between Los Mangos and La Moca, *Eggers* 13379 (C, F, GH, L, US); between San José al Río Chico and Barlovento, *Pittier* 6355 (NY, US); SW of Palo Quemado, *Liesner & González* 9188 (MO). Anzoategui: 16 km E of Boca de Uchire, *A. Gentry & Berry* 14867 (MO). Sucre: Cumana, *Mocquerys* 822 (BR, P, WAG). Amazonas: Palambra del Dr. Norte de San Carlos, *Lopez & Gutierrez* 662 (MO); near San Juan de Manapiare, *Agostini* 1502 (NY, U, US, Z). Bolívar: Serranía de Pijiguao, *Steyermark* et al. 131780 (WAG); Mun. Raul Leoní, Río Ariza, *Delgado* 201 (MO); La Prison, *L.l. Williams* 11540a (F, UC; other part as 11540 belongs to *T. cymosa*); Mun. Sucre, *Delgado* 1136 (MO); Represa Guri, *Liesner & González* 11282 (MO, WAG); Cerro Cordoba, *Wurdack* 362 (GH, K, P, US); La Paragua, *Ll. Williams* 12715 (F, MO, US, VEN); Río Paragua, near Dando y dando, *Stergios* 10862 (MO); Cerro Cotorra, *Steyermark* 86846 (MO); Caroni, *de Grosourdy* 13 (P); 17 km from Upata, *Blanco* 123 (MO, VEN); NW of El Manteco, *Delascio & Liesner* 7415 (VEN); 30 km S of El Manteco, *Steyermark* 86940 (MO, VEN); San Martín de Turumban, *Delascio & Reyes Lopez* 8859 (VEN).

GUYANA. N Rupununi, *Davis* 875 (K, NY); W extremity of Kanuku Mts, *A.C. Smith* 3241 (A, B, F, G, K, MO, NY, P, S, U, US), 3280 (A, B, F, G, K, MO, NY, P, S, U, US); Rupununi Savanna, Mt Point near ranch of Shirley Humphries, *Jansen-Jacobs* et al. 469 (CAY, K, U, US, WAG); Realla, Corantyne R., *im Thurn* Nov. 1879 (K). Sin. loc., *Robert Schomburgk* I 767 (BM, CGE, G-DC, K).

SURINAME. Iengie Kondre, near Coronie W, *Lanjouw* 1065 (K, NY, U); Coronie, *Lanjouw & Lindeman* 1461 (F, K, NY, U); SE feet of Stockumberg, *Dawson* LBB 13155 (U); S of road to Coppename punt, km 69, *Heyde & Lindeman* 321 (U, US); Saramacca R., 40 km S of Paramaribo, *Maguire & Stahel* 23585 (A, BR, F, G, K, NY, P, U, UC, US); near Paramaribo, *Kegel* 158 (GOET, P); ibid., *Wullschlaegel* 315 (BR, GOET, U, W); ibid., Kwattaberg, *Hekking* 832 (A, BR, C, MO, U, USF, Z); Para Distr., *Hostmann* 1565 (C, MO, P, S, UPS); between Meerzorg and Nieuw Amsterdam, *Florschütz* 956 (U); near Galibi, *Went* 420 (U); sin. loc., *Hostmann* 421 (BM, K, NY, P).

FRENCH GUIANA. Hattes, near St. Laurent, *Moretti* 1103 (CAY); Mana, *Sagot* 389 (BM, K, P, U).

BRAZIL. Roraima: Ilha de Santa Rosa, *Milliken* 589V (K); Ilha de Maracá, *Hopkins* et al. 528 (GH, K, US, WAG); Road BR 401 between Boa Vista and Rom Fim (Guyana), *J.M. Pires & Leite* 14681 (Z); Rio Branco, near São Marcos, *Ule* 7828 (F, G, K, L); Upper Rio Branco, Boa Vista, *Kuhlmann* RB 3649 (S, U, US).

Map 55. ●. *Stemmadenia grandiflora*, ▲. *S. tomentosa*.

CULT. Hawai'i, Oahu, Honolulu, H.L.Lyon Arboretum, *Nagata* 415 (L); ibid., *Osgood* 211 (BISH). Netherlands, Baarn, Cantonspark, *E.A. Mennega* 6300 (U). India, Hort. Bot. Calcutta 247 in herb. *Pierre* 1865 (G).

Note. *Stemmadenia grandiflora* is easily confused with *S. donnell-smithii* by the leaves, corolla and the size of the calyx. The latter, however, has the calyx subtended by bracteoles about 0.2-0.3 x as long as the sepals, which are deciduous instead of persistent. In *S. grandiflora* bracteoles are absent.

5. Stemmadenia litoralis (H.B.K.) Allorge in Mém. Mus. natn. Hist. Nat. II, 30: 48 (1985). – Type: Mexico, Campeche, *Bonpland* s.n. (holotype P-BO; phot. F, GH, US).
Fig. 104, p. 417; map 48, p. 363

Basionym and homotypic synonym:

Tabernaemontana litoralis H.B.K., Nov. Gen. 3: 228 (1819). *Peschiera litoralis* (H.B.K.) Miers, Apoc. S. Am. 45 (1878).

Heterotypic synonyms:

Echites bignoniiflora Schldl. in Linnaea 26: 372 (1853). *S. bignoniiflora* (Schldl.) Miers, op. cit. 77 (both as *bignoniaeflora*). – Type: Mexico, sin. loc., *Schaffner* 4 (holotype B†; lectotype G, designated here; isotype S).
Odontostigma galeottianum A. Rich. in Sagra, Hist. fisic. polit. y nat. Cuba 11: 86 (1853) = Fl. Fanerog. 3: 86 (1853). *S. galeottiana* (A. Rich.) Miers, op. cit. 76, **syn. nov.** – Type: Cult., Cuba, sin. loc. *Galeotti* 12 (holotype P).
S. insignis Miers, op. cit. 76, pl. 10, B. – Type: Cult., Mexico, Yucatan, Merida, *Schott* 430 (holotype BM; isotypes F, US; phot. of US sheet WAG).
S. bella Miers, op. cit. 77. – Type: Mexico, sin. loc., *Botteri* 884 (holotype BM; isotypes CGE, G, K, LISU, P).

Shrub or tree 2-18 m high. Trunk up to at least 40 cm in diameter; bark grey, smooth. Branches pale brown, with mostly protruding lenticels; branchlets terete, glabrous. *Leaves* of a pair equal or unequal (larger up to 3 x as long as other and similarly shaped), petiolate; petiole glabrous, 4-20 mm long (ocreae slightly widened into intrapetiolar stipules); blade herbaceous when fresh, membranaceous or thinly papery and often with paler venation on both sides when dried, elliptic, 2-2.5 x as long as wide, 5-27 x 2-11.5 cm, acuminate at the apex, cuneate at the base, glabrous on both sides or sometimes partly and minutely white-lanate or -pilose along the costa and sometimes with scattered black dots beneath, with 7-15 pairs of arcuate secondary veins forming an angle of 45-80° with the costa; tertiary venation reticulate, often impressed above, less clear than in S. obovata. *Inflorescence* pedunculate, 4-14 cm long excluding flowers, depending on age of inflorescence, 1-10-flowered. Peduncle glabrous, 15-40 mm long; pedicels glabrous, 7-15 mm long. Bracts small, like bracteoles, 3-4 mm long. *Flowers* fragrant, very showy. *Calyx* subtended by 1-2 bracteoles; bracteoles broadly ovate to triangular, 1-1.3 x as long as wide, 3-6 x 3-5 mm, obtuse or rounded, glabrous on both sides; sepals pale green, connate at the base for 1-2 mm, glabrous on both sides, not ciliate, subequal or unequal, inner 1.1-2 x as long as outer, obovate or ovate, 1.4-2 x as long as wide, (7-)10-16 x 7-9 mm, outer ovate and often comparatively wider, from 7 mm long, with 1-2 dense rows of approximately 15 colleters at the base forming a continuum in the entire calyx; colleters 0.3-1.1 x 0.1-0.2 mm. *Corolla* white, with a yellow or orange throat, with an ovoid obtuse or acute head in the mature bud, glabrous outside, not ciliate, with 5 longitudinal wings, 9-16 x 1-2 mm from the insertion of the stamens just above them, pubescent on the filament-ridges for 1-2 mm just below the anthers and on the wings up to 0-3 mm below their apices; tube (2.5-)4-6.5 x as long as the calyx, 1.1-2.2 x as long as the lobes, (40-)45-66 mm long, narrowly infundibuliform, with a short narrow basal portion 7-11 mm long, 4-10 mm wide at the base, narrowed to 3-6 mm wide below the insertion of the stamens and from there gradually widenened to (10-)12-20 mm wide at approximately 0.7-0.8 of the length and from there almost cylindrical, slightly narrowed at the throat or not, twisted 0.25-0.4 turn around the anthers; lobes obliquely obovate or ovate, 0.15-0.8 x as long as the tube, 1-2.4 x as long as wide, 26-45 x 16-31 mm, rounded, not undulate, spreading. *Stamens* deeply included, inserted 7-11 mm above the corolla base; anthers sessile just below the wings, 3-4 x as long as wide, 6-7 x 1.5-2 mm, apex sterile for 0-0.7 mm. *Pistil* glabrous, 9-14 mm long; ovary broadly ovoid, 3-3.5 x 3-4.5 x 3-4.5 mm, more or less gradually narrowed into the style, of two carpels connate at the base by a 1-1.5 mm high slightly 5-lobed disk-like thickening; style 4-8.5 mm long; pistil head approximately 2 x 1.5 mm, composed of a basal ring, a 5-lobed depressed globe on a cone and a stigmoid apex. Ovules approximately 100-200 in each carpel. *Fruit* of 2 separate mericarps; mericarps orange or yellow, obliquely ellipsoid, 40-65 x 20-40 x 20-40 mm, acuminate or apiculate at the recurved apex, with a lateral ridge at each side, smooth or warty and then mostly lenticellate; wall 3-4 mm thick in dried fruits; aril orange or red. *Seed* black or brown, obliquely ellipsoid, 9.5-10.5 x 5.5 x 4-5 mm; embryo 7-7.5 mm long; cotyledons ovate, 1.3-1.4 x as long as wide, 3.5-4 x 2.5-3 mm, rounded at the apex, cordate at the base; rootlet 3.5-4 x 0.9 mm.

DISTRIBUTION: From Mexico to Colombia.

ECOLOGY: Mostly deciduous or secondary forest. Alt. 0-1500 m. Flowering mainly from March to June. A clearcut fruiting season could not be deduced from the available collection data.

Geographical selection of the approximately 120 specimens examined:

MEXICO: Nayarit: near Acaponeta, *Rose* et al. 14484 (US). Colima: near Manzanillo, *Ferris* 6239 (US). Tlaxcala: near Texoloc, *Pringle* 8103 (A, BM, BR, E, F, G, GH, HBG, K, L, M, MEXU, MO, NY, P, S, UC, US, WU, Z). Veracruz: Mun. Tlapacoyan, El Arco, *Ventura* 1095 (F, MO, US); Mun. Jilotepec, La Concepción, *Ventura* 11346 (MEXU); Mun. Misantla, *Calzada* 4429 (F, WIS); near Misantla, *Hahn* 20 June 1865 (K, P); 1 km NW of Xico, *Nee & K. Taylor* 29402 (F, USF); 6 km NE of Xico, *Marquez & Gandara* 99 (BM, F, K, MO); Mirador, *Liebmann* 11908 (C,

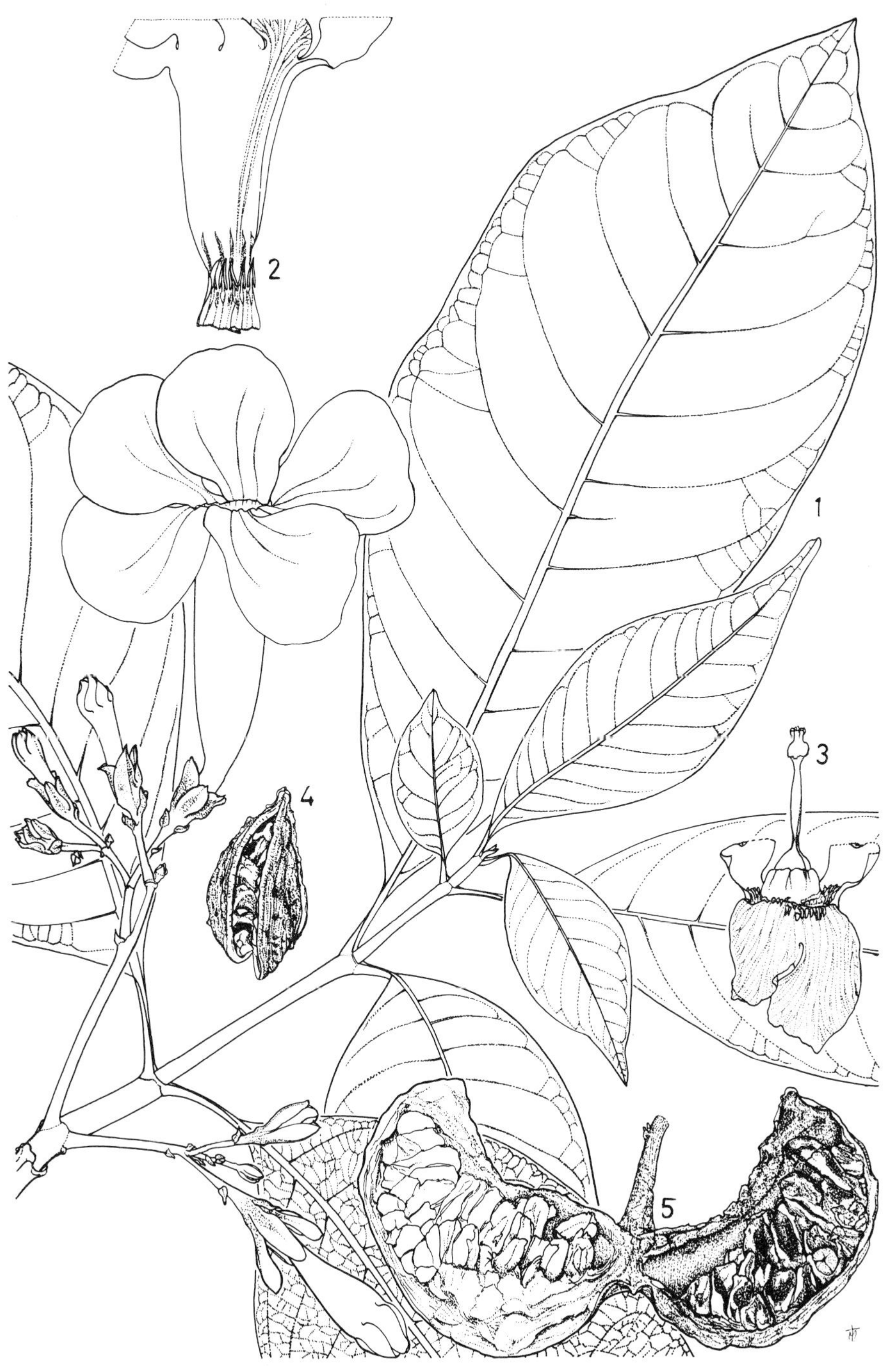

Fig. 104. *Stemmadenia litoralis.* **1,** habit (x 2/3); **2,** opened corolla (x 2/3); **3,** calyx with pistil (x 2); **4,** mericarp (x 2/3); **5,** section of fruit (x 2/3). 1-3 from Pringle 8103; 4 from Caum 5 Nov. 1941; 5 from Nee 29402.

F, UC, US); km 25-40 Paso del Toro-Alvarado Road, *Schubert & Rojas* 1844 (MEXU); Cordoba, *Matuda* 546 (MEXU, US); region of Orizaba, *Bourgeau* 2440 (BR, F, FI-W, G, GH, L, P, S, US); Orizaba, *Bilimeck* 269 (BM, BR, GH, K, P); Santa Ana Atzacan, *Rosas* 387 (A, K, P, U, UPS); Zacuapan, *Purpus* 7740 (A, GH, HBG, MO, NY, US); 13 km E of Tebanca, *Hansen & Nee* 7608 (F, USF, WAG); near Catemaco, *Cedillo* 226 (BR, F, MO); Est. Biol. Las Tuxtlas, *Sinaca* 603 (MEXU, MO); ca 14 km E of Lago Catemaco, *A. Gentry* et al. 32401 (LL, MO, U, WAG). Puebla: near Xicotepec Juárez (= Villa Juárez), *L. González* 784 (WIS); Mun. Cuetzalan, Acaxiloco, *Ventura* 1106 (F, MO). Oaxaca: between El Faro and Fentila, *Conzatti* 3824 (US); near Palantla, *Schultes & Reko* 643 (GH, MO); near Chiltepec, *Miranda* 8279 (MEXU); Cerro del Machete, Pochutla, *Reko* 6274 (F); sin. loc., *Galeotti* 1599 (F, G, P, US), 1605 (BR, NY, US). Chiapas: Mun. Angel Albino Corzo, above Finca Cuxtepec, *Breedlove* 51199 (LL, MO). Tabasco: Teapa, *Linden* 330 (K, P). Campeche: Campeche, *Bonpland* s.n. (P-BO; phot. F, GH, US, type). Yucatan: Merida, *Millspaugh* 27 (F). Sin. loc., *Botteri* 884 (BM, CGE, G, K, LISU, P, type of *S. bella*); *Schaffner* 4 (G, S, type of *S. bignoniiflora*).

GUATEMALA. Peten: La Libertad, *Lundell* 2406 (GH, MO, S, US); Poptun, *Lundell* 16430 (LL, MO). Alta Verapaz: Coban, *J.D. Smith* 1800 (US). Quezaltenango: San Carlos Miramar, *Tonduz & Rojas* 147 (BR, MO, US).

NICARAGUA. Matagalpa: Cerro El Pichacho, *Stevens & Moreno* 22125 (MO, WAG).

COSTA RICA. Guanacaste: Volcán Santa Maria, *Brenes* 27 Jan. 1901 (NY). Puntarenas: Monteverde, *Haber* et al. 4505 (MO). Alajuela: 5 km N of Alajuela, *Lent* 3476 (F, MO). San José: between San Pedro de Montes de Oca and Curridabat, *Standley* 32793 (US); near San José, *Tonduz* 7293 in CR 10123 (F, US); Cerro de San Isidro, *Brenes* CR 14274 (US). Cartago: near Cartago, *Standley* 35459 (US); San Jerónimo de Moravia, *Echeverria* 664 (F).

COLOMBIA. Antioquia: valle de Aburrá, Poblado, *Soejarto* 2075 (COL); Medellín, Bro. *Daniel* 2308 (US). Valle: Cali, *Forero & Hernández* 1546 (MO).

CULT. U.S.A., Florida, Oneco, *R.F. Martin* Aug. 1938 (MO); Hawai'i: Honolulu, Manoa Valley, *Meebold* Nov. 1940 (BISH, M); Manoa Campus, *Trax* 4 Jan. 1963 (MU); Old Pali Road, Nuuanu Valley, *Rock & Neal* 4 Feb. 1956 (AAU, BISH, MICH). Mexico, Veracruz, Tehuantepec, *Juan Evangelísta* s.n. (K); Yucatan, Merida, *Schott* 430 (BM, F, US; phot. of US sheet WAG, type of *S. insignis*). Guatemala, Alta Verapaz, near Cobán, *Standley* 91365 (F, G, MO, US). Costa Rica, Guanacaste, Nicoya, *Cook & Doyle* 686 (US). Cuba, sin. loc. *Galeotti* 12 (P, type of *S. galeottiana*). Jamaica, Hope Gardens, *Harris* 80 (K). Colombia, Valle, Cali, *Silverstone-Sopkin* 3367 (MO). South Africa, Natal, Durban, Bot. Garden, *Strey* 7059 (K). Sri Lanka, 5 km N of Peradeniya, *Wirawan* 600 (K, NY, US). Indonesia, Java, Bogor Bot. Garden XV J.A.XIV.8, *Leeuwenberg* 11873 (WAG).

Notes. *Stemmadenia litoralis* is closely allied to *S. obovata* by the large showy flowers. They differ from each other as is described in couplet 9 of the key to the species, p. 400. The corolla is white and not yellow. Its tube varies in length from 40 to 66 mm, not from 6 to 7 cm as supposed by Madame Allorge. The corolla tube of the type is 45 mm, not 38 mm long as was supposed by her and the colour of its corolla is not indicated. *S. galeottiana* is reduced to a synonym of *S. litoralis*, as it falls within the variation observed for the species.

6. Stemmadenia macrophylla Greenm. in Proc. Am. Acad. 35: 310 (1900); in Contrib. Gray Herb. II, 18: 310 (1900). – Type: Guatemala, Alta Verapaz, Pansamalá, *von Tuerckheim* 981 (holotype GH; isotypes G, K, MO, NY, P, US; phot. of US sheet WAG). Fig. 105, p. 420; map 56, p. 422

Heterotypic synonyms:

S. greenmannii Woods. in Ann. Miss. Bot. Gard. 15: 360, pl. 48, fig. 2 (1928). – Type: Costa Rica, Alajuela, near San Ramón, *Brenes* 14275 (holotype GH; isotypes BM, F, G, K, US; phot. of US sheet WAG).
S. eubracteata Woods. in op. cit. 368, pl. 49, fig. 2, **syn. nov.** – Type: Guatemala, Santa Rosa, Volcán Tecuamburro, *Heyde & Lux* 4538 (holotype GH; isotypes BM, F, K, M, US).
S. robinsonii Woods. in op. cit. 369, **syn. nov.** – Type: Costa Rica, Limón, Valley of Zhorquin, Talamanca Mts, *Tonduz* in *Pittier* 8617 (holotype US; phot. WAG; isotypes BR, G).
S. macrantha Standl. in Publ. Field Mus. Nat. Hist. Bot. Ser. 4: 253 (1929), **syn. nov.** – Type: Panama, Bocas del Toro, Region of Altamirante, *Cooper* 510 (holotype F; isotypes FHO, G, GH, K, NY, US).
S. allenii Woods. in op. cit. 28: 461 (1941), **syn. nov.** – Type: Panama, Coclé, N of El Valle de Antón, *Allen* 2187 (holotype MO; isotype US; phot. of isotype WAG).
S. lagunae Woods. in op. cit. 29: 364 (1942). – Type: Panama, Bocas del Toro, Careening Cay, *von Wedel* 570 (holotype MO; isotype GH).
S. minima A. Gentry in Ann. Miss. Bot. Gard. 64: 322 (1977), **syn. nov.** – Type: Panama, Panamá, Cerro Jefe, *A. Gentry* 6763 (holotype MO).

Shrub or small tree, 1.5-6 m high. Trunk up to at least 5 cm in diameter; bark smooth, pale grey-brown, lenticellate. Branches pale brown, lenticellate; branchlets terete, glabrous. *Leaves* of a pair equal or unequal (the larger one up to 3 x as long as the smaller one and similarly shaped, or the smaller one comparatively wider), petiolate; petiole glabrous, 1-6 mm long; (ocreae not or slightly widened into intrapetiolar stipules); blade membranaceous when dried, elliptic or narrowly elliptic, 1.5-4 x as long as wide, 2.5-16 x 1-7 cm, acuminate or caudate at the apex, cuneate and often unequal-sided or in smaller leaf of a pair often rounded at the base, glabrous on both sides, with scattered black dots beneath or not, with 4-12 pairs of upcurved secondary veins forming an angle of 60-80° with the costa; tertiary venation reticulate and conspicuous or not. *Inflorescence* shortly pedunculate, 1-4-flowered. Peduncle glabrous, 1-2 mm long; pedicels glabrous, 2-6 mm long. Bracts often numerous, like bracteoles, but usually smaller. *Flowers* very variable in size, fragrant. *Calyx* subtended by a small subtriangular bracteole, about as long as wide, 2.4-4 x 2.5-4 mm, or not; sepals white or whitish, oblong or narrowly oblong, less often ovate, inner up to 2.3 x as long as outer and usually narrower, (2-)2.5-10 x as long as wide,(3-) 5-30 x 2-10 mm, obtuse or sometimes rounded at the apex, entire, glabrous or occasionally pilose on both sides or only outside at the base, veined, outer from 3 x 1 mm and therefore easily confused with bracteole, with 2-5 colleters in the middle at the base; colleters 0.2-0.3 x 0.1 mm. *Corolla* yellow and often with a darker throat, or white and often with a yellow throat, reversed flask-shaped in the mature bud, with a large ellipsoid or ovoid mostly obtuse head approximately 0.3 of the bud length, glabrous outside, not ciliate at lobes, with 5 longitudinal wings 7-8 x 1 mm from the insertion of the stamens, pubescent in a belt from 3-11 mm below the insertion of the stamens which may be interrupted between the filament ridges, and on the wings which may be apically or entirely glabrous; tube 1.4-4.6 x as long as the calyx, 1-2.3 x as long as the lobes, 20-67 mm long, infundibuliform or narrowly so, 3-8 mm wide at the base,

Fig. 105. *Stemmadenia macrophylla.* **1-2,** habit (x 2/3); **3,** flower (x 1); **4,** opened corolla (x 2); **5,** pistil with one sepal in old inflorescence (x 2); **6,** sepal inside (x 6); **7,** fruit (x 2/3). 1 and 3-6 from Knapp et al. 5465; 2 from Hammel 5597; 7 from Antonio 2141.

narrowed below and/or above the anthers to 2-4 mm wide, from above the stamens gradually widenened towards the throat to 5-14 mm wide, not contracted at the throat, twisted 0.25-0.7 turn around the anthers; lobes obliquely ovate, obovate or slightly dolabriform and with an obtuse lateral tooth, 0.4-1 x as long as the tube, 1.25-2.5 x as long as wide, 15-45 x 8-20 mm, rounded, not undulate, spreading. *Stamens* deeply included, inserted 0.2-0.5 of the length of the corolla tube, (at 4-26 mm from the base); anthers sessile, just below the wings, narrowly triangular, 3-4 x as long as wide, 4-5.2 x 1.3-1.5 mm, apex sterile for 0.3-0.5 mm. *Pistil* glabrous, 11-28 mm long; ovary broadly ovoid, 2 x 2 x 2 mm, gradually narrowed into the style, of two carpels, connate at the base with a slightly 5-lobed disk-like thickening 1 mm high; style 5.5-24 mm long; pistil head 1.5 x 1.5 mm, of a basal ring 0.3 x 1.5 mm, a stipitate 5-lobed depressed globe 0.5 x 0.7 mm and a stigmoid apex about 0.2 x 0.2 mm. Ovules approximately 50 in each carpel. *Fruit* of 2 separate mericarps; mericarps green or yellow, obliquely ellipsoid or almost pod-like, 25-50 x 15-25 x 15-25 mm, recurved, rounded, apiculate or acuminate at the apex, without lateral ridges, smooth; wall approximately 2 mm thick in dried fruits; aril orange(?). *Seed* dark brown or black, obliquely ellipsoid, 10-11 x 4-5 x 3-4 mm; embryo 7-7.5 mm long; cotyledons ovate, 1.2-1.4 x as long as wide, 3-3.5 x 2.5 mm, rounded at the apex, cordate at the base; rootlet 1.3-1.6 x as long as the cotyledons, 4.5-4.7 x 0.6-0.7 mm.

DISTRIBUTION: From Mexico to Colombia.

ECOLOGY: Wet forest. 0-1000 m. Flowering from March to August and fruiting mainly in August in Mexico and scattered over the year in other countries.

Geographical selection of the c. 180 specimens examined:

MEXICO. Guerrero: Acapulco, *Boege* 2451 (MEXU); Xochistlahuaca, *B. Lopez* 86-5-10 (USF). Veracruz: Mun. Hidalgotitlan, km 8-12 Plan de Arroyos-Alvaro Obregon Road, *Dorantes* et al. 2872 (F); Mun. Minatitlán, 6.6 km N of Laguna Grande-Río Grande, *Wendt & Villalobos* 2554 (MEXU). Oaxaca: 2.5 km W of Río Corte, Sarabia-Uxpanapa Road, *Croat & Hannon* 65421 (WAG). Chiapas: 8-10 km SE of Chiapa de Corzo, *Thorne & Lathrop* 40230 (LL, MO, NY); Zinacantán, *Laughlin* 1613 (F); Venustiano Carranza, *Breedlove* 5194 (LL, MO); 6-8 km E of Frontero Comalapa, *Breedlove* 39117 (MO); La Grandeza, *Matuda* 5553 (F, LL, MEXU).

GUATEMALA. Alta Verapaz: Pansamalá, *von Tuerckheim* 981 (G, GH, K, MO, NY, P, US; phot. of US sheet WAG, type). Huehuetenango: between Democracia and Cañon Chamushú, *Steyermark* 51279 (F, MO). Santa Rosa: Volcán Tecuamburro, *Heyde & Lux* 4538 (BM, F, GH, K, M, US, type of *S. eubracteata*).

EL SALVADOR. Ahuachapán: Cerro Campana, *Witsberger* 595 (MO), 752 (MO). Santa Ana: Volcán Santa Ana, *L.O. Williams* 13569 (F, GH). La Union: near La Union, *Grant* 725 (A).

HONDURAS. Copan: N of Copan, Poole & Watson 1151 (LL).

NICARAGUA. Zelaya: E of Caño Angustura, *Sandino* 4635 (MO, WAG); Caño Cedro Macho, *Moreno* 23932 (MO, WAG); 3.6 km SE of Cerro Isidro, *Proctor* et al. 27030 (NY); Bluefields Airport, *Neill* 2588 (MO); El Zapote, *Sandino* 4803 (MO, WAG). Río San Juan: 5 km NE of Sábalos, *Moreno* 23382 (MO, WAG).

COSTA RICA. Guanacaste: Volcán Orosi, *Wilbur & Stone* 10208 (F, GH, MO, US); Rincón de la Vieja National Park, *Herrera & Robles* 802 (WAG). Puntarenas: N of Clarita, Burica Peninsula, *Lent* 3066 (F). Alajuela: ca 11 km SE of Upala, *Grayum* 9057 (MO); Santa Maria National Park, *Liesner* 4765 (MO, WAG); near San Ramón, *Brenes* 14275 (BM, F, G, GH, K, US; phot. of US sheet WAG, type of *S. greenmannii*). Heredia: Finca La Selva, *Hammel* 9605 (MO); W of San José, *Garwood* et al. 1032 (BM, MO). Cartago: 24 km NE of Turrialba, *Liesner* et al. 15392 (MO, WAG, WIS). Limón: Cerro Coronel, *Stevens* 24380 (MO, WAG); Finca Hamburg, below

Cairo, *Standley & Valerio* 48776 (US); Cordillera de Talamanca, between Quebrada Camagre and Río Barbilla, *Grayum* et al. 8932 (MO, WAG); Valley of Zhorquin, Talamanca Mts, *Tonduz* in *Pittier* 8617 (BR, G, US; phot. of US sheet WAG, type of *S. robinsonii*); between headwaters of Quebrada Mata de Limón and Quebrada Tigre, *Grayum* et al. 4460 (MO).

PANAMA. Bocas del Toro: Kankintoe, Region of Altamirante, *Cooper* 510 (F, FHO, G, GH, K, NY, US, type of *S. macrantha*); Fish Creek Mts, near Chiriqui Lagoon, von Wedel 2247 (GH, MO, NY, US); Careening Cay, *von Wedel* 570 (GH, MO, type of *S. lagunae*); above Chiriquí Grande, *McPherson* 12297 (WAG); Water Valley (= Cayo de Agua), *von Wedel* 587 (GH, MO, paratype of *S. lagunae*). Veraguas: near Santa Fé, *Antonio* 3986 (MO, WAG); Isla de Coiba, *Antonio* 2370 (MO, WAG); mouth of Río Concepción, *Lewis* et al. 2853 (AAU, GH, MO). Coclé: El Cope, *Antonio* 2141 (WAG); N of El Valle de Antón, *Allen* 2187 (MO, US; phot. of US sheet WAG, type of *S. allenii*), 2239 (MO, US, paratype of *S. allenii*); Cerro Pilon, *Liesner* 765 (C, F, MO, NY, US, USF, WAG); Alto Calvario, *Folsom* 3414 (MO, WAG). Canal Zone: Madden Dam area, *Dwyer & Lallathin* 8827A (F, MO). Colón: Santa Rita Ridge, 6-8 km from Transisthmian Highway, *A. Gentry* 6090 (F, MO, Z). Panamá: Cerro Campana, *Hammel* 5597 (WAG); near Arraiján, *D'Arcy* 10664 (MO, NY); Cerro Jefe, *A. Gentry* 6763 (MO, type of *S. minima*); ibid., *Knapp* 3522 (MO, WAG); between Cerro Azul and Cerro Jefe, *Dressler* 3333 (AAU, GH, MO); El LLano-Cartí Road, 8.7 km from Panam. Highway, *Folsom* et al. 6164 (MO, USF, WAG). San Blas: 34-38 km from Panam. Highway, on El Llano-Cartí Road, *Knapp & Schmalzel* 5465 (WAG). Darién: Camp Summit, near San Blas border, Sea Level Canal, *Oliver* et al. 3695 (GH, MO); 5 km W of Cocalito, *Garwood* 743 (F).

COLOMBIA. Guajira: Serrania de Macuira, *Sugden* 51A (COL); Isamana Region, *Bernal & Sugden* 92 p.p. (FHO). Valle: Puerto Merizalde, *A,. Gentry & Juncosa* 40568 (MO). Cauca: near El Pastico, *A. Gentry & Juncosa* 40639 (MO).

Map 56. *Stemmadenia macrophylla.*

Notes. *Stemmadenia macrophylla* varies greatly in the size and shape of the sepals and the size of the corolla. Small sepals have been observed in e.g. the types of *S. allenii, S. minima* and *S. macrophylla*. Large sepals are present in the types of *S. eubracteata* and *S. robinsonii*. The corollas of the types of *S. eubracteata, S. macrophylla* and *S. robinsonii* form together with many other specimens a continuous range of variation. When some of the extremes among the specimens examined are seen, it is hardly understandable that they all belong to a single species. When the sepals are approximately 5-6 mm long, they are approximately 1.5-2 x as long as wide. When they are longer, they are at least 3 x as long as wide. In the first case they are usually rounded and in the second mostly obtuse, even seemingly acute in dried flowers. The variation in size of the corolla is more or less continuous. The species is distinguished with difficulty from the related *S. alfari* and *S. grandiflora*. Fruiting specimens in which the sepals have been shed cannot be named with certainty.

The three species are distinguished here as follows:

1. Sepals pale green, orbicular or broadly ovate, 0.8-1.3 x as long as wide, 2-4 x 2-4 mm, rounded, equal or subequal; corolla with a depressed-subglobose head in the mature bud, approximately 0.1-0.2 x as long as the bud, rounded at the apex, tube 8-13 x as long as the longest sepals, 22-32 mm long. Costa Rica, Panama. **.S. alfari**

   Sepals white, whitish or pale green; corolla with an ovoid or ellipsoid head in the mature bud, obtuse or acute at the apex; tube 1.4-4.6 x as long as the longest sepals, 20-67 mm long. 2

2. Sepals white or whitish, mostly strongly unequal, oblong, narrowly oblong or less often ovate and then 5-6 mm long, otherwise 2-10 x as long as wide, 5-30 x 2-10 mm, mostly obtuse, less often rounded; corolla tube 20-67 mm long, infundibuliform or narrowly so. Mexico to Colombia ....**S. macrophylla**

   Sepals pale green, unequal, ovate, broadly ovate or elliptic, 1-6 x as long as wide, but outer mostly less than 1.5 x as long as wide, 11-23 x 5-20 mm, outer from 8 x 5 mm, all rounded to acuminate; corolla tube 23-35 mm long, almost cylindrical, Costa Rica to northern S. America ....**S. grandiflora**

7. Stemmadenia obovata (Hook. & Arn.) K. Schum. in Engler & Prantl, Nat. Pflanzenf. 4, 2: 149 (1895). – Type: Mexico, San Luis Potosi, Realejo, *Sinclair* s.n. (holotype K). Fig. 106, p. 425; map 57, p. 428

Basionym and homotypic synonym:

Bignonia(?) *obovata* Hook. & Arn., Bot. Beech. Voy. 439 (1841). *S. pubescens* Benth., Bot. Voy. Sulph. 125 (1845).

Heterotypic synonyms:

S. glabra Benth., op. cit. 124, t. 44, **syn. nov.** – Type: Honduras, Choluteca, Gulf of Fonseca, *Sinclair* s.n. (holotype K).

S mollis Benth., op. cit. 125. *S. obovata* var. *mollis* (Benth.) Woods. in Ann. Miss. Bot. Gard. 15: 358 (1928). – Type: Ecuador, Guayas, Guayaquil, *Sinclair* s.n. (holotype K).

S. calycina Brandeg. in Univ. Calif. Publ. Bot. 10: 188 (1922). – Type: Mexico, Veracruz, Remulatero, *Purpus* 8771 (holotype UC; isotypes GH, MO, NY, US; phot. of US sheet WAG).

Deciduous shrub or small tree 1-9 m high. Trunk 1-15 cm in diameter. Branches pale grey-brown, with scattered lenticels; branchlets terete, pubescent or sometimes glabrous. *Leaves* of a pair equal of unequal (larger upt to 5 x as long as other and similarly shaped), shortly petiolate; petiole pubescent or sometimes glabrous, 2-15 mm long; (ocreae not or slightly widened into intrapetiolar stipules); blade herbaceous when fresh, membranaceous and often with pale venation on both sides when dried, elliptic or obovate, or narrowly so, 1.5-2(-3) x as long as wide, 6-23 x 2.5-14 cm (in smaller leaves of a pair sometimes smaller), acuminate or apiculate at the apex, cuneate or rounded and at the same time often unequal-sided at the base, sparsely pubescent, sometimes especially on the costa, to glabrous, pubescent or sometimes glabrous beneath, with 12-22 pairs of rather straight secondary veins forming an angle of (45-)60-80° with the costa; tertiary venation reticulate, often dark beneath and often impressed above. *Inflorescence* 1-4-flowered, shortly pedunculate, 1-5 cm long excluding flowers. Peduncle pubescent or sometimes glabrous, 5-30 mm long; pedicels pubescent or sometimes glabrous, 5-10 mm long. Bracts triangular or nearly so, 2-3 mm long, acute or obtuse at the apex, pubescent or sometimes glabrous outside. *Flowers* fragrant, very showy. *Calyx* subtended by (1-)2(-3) bracteoles, the outer of which broadly ovate, 0.6-1.2 x as long as wide, 3-7 x 3-7 mm, obtuse or rounded, usually pubescent outside, rarely ciliate, glabrous inside, second bracteole mostly 1.5-2 x as large and less hairy; sepals light green, free from each other, glabrous inside, unequal, inner one or two approximately 1.5-2 x as long as other, obovate or elliptic, 1.2-3 x as long as wide, 15-34 x 6-14 mm, retuse or rounded, glabrous outside, not ciliate, other sepals elliptic or ovate, glabrous or partly pubescent outside, especially at the base, with a continuous ring of colleters in 2-3 irregular rows inside at the base, approximately 30 with each sepal; colleters 0.7-1.2 x 0.2-0.3 mm. *Corolla* bright yellow, with an ovoid obtuse head in the mature bud, glabrous outside, not ciliate, with 5 longitudinal wings 8-12 mm long and approximately 1 mm wide, the wings pubescent all over or only at the base, from 2-3 mm above the anther bases, remainder of tube with a belt of pubescence 7-9 mm wide from there down which is often interrupted among the filament ridges or sometimes absent, otherwise glabrous; tube 1.4-3 x as long as the calyx, 1-1.5 x as long as the lobes, 35-55 mm long, infundibuliform, 6-12 mm wide above the base, below the insertion of the stamens narrowed to 3-7 mm wide and from there gradually widened to 14-23 mm at the mouth which is not contracted; lobes dolabriform, 0.54-1 x as long as the tube, 1.1-2.1 x as long as wide, 25-50 x 19-31 mm, with two rounded apices or obliquely obovate and rounded at the apex (curved to the right), obscurely auriculate at the left side of the base (seen from outside), entire, not undulate, spreading. *Stamens* deeply included, inserted approximately 2 mm above anther bases at 15-19 mm from the base of the corolla tube, 0.29-0.44 of the length of the corolla tube, sessile, 3-4 x as long as wide, 5.5-7 x 1.5-2 mm, apex sterile for 0.3-1 mm. *Pistil* glabrous or pubescent only on the ovary, 17-21 mm long; ovary broadly ovoid or almost conical, 3-5 x 3-5 x 3-5 mm, glabrous or pubescent, of two separate carpels, connate at the base with a slightly 5-lobed annular thickening, rather abruptly narrowed into the style; style cylindrical, 1 mm thick, 11.5-15 mm long; pistil head 1.5-2 mm high, composed of a basal 5-lobed ring 0.3 x 1.5-2 mm with acute horizontal lobes, a short cone topped by a 5-lobed depressed globe 0.5-0.6 x 0.9-1 mm and a stigmoid apex approximately 0.2 x 0.2 mm. Ovules approximately 100 in each carpel. *Fruit* of 2 separate mericarps; mericarps green, obliquely ellipsoid, 40-65 x 22-40 x 20-40 mm, acuminate or caudate at the recurved apex, acumen 4-15 mm long, not ridged, smooth, glabrous or pubescent; wall 6 mm thick (in dried fruits 3-4 mm); aril orange or red. *Seed* black, obliquely ellipsoid, 8-11 x 3-5 x 3-4 mm; embryo 5-7.5 mm long, cotyledons ovate or nearly so, 1-1.5 x as long as wide, 2-3.5 x 2-2.5 mm, rounded at the apex, rounded or truncate at the base; rootlet 1.1-1.5 x as long as the cotyledons, 3-4 x 0.6-0.8 mm.

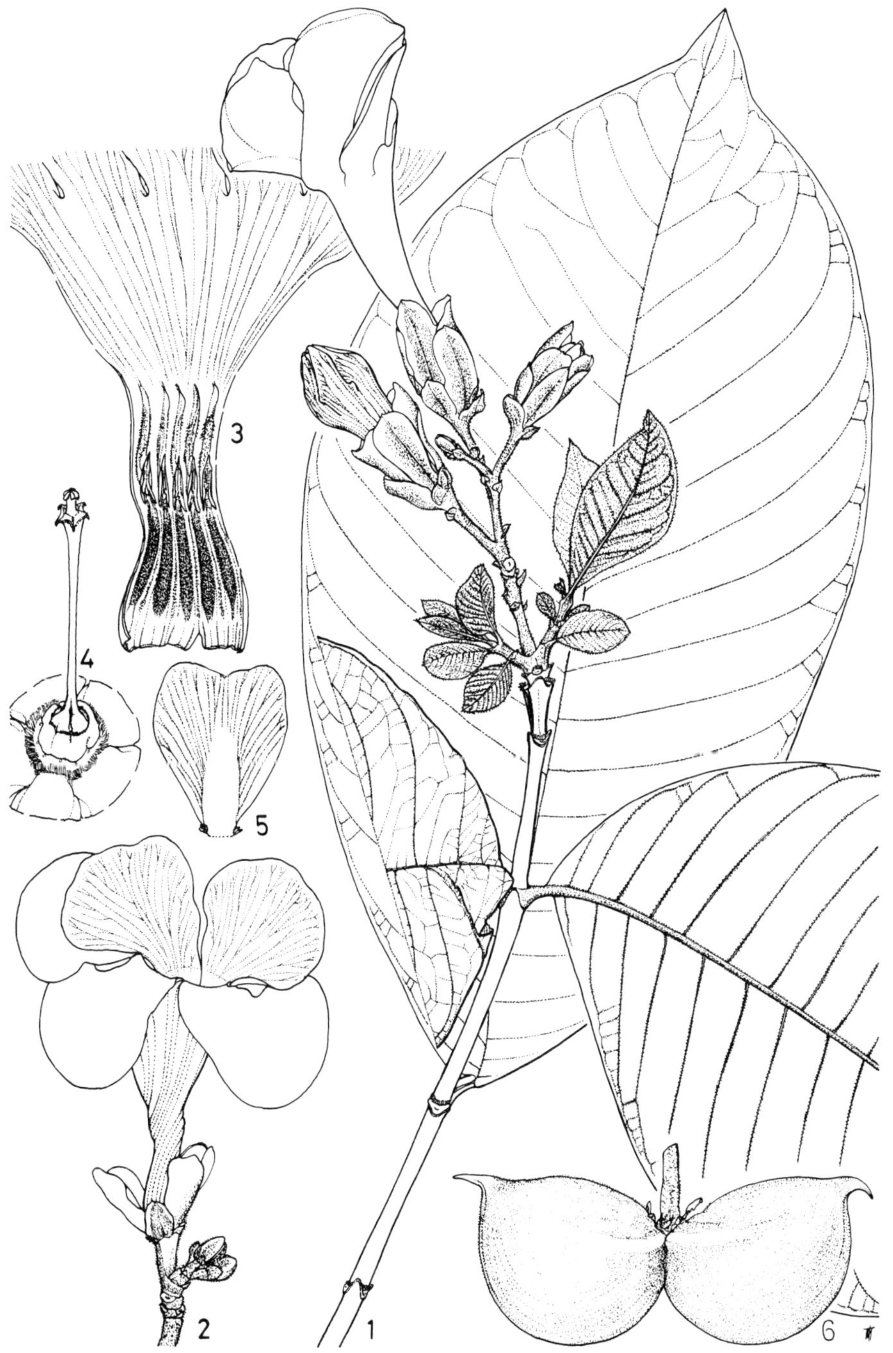

Fig. 106. *Stemmadenia obovata.* **1,** habit (x 2/3); **2,** flower (x 2/3); **3,** opened corolla (x 1.2); **4,** pistil with sepal bases (x 2); **5,** sepal inside (x 1); **6,** fruit (x 2/3). 1 from Ochoa et al. 202 and Stevens et al. 2724; 2-5 from Ochoa et al. 202; 6 from de Jong 120.

DISTRIBUTION: Tropical America from Mexico to Bolivia.

ECOLOGY: Dry thickets or low often thorny forests. Alt. 0-900 m. In Mexico flowering mainly but not exclusively from April to July and fruiting from June to November. In Nicaragua flowering from May to October and fruiting from July to December. In Costa Rica flowering usually from April to August. In other countries scattered over the year.

Geographical selection of the approximately 530 specimens examined:

MEXICO. Sinaloa: Mun. Rosario, 3 km N of Chilillos, *Martínez* et al. 3996 (MO). San Luis Potosi: Realejo, *Sinclair* s.n. (K, type). Mexico: Distr. Temascaltepec, Bejucos, *Hinton* 722 (GH, MO, NY, US); Durango, *Soule* 2151 (MO). Veracruz: Baños del Carrizal, *Purpus* 6230 (BM, F, GH, MO, NY, UC, US); Cerro Monte de Oro, *Vazquez* 578 (F); near Veracruz, *Hahn* 281 (FI-W, P); Paso de Ovejas, *Ventura* 5839 (IBSC, MO); Puente Nacional, *Linden* 355 (FI-W, K); 2-6 km SE of Emiliano Zapata, *Hansen & Nee* 7465 (F, USF, WAG); Remulatero, *Purpus* 8771 (GH, MO, NY, UC, US; phot. of US sheet WAG, type of *S. calycina*); 11 km S of Palma Sola, *Dorantes* et al. 1188 (F). Jalisco: Est. Biol. Chamela, *Solís* 2452 (MO). Michoacan: Correo, *Langlassé* 1029 (G, GH, K, P, US); 3 km NW of Turicato, *Soto* 4839 (MEXU); 2 km SE of Carácuaro, *Soto* 3943 (MO); 16 km SW of Melchor Ocampo, *Soto* 802 (IBSC); Tzezénguaro, *Soto* 3285 (MEXU); Mun. Tzitzio, Temazcal-Huetamo Road, *Soto* 432 (MEXU). Morelos: 10 km S of Yautepec, *Hernández & Chacón* 117 (MO, WAG); Huajitlán, *Cox & Guzman* MCG 633 (ECON); 46 km SW of Cuernavaca, *Frye* 2561 (BH, GH, MO, NY, UC, US); 8 km S of Cuernavaca, *L. González* 2613 (BH, F, LL). Guerrero: Temisco, *Mexia* 8750 (B, F, G, GH, K, MO, NY, S, U, UC, US); near Balsas, *Halbinger* 23 (GH); Iguala, *Rose* et al. 9274 (C, F, MO, NY, P, S, US); Mun. Ometepec, 25 km NW of Cuajinicuilapa, *Martínez & Téllez* 126 (MO); Distr. Coyuca, Jaripo, *Hinton* et al. 5983 (F, GH, K, MO, NY, US); near Trapiche Viejo, 40 km NE of Chilapa, *Acosta & R. López* 142 (MEXU); 21 km N of Chilpancingo, *J. García & A. Delgado* 1012 (F, L, LG, LL, WIS); El Cuindacito, *Martínez* et al. 3666 (KUN, MO); Costa Chica, Mun. Cuautepec, *Herrera* 12 (MEXU). Oaxaca: Pueblo Nuevo, *Alexander* 239 (NY, US); 7 km N of San Pedro Cajones, *Torres* 2900 (MO); S of Oaxaca, *Jürgensen* 21 (G, K); S of Matias Romero, *Cedillo* 729 (MO); 7 km E of Salina Cruz, *R.B. King* 1296 (LL, NY, US); Tehuantepec, *Andrieux* 252 (FI-W, G, K, M, P); 15 km NW of Tehuantepec, *A. García & Torres* 1677 (F); 17.2 km NW of Petuantepec, *Torres & Martínez* 5568 (MEXU); 2-4 km E of Tehuantepec, *R.M. King* 1210 (LL, NY, UC, US); km 17 Tapanatepec-Tuxtla Gutiérrez Road, *Hernández & Torres* 357 (MO); 5 km E of Pinotepa Nacional, *Roe* et al.495 (F, WIS). Chiapas: Petapa, *Goldman* 1027 (US); Distr. Pochutla, road to Tonameca, *Conzatti* et al. 3258 (US); road to Bahía de Santa Cruz, *Cedillo* et al. 1693 (LL, MO); Nuevo Amatenango, *Matuda* 4786 (LL); Las Arenas, 15 km NW of Arriaga, *Breedlove* 36825 (MO); SE of Chiapa de Corzo, *Saunders* 7 (LL); Paderon, Tonalá, *Matuda* 16322 (MO, US); Chicomuselo, *Matuda* 5645 (LL); SE of Comitán, *Carlson* 1941 (F, NY); El Chorreadero, 9 km E of Chiapa, *Breedlove* 9562 (F, LL); Mun. Venustiano Carranza, above Finca Carmen, *Ton* 3005 (WIS); Bosque San Diego, Dept. Metapan, *Current* 10 (MO).

GUATEMALA. Huehuetenango: near Cuilco, *Steyermark* 50755 (F); Sierra de los Cuchumatanes, *Steyermark* 51419 (F, MO). Retalhuleu: between Nueva Linda and Champerico, *Standley* 87619 (F); Champerico, *Standley* 66632 (F). Quiché: *Aguilar* 1385 (F). Baja Verapaz: 8 km SW of Granados, *Harmon & Dwyer* 3024 (MO). Guatemala: between Chiquín and Trapiche Grande, *Pittier* 133 (US). Santa Rosa: SE of Chiquimulilla, *Standley* 78734 (F). El Progreso: near Barranquillo, *Steyermark* 46445 (F, MO, US). Jalapa: El Rancho, *Maxon & Hay* 3766 (US). Zacapa: Sierra de las Minas, *Standley* 73977 (F); Zacapa, *Steyermark* 29000 (F, MO). Chiquimula: near Chiquimula, *Standley* 74329 (F). Jutiapa: 13 km SW of San Cristóbal, *Dunn* et al. 19 (G, K, MO).

EL SALVADOR. Sonsonate: near Sonsonate, *Standley* 22372 (GH, NY, US). San Salvador: E of La Libertad, *Weberling* 901 (ULM, WAG); near La Cebadilla, *Calderón* 1230 (GH, US). San Vincente: near San Vincente, *Standley & Padilla* 3523 (F). San Miguel: 2 km SW of Montecristo, *Tucker* 595 (BH, G, K, LL, NY, UC, US). La Unión: near La Unión, *Standley* 20686 (GH, NY, US).

HONDURAS. Santa Bárbara: between Ilama and Gualala, *Molina* 22029 (F, NY). Comayagua: El Banco, *J.V. Rodriguez* 2367 (F); Río Selguapa, *Burch* 6029 (MO, USF); San Jeronimo, *Molina* 6988 (F, LL); Agua Salada, *L.O. Williams & Molina* 12578 (F, GH). Morazán: near El Zambrano, *Standley* 21729 (P); Pedregal, *Molina* 118 (F, GH, UC, US); Río Yeguare, *L.O. Williams* 14081 (BM, F, GH). Valle: El Coyolito, Fonseca Gulf, *Molina* 25970 (BM, F, NY, US). Choluteca: Fonseca Gulf, *Sinclair* s.n. (K, type of *S. glabra*); 2 km from Choluteca, *Molina* 5490 (US).

NICARAGUA. Nueva Segovia: ca 6 km N of N edge of Ocotal, *Stevens* 3058 (MO, U); 4 km NE of El Jícaro, *Araquistain & Moreno* 2227 (MO, WAG). Madriz: E of Somoto, *L.O. Williams & Molina* 10891a (F, MO). Esteli: between Quebrada Jamaili and Cerro El Pedrero, *Stevens* 2589 (MO, WAG); Mun. San Juan de Limay, *Moreno* 1995 (MO, WAG); 3 km N of La Trinidad, *Moreno* 22429 (MO). Chinandega: 2 km from Cinco Pinos, *Moreno* 11760 (MO); SE of Volcán Casita, *Moreno* 1527 (MO, WAG); NW of Volcán Cosigüina, *Sandino & Aldubin* 4392 (MO, WAG). Leon: Llano San Lorenzo, *Moreno* 2318 (MO); Volcán Casita, *Grijalva & Estrada* 2529 (MO, WAG); Mun. Santa Rosa del Peñon, Soledad de la Cruz, *Moreno* 2543 (MO); La Paz, *C.F .Baker* 2270 (ECON, FI, G, GH, K, US); Volcán Momotombo, *Robbins* 6121 (BM, F, GH, MO, NY, UC). Matagalpa: old road to Jinotega, *Moreno* 22924 (WAG); Cerro El Pilon, *Stevens* 9404 (BM, MO). Boaco: Hacienda San Antonio, road to Boaquito, *Moreno* 21532 (MO, WAG); 1.5 km S of Santa Lucía, *Moreno* 10150 (MO, WAG); El Papayal, *Moreno* 3683 (MO). Managua: El Portillo, *Moreno* 21496 (MO, WAG); N shore of Peninsula de Chiltepec, *Stevens* 3545 (BM, MO). Granada: near Granada, *Levy* 72 (G, P); road from Casa Tejas to Finca San José del Mombacho, *Moreno* 16651 (MO, WAG). Masaya: near Vera Cruz, Laguna Verde, *J.S. Hall & Bockus* 7860 (B, BM, MO, NY, UC). Carazo: ca 3.3 km SE of Casares, *Stevens* 2724 (MO, U, WAG); Chococente, *Ocha* 202 (WAG). Rivas: El Jocote, *de Jong* et al. 120 (WAG); Isla Ometepe, *Robleto* 196 (MO, WAG). Chontales: near Juigalpa, *Standley* 9192 (F). San Francisco, *Moreno* 10078 (WAG). Río San Juan: Río Sábalos, *Moreno* 23058 (MO, WAG).

COSTA RICA. Guanacaste: La Cruz, *Stevens* 13603 (MO, WAG). Bahia El Coco, *Burger* 7724 (F, MO, UC, US); 5 km NW of Bagaces, *Opler* 1736 (F, MO, UC); near Río Cañas, road of Tilarán, *Brenes* 12634 (F); W of Santa Cruz, *Jimenez* 533 (F); Nicoya, *Tonduz* 13900 (BM, G, GH, K, NY, P, US). Punarenas: Isla San Lucas, *Grayum* 4189 (WAG); Monteverde, *Haber* 1058 (WAG). San José: San José, *Umana* 31 (F). Cartago: Guijarro, *Echeverría* 214 (F).

PANAMA: Los Santos: ca 8 km SE of Los Santos, *Lewis* et al. 1679 (E, LD, MO, NY, WAG); Salinas de Chitre, *D'Arcy & Croat* 4160 (C, GH, MO); Las Tablas, *Allen* 812 (F, GH, MO, NY, US).

ECUADOR. Guayas: Guayaquil, *Sinclair* s.n. (K, type of *S. mollis*); Durán, *Asplund* 14831 (LD, S, UPS); Rí Daule, near Guayaquil, *Spruce* 6337 (AWH, BM, CGE, E, F, G, K, P; phot. of P sheet F, GH, US); Pedro Carbo, *Haught* 3002 (MO, US); Cerro de Santana, *Jameson* 513 (BM, E, G, K, TCD); Chongón, *Asplund* 5875 (G, S, UPS, US); Guayaquil-Salinas Road, km 20, *Harling & Andersson* 21044 (QCA, USF); 26 km W of Guayaquil, *A. Gentry* 9998 (GB, MO, S); Cerro Santa Ana, *Asplund*

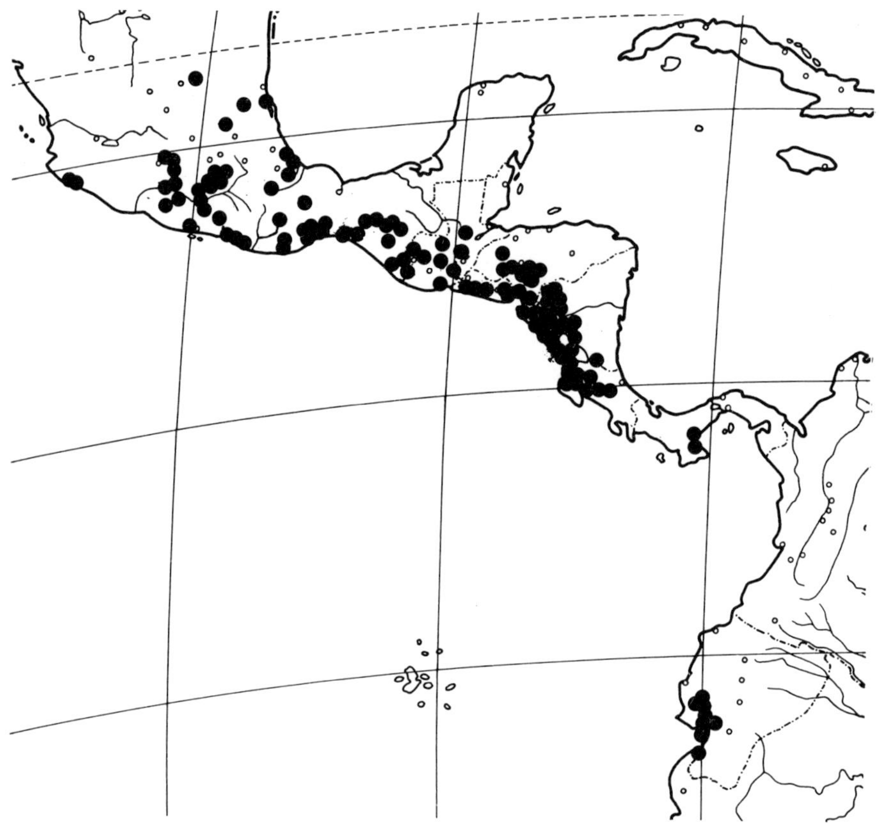

Map 57. *Stemmadenia obovata.*

15203 (B, K, LD, S, UPS); near Estero Salado, *Harling* 3047 (S).

PERU. Tumbes: Prov. Zarumilla, between El Caucho and Condor Flores, *Díaz & Vásquez* 3052 (WAG); Bosque Nacional de Tumbes, *Simpson & Schunke* 531 (F, G, NY, US).

BOLIVIA. near Yungas (presumably E slopes of Cordillera in Cochabamba), *Rusby* 1163 (NY, US).

CULT. Hawai'i, Honolulu, Forster Bot. Garden, *Lau* 1576 (WAG), ibid. *Potter* 19 June 1957 (BISH).

Notes. The indumentum of *Stemmadenia obovata* varies much in density. In a few cases the plants are glabrous. *S. obovata* is closely allied with *S. litoralis* (q.v.).

8. Stemmadenia pauli Leeuwenberg, **sp. nov.**

Fig. 107, p. 429; phot. 12, 13; map 50, p. 376

Frutex vel arbor parva ramis furcatis. Folia elliptica vel anguste elliptica subcoriacea. Inflorescentiae foliis multo breviores pauciflorae. Sepala alba ovata vel oblonga inter se inaequalia libera. Corolla fere alba tubo fere cylindrico intus alis quinque sparse pilosis supra stamina limbo patente lobis oblique obovatis. Stamina inclusa antheris anguste triangularibus. Ovarium glabrum basi incrassatum.

Fig. 107. *Stemmadenia pauli.* **1,** habit (x 2/3); **2,** flower (x 1.6); **3,** sepal inside (x 1.6); **4,** opened corolla (x 1.6); **5,** pistil (x 4); **6,** immature fruit (x 2/3). 1-3 from Maas 7850; 4-5 from Allen 5965; 6 from Liesner 3255.

Typus: Costa Rica, Puntarenas, Golfo Dulce F.R., Peninsula de Osa, *Paul J.M. Maas* 7850 (holotypus U; isotypus WAG).

Shrub or small tree 1-6 m high. Branches medium brown, lenticellate; branchlets terete, glabrous. *Leaves* shortly petiolate; petiole 3-7 mm long, glabrous; (ocreae widened into intrapetiolar stipules); blade subcoriaceous when dried, elliptic or narrowly elliptic, 2-2.5 x as long as wide, 7.5-20.5 x 3.5-10 cm, acuminate to caudate at the apex, cuneate or rounded at the base, glabrous on both sides, with 7-11 pairs of upcurved secondary veins forming an angle of 50-60° with the costa, with or without scattered black dots on both sides; tertiary venation reticulate. *Inflorescence* pedunculate, 4-5 x 4-5 cm, 1-4-flowered, lax or nearly so. Peduncle 2-13 mm long, glabrous, rather thin; pedicels glabrous, 7-13 mm long, thin. Bracts scale-like, about 0.1 x as long as the sepals. *Flowers* sweet-scented, open during the day. *Calyx* subtended by one sepal-like bracteole 0.2-0.25 x as long as the sepals; sepals white or when in bud often yellow, free, clasping the corolla base, ovate or oblong, 1.5-2 x as long as wide, unequal, larger 17 x 8-10 mm, smaller from 11 x 7 mm, rounded, glabrous outside, not ciliate, glabrous inside and with 5-7 colleters in one row at the base; colleters 0.2-0.3 x 0.2 mm. *Corolla* creamy, turning yellow at anthesis, 30-32 mm long in the mature bud and forming a broadly ovoid head 0.16-0.27 of the bud length, (5-7 x 5.7 mm), with an obtuse apex, glabrous outside, with 5 sparsely pilose wings 1 mm wide inside from the insertion of the stamens to 3 mm above the apices of the anthers; tube 1.53-1.6 x as long as the calyx, 1.7-1.8 x as long as the lobes, 26-27 mm long, almost cylindrical, 5-6 mm wide above the base, narrowed below the insertion of the stamens to 3-4 mm wide, 4-5 mm wide around the anthers, narrowed again above to 2.5-3 mm wide, 3-5 mm wide at the mouth, twisted 0.7-0.9 turn around the anthers; lobes obliquely obovate, 0.55-0.58 x as long as the tube, 1.15 x as long as wide, 15 x 13 mm, rounded, not undulate, spreading. *Stamens* with apex 9-10 mm below mouth of corolla tube, inserted 0.4-0.46 of the length of the corolla tube (at 10.5-12.5 mm from the base); anthers with tails 0.5 mm below the insertion, narrowly triangular, 3.7-4 x as long as wide, 5.5-6 x 1.5 mm, apex apiculate, sterile for 0.2-0.5 mm, sagittate at the base, with tails curved towards each other, glabrous. *Pistil* glabrous, 13-15 mm long; ovary broadly ovoid, 2.5-3 x 2-3 x 2 mm, of 2 separate carpels connate at the base with a disk-like ring 1.5 mm high; style filiform, 9-11 mm long; pistil head 1.2 mm high, composed of an undulate ring 0.2 x 1.5 mm, a stipitate 5-lobed depressed globe 0.3 x 0.7 mm and a stigmoid apical part 0.2 x 0.2 mm. *Fruit* of 2 separate mericarps; only immature mericarps known, reniform, 30-32 x 22-24 mm, rounded or minutely apiculate at the apex, not ridged, smooth.

DISTRIBUTION: Costa Rica, only known from Puntarenas Province.

ECOLOGY: Forest understorey. Alt. 0-550 m. Flowering from end of February to March. Immature fruits found May-July.

Paratypes:

COSTA RICA. Puntarenas: Río Sonador, 37 km E of San Isidro, *Webster* et al. 12409 (F); between Río Esquinas and Palmar Sur, *Allen* 5965 (BM, F, GH, US); between Rincón de Osa and Rancho Quemado, *Croat & Grayum* 59774 (MO); Peninsula de Osa, near Airport at Rincón, *Duke* 16112 (MO, Z); ibid., *Liesner* 1748 (MO, Z); Corcovado National Park, *Kernan* 185 (MO); ibid., *Liesner* 3255 (MO); 6 km N of Golfito, *Utley* 4893 (MO).

Fig. 108. *Stemmadenia stenoptera.* **1,** habit (x 2/3); **2,** leaf beneath (x 2/3); **3,** flower (x 2); **4,** calyx with pistil (x 2); **5,** sepal inside (x 4); **6,** opened corolla (x 6); **7,** pistil (x 8); **8,** pistil head (x 18). 1-8 from Lott et al. 359.

9. Stemmadenia stenoptera Leeuwenberg, sp. nov.

Fig. 108, p. 431; map 54, p. 403

Frutex ramis furcatis. Folia anguste elliptica membranacea juvenilia ciliata. Inflorescentiae foliis multo breviores pauciflorae pedunculo bracteis suffulto. Sepala inter se inaequalia ovata fere libera. Corolla parva tubo fere cylindrico intus supra stamina alis angustis quinque limbo patente lobis anguste oblongis inaequalilateralibus. Stamina inclusa antheris oblongis. Ovarium basi incrassatum glabrum. Fructus ignotus.

Typus: Mexico, Colima, Mun. Manzanillo, Playa Miramar, 19 km NW of Manzanillo, 9 March 1981, *Lott* et al. 359 (holotypus USF; isotypus WAG).

Etymology: στενός, narrow; πτερόν, wing; stenoptera = with narrow wings. The wings inside the corolla tube are narrow.

Shrub 2 m high. Branches pale brown, lenticellate; branchlets terete, glabrous. *Leaves* petiolate; petiole glabrous, 2-10 mm long; (ocreae slightly widened into intrapetiolar stipules); blade membranaceous when dried, narrowly elliptic, 3-4.5 x as long as wide, 2-12.5 x 0.5-3.7 cm, acuminate at the apex, cuneate at the base, entire or shallowly sinuate, glabrous on both sides, ciliate when young, with 5-7 pairs of upcurved secondary veins; tertiary venation reticulate. *Inflorescence* pedunculate, 15-25 x 15-25 mm, 3-5-flowered, lax. Peduncle 2-11 mm long, glabrous, with several bracts; pedicels 2-10 mm long, rather slender, glabrous. Bracts scale-like, 1-2 mm long. Flowers open during the day. *Calyx* subtended by one bracteole 1-2 mm long; sepals green(?), connate at the base for 0.5 mm, erect, ovate, unequal, 1.7-2.3 x as long as wide, 4-7 x 2-3 mm, obtuse, glabrous outside, not ciliate, glabrous inside and with 10 colleters in 1-2 rows 0.5 mm above the base; colleters 0.4-0.8 x 0.1-0.2 mm. *Corolla* yellow, 17 mm long in the mature bud and forming a comparatively rather large ovoid head, 7 x 4 mm, with an acute apex, glabrous outside; with 5 narrow wings 0.5 mm wide inside from the insertion of the stamens and from there pilose almost to the mouth; tube 1.8-3 x as long as the calyx, about as long as the lobes, 11-13 mm long, almost cylindrical, 4 mm wide above the base, narrowed just below the insertion of the stamens to 2.4 mm wide, widened again to 3 mm wide around the stamens, twisted 0,5 turn around the anthers; lobes obliquely oblong, about twice as long as wide, 12 x 5.5 mm, obtuse, with an obscure lateral lobe, spreading. *Stamens* with apex 2-2.5 mm below mouth of corolla tube, inserted 0.6-0.7 of the length of the corolla tube, (at 7.5-8 mm from the base); anthers narrowly oblong, 3 x 1 mm, apex acuminate, sterile for 0.3 mm, sagittate at the base, with tails 0.5 mm below the insertion. *Pistil* glabrous, 7-9 mm long; ovary ovoid, 3 x 1 x 1 mm, with a disk-like ring-shaped thickening 1 mm high; style 3-5 mm long; pistil head composed of a basal ring 0.2 x 0.8 mm, a stipitate 5-lobed depressed globe 0.5 x 0.6 mm and a stigmoid apex 0.2 x 0.2 mm. *Fruit* unknown.

Only known from the type.

10. Stemmadenia tomentosa Greenm. in Proc. Am. Acad. 35: 310 (1900); in Contrib. Gray Herb. II, 18: 310 (1900). – Type: Mexico, Jalisco, near Zapotlan, *Pringle* 4370 (holotype GH; isotypes BM, BR, E, F, G, HBG, K, M, MO, MU, NY, P, S, UC, US, WU, Z; phot. of US sheet WAG).

Fig. 109, p. 434; map 55, p. 415

Heterotypic synonyms:

S. palmeri Rose ex Greenm. in op. cit. 311. *S. tomentosa* var. *palmeri* (Rose ex Greenm.) Woods. in Ann. Miss. Bot. Gard. 15: 354, pl. 47, fig. 5 (1928). – Type: Mexico, Sinaloa, Imala, *Palmer* 1470 (holotype US; phot. WAG; isotype GH).
S. sinaloana Woods. in op. cit. 356, pl. 48, fig. 1, **syn. nov.** – Type: Mexico, Sinaloa, Rosario, *Lamb* 467 (holotype GH).
S. decipiens Woods. in op. cit. 363, **syn. nov.** – Type: Mexico, Sinaloa, between Rosario and Colomas, *Rose* 1614 (holotype US; phot. WAG).

Deciduous shrub or small tree, 1.50-10 m high. Trunk up to at least 15 cm in diameter; bark pale grey, shallowly and longitudinally fissured, lenticellate. Branches pale brown, lenticellate; branchlets terete, glabrous. *Leaves* of a pair equal or unequal (larger up to 3 x as long as other and similarly shaped), petiolate; petipole glabrous, 5-15 mm long; (ocreae slightly widened into intrapetiolar stipules or not); blade herbaceous when fresh, membranaceous when dried, elliptic, or narrowly elliptic, 1.7-3.5 x as long as wide, 4-21 x 1.5-10.5 cm, acuminate at the apex, cuneate at the base, glabrous above, white-pilose or -lanate, or with a white cobweb-like indumentum along the costa, entirely glabrous or sometimes (only seen in the type) entirely tomentose beneath, with 10-20 pairs of rather straight or upcurved secondary veins forming an angle of 50-70° with the costa; tertiary venation reticulate, often impressed above. *Inflorescence* pedunculate, 5-55 mm long excluding flowers, 1-10-flowered. Peduncle glabrous, 2-35 mm long; pedicels glabrous, 3-10 mm long. Bracts small, like bracteoles. *Flowers* fragrant, very showy. *Calyx* subtended by 0-2 sepal-like bracteoles, 1-1.3 x as long as wide, 2-4 x 2-3 mm; sepals pale green, subequal or unequal and larger up to 1.5 x as long as smaller, connate at the extreme base, ovate or broadly ovate, (0.5-)1-1.5 x as long as wide, the outer of which the broader, (2-)3-6 x 3-4 mm, obtuse or sometimes rounded or acute at the apex, glabrous outside, not ciliate, glabrous inside and with a row of approximately 15 colleters at the base, forming a continuous ring; colleters 0.5-1 x 0.1-0.2 mm. *Corolla* pale yellow, with an ovoid obtuse or acute head in the mature bud, glabrous outside, not ciliate on the lobes, with 5 longitudinal wings, 8-14 x 1-2 mm inside from just above the anthers, and pubescent on the filament ridges just below the anthers for 1-3 mm and on the wings up to 0-2 mm below their apices, otherwise glabrous; tube 5.5-14 x as long as the calyx, 1-1.5 x as long as the lobes, 33-45 mm long, 6-10 mm wide just above the base, narrowed to 3-6 mm wide below the insertion of the stamens and from there gradually widened to the throat and at the throat 10-21 mm wide, twisted 0.25-0.7 turn around the anthers; lobes obliquely obovate or nearly so, 0.65-1 x as long as the tube, 1.2-1.5 x as long as wide, 24-35 x 18-30 mm, rounded, not undulate, spreading. *Stamens* deeply included, inserted about 1-2 mm above anther tail ends, with tails 0.25-0.4 of the length of the corolla tube, (at 9-16 mm from the base); anthers sessile, just below the wings, narrowly triangular, 3-4 x as long as wide, 5.5-7 x 1.5-2 mm, apex sterile for 0.3-0.5 mm. *Pistil* glabrous, 14-19 mm long; ovary broadly ovoid, 3-3.5 x 3 x 2-3 mm, gradually narrowed into the style, of two carpels connate at the base with a disk-like thickening 1-2 mm high and mostly 5-lobed; style 9-14 mm long; pistil head 1.5-2.2 x 1.5-2 mm, composed of ring 0.5 x 1.5-2 mm, a 5-lobed mostly depressed globe 0.7 x 0.7-0.9 mm and a stigmoid apex about 0.2 x 0.2 mm. Ovules approximately 100 in each carpel. *Fruit* of two separate mericarps; mericarps green, obliquely ellipsoid, 30-43 x 25-30 x 20-25 mm, apiculate or acuminate at the recurved apex, with 2 lateral ridges, one at each side, lenticellate or smooth; wall about 3 mm thick in dried fruits; aril orange or red. *Seed* dark brown, obliquely ellipsoid, 9-11 x 3.5-5 x 3-4 mm; embryo 6.5-9 mm long; cotyledons ovate, 1.4-1.7 x as long as wide, 3.5-5 x 2.5-3 mm, rounded at the apex, cordate at the base; rootlet slightly shorter than cotyledons, 3-4 x 0.8-0.9 mm.

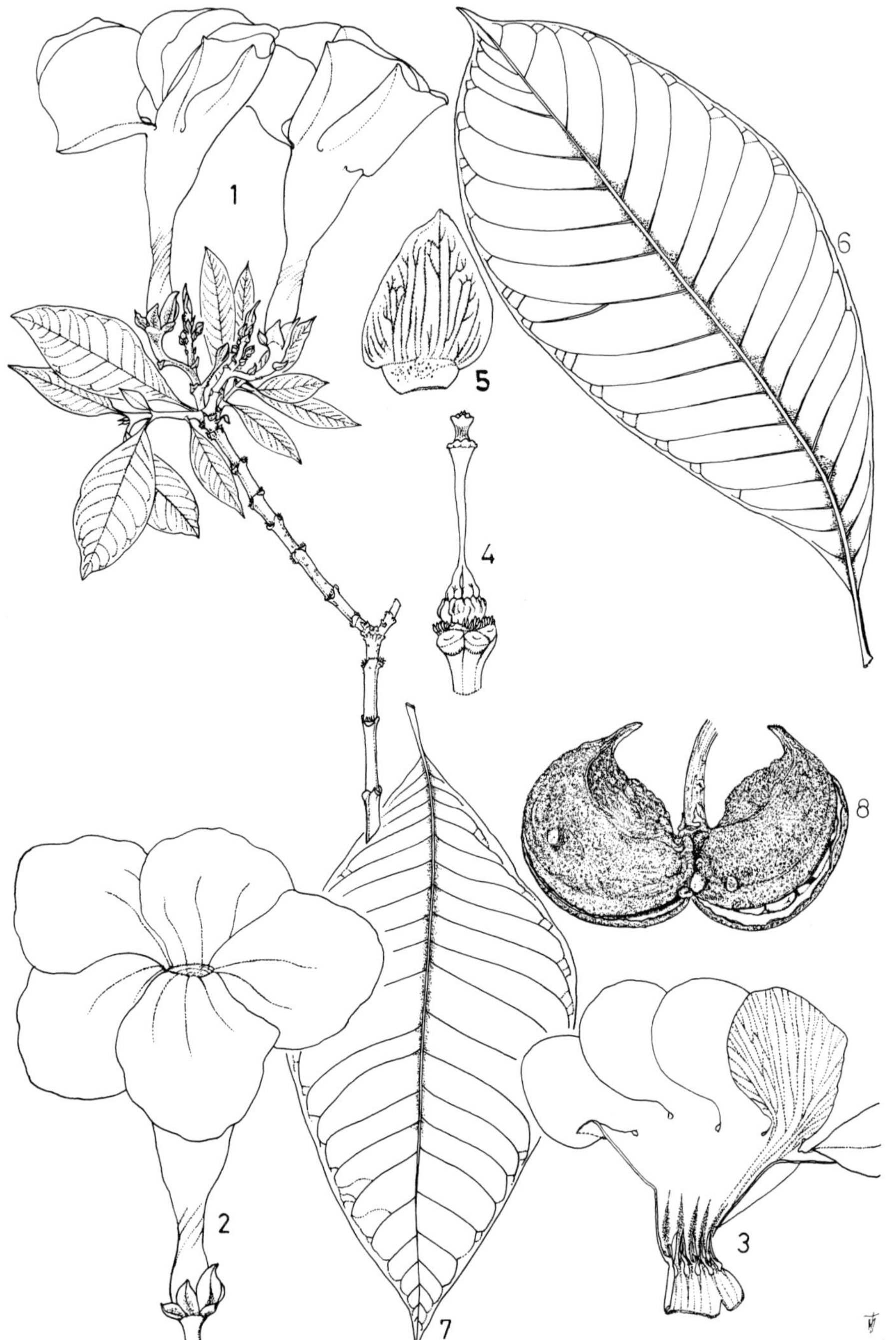

Fig. 109. *Stemmadenia tomentosa.* **1,** habit (x 2/3); **2,** flower (x 1); **3,** opened corolla (x 2/3); **4,** pistil (x 2); **5,** sepal inside (x 4); 6-7 leaves (x 2/3); **8,** fruit (x 2/3). 1 from Pringle 4370; 2 and 7 from Marcks 1162; 3-5 from McVaugh 10217; 6 and 8 from Parks et al 225.

Distribution: Mexico.

Ecology: Forest, often low, or thickets. Alt. 1100-1500 m (or less?). Flowering mainly from April to June. Fruits found throughout the year.

Geographical selection of the approximately 120 specimens examined:

Mexico. Sonora: Caramechi, *H.S. Gentry* 1175 (A, F, MO). Chihuahua: W of La Bufa, *Bye* 4017 (ECON, LL); near La Bufa, Río Batopilas, *Bye* 2889 (MEXU); near Guadelupe, *Rose* et al. 14675 (GH, US); near Batopilas, *Palmer* 11"M" (GH, K, LE, NY, US); Los Nogales, *Pennington* 1 (LL). Sinaloa: 4 km NE of El Cajon, *Nash* et al. L19587 (NY); El Andable, *Pérez* 57 (UC, WIS); Imala, *Palmer* 1470 (GH, US; phot. of US sheet WAG, type of *S. palmeri*); Colomas, foothills of Sierra Madre, *Rose* 1688 (K, MO, NY, US); SW of Altamirando, *Martínez* 3666 (KUN); Capradia, near Culiancan, *Brandegree* 23 Oct. 1904 (UC); S of Higuer de Abuya, *Breckon & Christman* 532 (GH, LL); near junction of road to La Cruz, 144 km NW of Mazatlan, *Waterfall* 12791 (GH, MO, UC, US); San Ignacio, *Narvaez & Salazar* 268 (US); 32 km N of Mazatlan, *Freeland & Spetzman* 8 (NY); La Noria, *Mexia* 164 (MO, UC); Rincón de Urias, *Ortega* 7068 (BR, F, MEXU); near Villa Unión, *Rose* et al. 13874 (F, US); 14 km E of Concordia, *Parks* et al. 225 (UC); Rosario, *Lamb* 467 (GH, type of *S. sinaloana*); ibid., *Rose* 1573 (MO, NY, US); between Rosario and Colomas, *Rose* 1614 (US; phot. WAG, type of *S. decipiens*). Nayarit: near Ixtlan, *Mexia* 733 (A, UC, US). Jalisco: Mt ridge 3 km E of Nayarit border, km 800 Guadalajara-Tepic Road, *Marcks* 1162 (LL, WIS); Bolaños, *Rose* 2888 (US); San Sebastian, *Mexia* 1490 (A, BM, F, G, GH, MO, NY, UC, US); Chiquilistlán, *Jones* 335 (MO, US); Est. Biol. Chamela, *Pérez* 1671 (MEXU); Ameca, *Lomelí* 18662 (MO); 10-12 km from Autlán, *McVaugh* 10217 (BM, G, LL, MO, NY, US); Tequila, *Rose & Hough* 4777 (US); Barranca of Guadalajara, *Palmer* 132 (BM, G, GH, K, NY, US); ibid., *Pringle* 6872 (BM, E, F, G, GH, K, MO, NY, P, UC, US, WU); 24 km S of Tapalpa, *Lott & Magellanes* 419 (USF); 9 km S of El Chante, *Iltis* et al. 1146 (GH); Barranca de Beltran, *Seler* 3436 (GH); San Cristobal de la Barranca, *C.L. Díaz* 3238 (MO, USF); near Zapotlan, *Pringle* 4370 (BM, BR, E, F, G, GH, HBG, K, M, MO, MU, NY, P, S, UC, US, WU, Z; phot. of US sheet WAG, type); Mun. Tonila, ca 19 km SW of Atenquique, *Lott* et al. 987 (USF). Colima: Sierra de Manantlán, *Cochrane* et al. 11763 (USF); Mun. Ixtlahuacán, 6.3 km SW of Agua de la Virgen, *Lott & Magellanes* 928 (USF). Mexico: Tejupilco, *Hinton* 7625 (F, GH, K, MO, NY, US). Morelos: near Cuernavaca, *Pringle* 6847 (A, BR, GH, HBG, K, M, US, Z); Barranca Honda, *Gutierrez* 256 (MEXU). Michoacan: Hacienda Coahuayula, *Emrick* comm. Sept. 1906 (F); Coalcomán, *Hinton* 12982 (MO, NY, US). Guerrero: Cuachicle, Tepechicle, *Cox & Guzman* MCG 617 (ECON); Aguazarca, *Hinton* et al. 10455 (G, K, NY, P, S, US, USF, WAG); 12 km from Zihuatanejo, *German* et al. 264 (MEXU, MO, NY); Mun. Alcozauca, Ixcuinatoyac, *Viveros & Casas* 238 (MEXU); La Palma, *Hinton* 10111 (K, LL, NY, UC, US, USF). Oaxaca: Distr. Pochutla, Caf. Alianza, *Conzatti* et al. 3172 (MO). Veracruz: Misantla, *Calzada* 4106 (F).

Cult. Indonesia, Java, Bogor Bot. Garden XV. J. B. XIV. 13, *Leeuwenberg* 13193 (WAG).

Notes. Fruiting specimens of *Stemmadenia litoralis* and *S. tomentosa* cannot be distinguished when they do not also bear flowers. For the time being they can be recognized, if their collecting locality is known. All specimens of *S. tomentosa* have been collected in northern and central Mexico, *S. litoralis* is known from southern Mexico to Colombia.

A NEW SYNONYM FOR AN OLD WORLD SPECIES

44. Tabernaemontana divaricata (L.) R. Br. ex Roem. & Schult., Syst. 4: 427 (1819); Leeuwenberg in Journ. Ethnopharmacology 10: 11 (1984); Tabernaemontana One, The Old World Species, Kew 153 (1991). – Type: Cult., Sri Lanka, sin. loc., "Apocynum zeylanicum arborescens", with 2 buds of double flowers, herb. Hermann Vol. 1: 7 (lectotype BM, designated by Leeuwenberg in 1984, after consultation with C. Jarvis).

Heterotypic synonym:

T. discolor Sw., Nov. gen. et sp. pl. seu Prod. Veg. Ind. Occ. 52 (1788), **syn. nov.** *Taberna discolor* (Sw.) Miers, Apoc. S. Am. 62 (1878). Type: Jamaica, sin. loc., apparently cultivated, *Swartz* s.n. (holotype S; isotypes B-W 5197, BM, C, M, P and possibly also *Wiles* s.n. in herb. *Hooker* s.n. (K, P, U)).

The above cited type is a poor, single-flowered specimen of *T. divaricata*. It could be recognized only as such after a long and intimate association with the taxa leading to this monograph of the genus. Dr. C.D. Adams informed the author personally that it never has been found in Jamaica.

NOMEN REJICIENDUM

Tabernaemontana echinata Aubl., Pl. Guian. 1: 263 and 3: pl. 103 (1775). *Peschiera echinata* (Aubl.) A. DC., Prod. 8: 360 (1844), who had a specimen of *T. heterophylla* under this name. *Anacampta echinata* (Aubl.) Mgf. in Notizbl. Bot. Gart. Berlin 14: 163 (1938). The name has been based on a number of different elements. See notes with *T. lagenaria*.

DOUBTFUL SPECIES

Tabernaemontana ovalifolia Glaziou in Bull. Soc. Bot. Fr. 57 Mém. 3: 452 (1910), non Urb. (1908), description inadequate. Brazil, Minas Gerais, Gandarela, *Glaziou* 14801 (not seen, not in P, probably excluded and specimen placed elsewhere).

NOMINA NUDA

Tabernaemontana(?) *bicolor* Klotzsch ex Schomb., Reisen Br. Guian. 3: 952 (1848) = *T. lorifera* (Miers) Leeuwenberg.

T. plumieri Krause in Beih. Bot. Centralbl. 32, 2: 344 (1914) = *T. citrifolia L.*

T. sessilifolia Klotzsch ex Schomb., l.c., non Bak. (1883) = *T. rupicola* Benth.

T. ulei K. Schum. in Engler, Bot. Jahrb. 40: 167 (1907) = *T. coriacea* Link ex Roem. & Schult.

EXCLUDED SPECIES (NEW WORLD)

Anartia bogotensis (H.B.K.) Miers, Apoc. S. Am. 82 (1878) = *Echites bogotensis* H.B.K. = *Mandevilla bogotensis* (H.B.K.) Woods.

A. flavescens (Willd. ex Roem. & Schult.) Miers, l.c., see *T. flavescens* Willd. ex Roem. & Schult. in this list.

Merizadenia amplifolia Miers, op. cit. 79 = *Macoubea guianensis* Aubl.

Peschiera praeclara Miers, op. cit. 47. *Tabernaemontana praeclara* (Miers) Kunth in Fedde, Repert. Beih. 43: 571 (1927). – Type: Venezuela, Caracas, Cockburn s.n.(holotype BM). = **Chionanthus guianensis** (Aubl.) P. Green, **comb. nov.** See also list of new names, p. vii. Basionym *Mayepea guianensis* Aubl., Pl. Guian. 1: 81 and 3 pl. 31 (1775). (*Oleaceae*).

Stemmadenia cerea Woods. in Bull. Torr. Bot. Cl. 75: 557 (1948) = *Tabernaemontana cerea* (Woods.) Leeuwenberg.

S. guatemalensis Muell. Arg. in Linnaea 30: 410 (1860) = *Malouetia guatemalensis* (Muell. Arg.) Standl.

S. nervosa Standl. & L.Wms. in Ceiba 3: 126 (1952) = *Tabernaemontana undulata* Vahl.

S. rugosa M.E. Jones in Contrib. West. Bot. 18: 67 (1933) = *Ipomoea intrapilosa* Rose (Convolv.).

Tabernaemontana alfari Donn. Sm. in Bot. Gaz. 24: 396 (1897) = *Stemmadenia alfari* (Donn. Sm.) Woods.

T. amsonia L., Sp. Pl. ed. 2: 308 (1762) = *Amsonia tabernaemontana* Walt.

T. angustifolia Ait., Hort. Kew. ed. 1, 1: 300 (1789) = *Amsonia ciliata* Walt.

T. aphlebia Standl. in Publ. Field Mus. Nat. Hist. Bot. Ser. 18: 946 (1938) = *Rauvolfia aphlebia* (Standl.) A. Gentry.

T. aubletii Pulle in Rec. Trav. Bot. Néerl. 9: 157, t. 3 (1912) = *Macoubea guianensis* Aubl.

T. cestroides Nees & Mart. in Nov. Act. Nat. Cur. 11, 1: 83 (1823) = *Malouetia cestroides* (Nees. & Mart.) Muell. Arg.

T. didyma Vell., Fl. Flum. 106, 3: t. 27 (on plate as *Echites didyma*) (1829) = *Prestonia didyma* (Vell.) Woods.
T. donnell-smithii Rose ex Donn. Sm. in Bot. Gaz. 18: 206 (1893) = *Stemmadenia donnell-smithii* (Rose ex Donn. Sm.) Woods.
T. echites L., Syst. ed. 10: 945 (1759) = *Echites umbellata* Jacq.
T. fasciculata Lam., Tableau Enc. 2: 300 (1792); Poiret in Lamarck, Enc. 7: 53 (1806) = *Parahancornia fasciculata* (Lam.) R.Benoist.
T. flavescens Willd. ex Roem. & Schult., Syst. 4: 797 (1819) = *Malouetia flavescens* (Willd. ex Roem. & Schult.) Muell.Arg.
T. funiformis Muell. Arg. ex Miers, Apoc. S. Am. 59 (1878) = *Mandevilla funiformis* (Vell.) K. Schum.
T. funiformis var. *peduncularis* Muell. Arg. ex Miers, l.c. = *M. funiformis* (Vell.) K. Schum.
T. gracilis Benth. in Hooker, Journ. Bot. 3: 244 (1841), not Muell. Arg. (1860) = *Malouetia gracilis* (Benth.) A. DC.
T. grandiflora Jacq., Enum. Pl. Carib. 14 (1760) = *Stemmadenia grandiflora* (Jacq.) Miers.
T. laevis Vell., Fl. Flum. 105, icon. 3: t. 18 (1829) = *Geissospermum laeve* (Vell.) Miers.
T. latifolia (Mich.) Parmentier, Cat. Jard. Bot. St. Seb. 77 (1818) = *Amsonia tabernaemontana* Walt.
T. laxa Benth. in Hooker, Journ. Bot. 3: 244 (1841) = *Molongum laxum* (Benth.) Pichon.
T. litoralis H.B.K., Nov. Gen. 3: 228 (1819) = *Stemmadenia litoralis* (H.B.K.) Allorge.
T. longiflora Rusby, Descr. 300 New Sp. S. Am. Pl. 82 (1920), non Benth. (1849). – Type: Venezuela, Bolívar, *Rusby* April 1896 (holotype NY), garden escape or label exchanged(?) = *Kopsia fruticosa* (Ker) A.DC.
T. lucida H.B.K., Nov. Gen. 7: 162 [folio], 209 [quarto] (1825) = *Molongum lucidum* (H.B.K.) Zarucchi.
T. macrophylla Lam., Tableau Enc. 2: 299 (1792); Poiret in Lamarck, Enc. Supl. 5: 276 (1817), not Muell. Arg.(1860) = *Macoubea guianensis* Aubl.
T. montana Oken in Allgem. Naturgesch. 3, 2: 1039 (1841) = *Lacmellea utilis* (Arn.) Mgf.
T. odorata Vahl, Eclog. Am. 2: 22 (1798) = *Malouetia odorata* (Vahl) Miers. = *M. tamaquarina* (Aubl.) A. DC.
T. pauciflora Spruce ex Muell. Arg. in Martius, Fl. Bras. 6, 1: 87 (1860), not Blume (1826), nor Roxb. ex Wight (1840) = *Macoubea sprucei* (Muell. Arg.) Mgf.
T. praeclara (Miers) Kunth, see *Peschiera praeclara* Miers in this list.
T. quadrangularis Achenbach in Zeit. für Naturforschung 35b: 219-225 (1980) = *Rhigospira quadrangularis* (Muell. Arg.) Miers ?
T reticulata A. DC., Prod. 8: 366 (1844) = *Macoubea guianensis* Aubl.
T. riparia H.B.K, l.c. = *Stemmadenia grandiflora* (Jacq.) Miers.
T. rosea Tenore, Cat. Orto Bot. Napoli 97 (1845). – Type: Cult, Italy. Naples, *Tenore* s.n.(holotype NAP, not seen, phot. WAG). = *Kopsia fruticosa* (Ker) A.DC.
T. sessilis Vell., Fl. Flum. 106, 3: t. 35 (on plate as *Echites sessilis*) (1829) = *Malouetia sessilis* (Vell.) Muell. Arg.
T. sprucei Muell. Arg. in Martius, Fl. Bras. 6,1: 86 (1860) = *Macoubes sprucei* (Muell. Arg.) Markgr. in Notizbl. Bot. Gart. Berlin 14: 179 (1938).
T. ternstroemiacea Muell. Arg. in op. cit. 88. = *Neocouma ternstroemiacea* (Muell. Arg.) Pierre.
T. utilis Arn. in Edinb. N. Phil. Journ. 8: 319 (1830) = *Lacmellea utilis* (Arn.) Mgf.

ACKNOWLEDGEMENTS

The author is greatly indebted to the Directors and Curators of the following herbaria for putting material at his disposal:

A, AAU, AMD, AWH, B, B-W, BH, BISH, BM, BO, BP, BR, C, CAY, CGE, COL, CR, E, ECON, F, FHO, FI, FI-W, FMB, G, G-DC, GB, GENT, GH, GOET, GUA, HBG, IBSC, IJ, INPA, K, KUN, L, LD, LE, LG, LINN, LISU, LL, M, MA, MEL, MEXU, MG, MICH, MJG, MO, MPU, MU, MY, NA, NSW, NY, OXF, P, P-LA, PORT, QCA, RB, S, SING, SP, TCD, TL, U, UC, ULM, UPS, US, USF, VEN, W, WAG, WIS, WU, Z.

On behalf of the Director of the Herbarium Vadense (WAG) he expresses his gratitude that so many duplicates of Apocynaceae were sent for identification by the Directors and Curators of several of the above cited herbaria, and also of CEPEC, CVRD and MBM. These duplicates, now in WAG, increase the value of the collections of this family; the important contributions sent from MO and NY merit special mention.

Grateful thanks go to Miss Yuen Fang Tan for preparing the excellent drawings, to Professor L.J.G. van der Maesen and the reviewer Dr. D.J. Goyder for correcting the English text, to Dr. C.D.Adams, Dr. F.J. Breteler and Dr. P.J.M. Maas for the nice photographs, and to Dr. P.J.M. Maas for maps made for the Flora Neotropica. The map of Jamaica has been copied from the herbarium labels of Dr. W.T. Stearn.

Last but not least, the author would like to thank Professor G. Ll. Lucas, Keeper of the Herbarium, Royal Botanic Gardens, Kew, for offering facilities to publish both the first and the second part of the revision of *Tabernaemontana*.

INDEX OF SCIENTIFIC NAMES (both volumes)

New names in **bold face**, synonyms in *italics*. Page numbers of principal entries are in **bold face**, those of figures and maps in *italics*. Tabernaemontana two, the New World Species starts with p. 213.